T0100493

Introduction to Applied Linear Algebra

Vectors, Matrices, and Least Squares

Introduction to Applied Linear Algebra

Vectors, Matrices, and Least Squares

Stephen Boyd

Department of Electrical Engineering
Stanford University

Lieven Vandenberghe

Department of Electrical and Computer Engineering
University of California, Los Angeles

CAMBRIDGE
UNIVERSITY PRESS

CAMBRIDGE
UNIVERSITY PRESS

University Printing House, Cambridge CB2 8BS, United Kingdom

One Liberty Plaza, 20th Floor, New York, NY 10006, USA

477 Williamstown Road, Port Melbourne, VIC 3207, Australia

314–321, 3rd Floor, Plot 3, Splendor Forum, Jasola District Centre,
New Delhi – 110025, India

79 Anson Road, #06–04/06, Singapore 079906

Cambridge University Press is part of the University of Cambridge.

It furthers the University's mission by disseminating knowledge in the pursuit of
education, learning, and research at the highest international levels of excellence.

www.cambridge.org
Information on this title: www.cambridge.org/9781316518960
DOI: 10.1017/9781108583664

© Cambridge University Press 2018

First published 2018
Reprinted 2018

Printed and bound in Great Britain by Clays Ltd, Elcograf S.p.A.

A catalogue record for this publication is available from the British Library.

ISBN 978-1-316-51896-0 Hardback

Additional resources for this publication at www.cambridge.org/IntroAppLinAlg

For

Anna, Nicholas, and Nora

Daniël and Margriet

Contents

Preface

This book is meant to provide an introduction to vectors, matrices, and least squares methods, basic topics in applied linear algebra. Our goal is to give the beginning student, with little or no prior exposure to linear algebra, a good grounding in the basic ideas, as well as an appreciation for how they are used in many applications, including data fitting, machine learning and artificial intelligence, tomography, navigation, image processing, finance, and automatic control systems.

The background required of the reader is familiarity with basic mathematical notation. We use calculus in just a few places, but it does not play a critical role and is not a strict prerequisite. Even though the book covers many topics that are traditionally taught as part of probability and statistics, such as fitting mathematical models to data, no knowledge of or background in probability and statistics is needed.

The book covers less mathematics than a typical text on applied linear algebra. We use only one theoretical concept from linear algebra, linear independence, and only one computational tool, the QR factorization; our approach to most applications relies on only one method, least squares (or some extension). In this sense we aim for intellectual economy: With just a few basic mathematical ideas, concepts, and methods, we cover many applications. The mathematics we do present, however, is complete, in that we carefully justify every mathematical statement. In contrast to most introductory linear algebra texts, however, we describe many applications, including some that are typically considered advanced topics, like document classification, control, state estimation, and portfolio optimization.

The book does not require any knowledge of computer programming, and can be used as a conventional textbook, by reading the chapters and working the exercises that do not involve numerical computation. This approach however misses out on one of the most compelling reasons to learn the material: You can use the ideas and methods described in this book to do practical things like build a prediction model from data, enhance images, or optimize an investment portfolio. The growing power of computers, together with the development of high level computer languages and packages that support vector and matrix computation, have made it easy to use the methods described in this book for real applications. For this reason we hope that every student of this book will complement their study with computer programming exercises and projects, including some that involve real data. This book includes some generic exercises that require computation; additional ones, and the associated data files and language-specific resources, are available online.

If you read the whole book, work some of the exercises, and carry out computer exercises to implement or use the ideas and methods, you will learn a lot. While there will still be much for you to learn, you will have seen many of the basic ideas behind modern data science and other application areas. We hope you will be empowered to use the methods for your own applications.

The book is divided into three parts. Part I introduces the reader to vectors, and various vector operations and functions like addition, inner product, distance, and angle. We also describe how vectors are used in applications to represent word counts in a document, time series, attributes of a patient, sales of a product, an audio track, an image, or a portfolio of investments. Part II does the same for matrices, culminating with matrix inverses and methods for solving linear equations. Part III, on least squares, is the payoff, at least in terms of the applications. We show how the simple and natural idea of approximately solving a set of overdetermined equations, and a few extensions of this basic idea, can be used to solve many practical problems.

The whole book can be covered in a 15 week (semester) course; a 10 week (quarter) course can cover most of the material, by skipping a few applications and perhaps the last two chapters on nonlinear least squares. The book can also be used for self-study, complemented with material available online. By design, the pace of the book accelerates a bit, with many details and simple examples in parts I and II, and more advanced examples and applications in part III. A course for students with little or no background in linear algebra can focus on parts I and II, and cover just a few of the more advanced applications in part III. A more advanced course on applied linear algebra can quickly cover parts I and II as review, and then focus on the applications in part III, as well as additional topics.

We are grateful to many of our colleagues, teaching assistants, and students for helpful suggestions and discussions during the development of this book and the associated courses. We especially thank our colleagues Trevor Hastie, Rob Tibshirani, and Sanjay Lall, as well as Nick Boyd, for discussions about data fitting and classification, and Jenny Hong, Ahmed Bou-Rabee, Keegan Go, David Zeng, and Jaehyun Park, Stanford undergraduates who helped create and teach the course EE103. We thank David Tse, Alex Lemon, Neal Parikh, and Julie Lancashire for carefully reading drafts of this book and making many good suggestions.

Stephen Boyd   *Stanford, California*
Lieven Vandenberghe   *Los Angeles, California*

Part I

Vectors

Chapter 1

Vectors

In this chapter we introduce vectors and some common operations on them. We describe some settings in which vectors are used.

1.1 Vectors

A *vector* is an ordered finite list of numbers. Vectors are usually written as vertical arrays, surrounded by square or curved brackets, as in

$$\begin{bmatrix} -1.1 \\ 0.0 \\ 3.6 \\ -7.2 \end{bmatrix} \quad \text{or} \quad \begin{pmatrix} -1.1 \\ 0.0 \\ 3.6 \\ -7.2 \end{pmatrix}.$$

They can also be written as numbers separated by commas and surrounded by parentheses. In this notation style, the vector above is written as

$$(-1.1, 0.0, 3.6, -7.2).$$

The *elements* (or *entries*, *coefficients*, *components*) of a vector are the values in the array. The *size* (also called *dimension* or *length*) of the vector is the number of elements it contains. The vector above, for example, has size four; its third entry is 3.6. A vector of size n is called an *n-vector*. A 1-vector is considered to be the same as a number, *i.e.*, we do not distinguish between the 1-vector $[\, 1.3 \,]$ and the number 1.3.

We often use symbols to denote vectors. If we denote an n-vector using the symbol a, the ith element of the vector a is denoted a_i, where the subscript i is an integer index that runs from 1 to n, the size of the vector.

Two vectors a and b are *equal*, which we denote $a = b$, if they have the same size, and each of the corresponding entries is the same. If a and b are n-vectors, then $a = b$ means $a_1 = b_1, \ldots, a_n = b_n$.

The numbers or values of the elements in a vector are called *scalars*. We will focus on the case that arises in most applications, where the scalars are real numbers. In this case we refer to vectors as *real vectors*. (Occasionally other types of scalars arise, for example, complex numbers, in which case we refer to the vector as a *complex vector*.) The set of all real numbers is written as $\mathbf{R}$, and the set of all real n-vectors is denoted $\mathbf{R}^n$, so $a \in \mathbf{R}^n$ is another way to say that a is an n-vector with real entries. Here we use set notation: $a \in \mathbf{R}^n$ means that a is an element of the set $\mathbf{R}^n$; see appendix A.

Block or stacked vectors.   It is sometimes useful to define vectors by *concatenating* or *stacking* two or more vectors, as in

$$a = \begin{bmatrix} b \\ c \\ d \end{bmatrix},$$

where a, b, c, and d are vectors. If b is an m-vector, c is an n-vector, and d is a p-vector, this defines the $(m+n+p)$-vector

$$a = (b_1, b_2, \ldots, b_m, c_1, c_2, \ldots, c_n, d_1, d_2, \ldots, d_p).$$

The stacked vector a is also written as $a = (b, c, d)$.

Stacked vectors can include scalars (numbers). For example if a is a 3-vector, $(1, a)$ is the 4-vector $(1, a_1, a_2, a_3)$.

Subvectors.   In the equation above, we say that b, c, and d are *subvectors* or *slices* of a, with sizes m, n, and p, respectively. *Colon notation* is used to denote subvectors. If a is a vector, then $a_{r:s}$ is the vector of size $s - r + 1$, with entries $a_r, \ldots, a_s$:

$$a_{r:s} = (a_r, \ldots, a_s).$$

The subscript $r:s$ is called the *index range*. Thus, in our example above, we have

$$b = a_{1:m}, \qquad c = a_{(m+1):(m+n)}, \qquad d = a_{(m+n+1):(m+n+p)}.$$

As a more concrete example, if z is the 4-vector $(1, -1, 2, 0)$, the slice $z_{2:3}$ is $z_{2:3} = (-1, 2)$. Colon notation is not completely standard, but it is growing in popularity.

Notational conventions.   Some authors try to use notation that helps the reader distinguish between vectors and scalars (numbers). For example, Greek letters (α, β, ...) might be used for numbers, and lower-case letters (a, x, f, ...) for vectors. Other notational conventions include vectors given in bold font ($\mathbf{g}$), or vectors written with arrows above them ($\vec{a}$). These notational conventions are not standardized, so you should be prepared to figure out what things are (*i.e.*, scalars or vectors) despite the author's notational scheme (if any exists).

Indexing.   We should give a couple of warnings concerning the subscripted index notation a_i. The first warning concerns the range of the index. In many computer languages, arrays of length n are indexed from $i = 0$ to $i = n - 1$. But in standard mathematical notation, n-vectors are indexed from $i = 1$ to $i = n$, so in this book, vectors will be indexed from $i = 1$ to $i = n$.

The next warning concerns an ambiguity in the notation a_i, used for the ith element of a vector a. The same notation will occasionally refer to the ith vector in a collection or list of k vectors $a_1, \ldots, a_k$. Whether a_3 means the third element of a vector a (in which case a_3 is a number), or the third vector in some list of vectors (in which case a_3 is a vector) should be clear from the context. When we need to refer to an element of a vector that is in an indexed collection of vectors, we can write $(a_i)_j$ to refer to the jth entry of a_i, the ith vector in our list.

Zero vectors.   A *zero vector* is a vector with all elements equal to zero. Sometimes the zero vector of size n is written as 0_n, where the subscript denotes the size. But usually a zero vector is denoted just 0, the same symbol used to denote the number 0. In this case you have to figure out the size of the zero vector from the context. As a simple example, if a is a 9-vector, and we are told that $a = 0$, the 0 vector on the right-hand side must be the one of size 9.

Even though zero vectors of different sizes are different vectors, we use the same symbol 0 to denote them. In computer programming this is called *overloading*: The symbol 0 is overloaded because it can mean different things depending on the context (*e.g.*, the equation it appears in).

Unit vectors.   A (standard) *unit vector* is a vector with all elements equal to zero, except one element which is equal to one. The ith unit vector (of size n) is the unit vector with ith element one, and denoted e_i. For example, the vectors

$$e_1 = \begin{bmatrix} 1 \\ 0 \\ 0 \end{bmatrix}, \qquad e_2 = \begin{bmatrix} 0 \\ 1 \\ 0 \end{bmatrix}, \qquad e_3 = \begin{bmatrix} 0 \\ 0 \\ 1 \end{bmatrix}$$

are the three unit vectors of size 3. The notation for unit vectors is an example of the ambiguity in notation noted above. Here, e_i denotes the ith unit vector, and not the ith element of a vector e. Thus we can describe the ith unit n-vector e_i as

$$(e_i)_j = \begin{cases} 1 & j = i \\ 0 & j \neq i, \end{cases}$$

for $i, j = 1, \ldots, n$. On the left-hand side e_i is an n-vector; $(e_i)_j$ is a number, its jth entry. As with zero vectors, the size of e_i is usually determined from the context.

Ones vector.   We use the notation $\mathbf{1}_n$ for the n-vector with all its elements equal to one. We also write $\mathbf{1}$ if the size of the vector can be determined from the context. (Some authors use e to denote a vector of all ones, but we will not use this notation.) The vector $\mathbf{1}$ is sometimes called the *ones vector*.

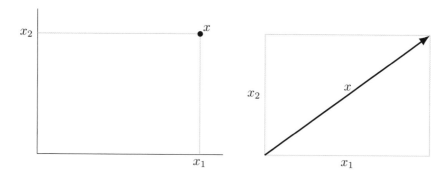

Figure 1.1 *Left.* The 2-vector x specifies the position (shown as a dot) with coordinates x_1 and x_2 in a plane. *Right.* The 2-vector x represents a displacement in the plane (shown as an arrow) by x_1 in the first axis and x_2 in the second.

Sparsity.   A vector is said to be *sparse* if many of its entries are zero; its *sparsity pattern* is the set of indices of nonzero entries. The number of the nonzero entries of an n-vector x is denoted $\mathbf{nnz}(x)$. Unit vectors are sparse, since they have only one nonzero entry. The zero vector is the sparsest possible vector, since it has no nonzero entries. Sparse vectors arise in many applications.

Examples

An n-vector can be used to represent n quantities or values in an application. In some cases the values are similar in nature (for example, they are given in the same physical units); in others, the quantities represented by the entries of the vector are quite different from each other. We briefly describe below some typical examples, many of which we will see throughout the book.

Location and displacement.   A 2-vector can be used to represent a position or location in a 2-dimensional (2-D) space, *i.e.*, a plane, as shown in figure 1.1. A 3-vector is used to represent a location or position of some point in 3-dimensional (3-D) space. The entries of the vector give the coordinates of the position or location.

A vector can also be used to represent a displacement in a plane or 3-D space, in which case it is typically drawn as an arrow, as shown in figure 1.1. A vector can also be used to represent the velocity or acceleration, at a given time, of a point that moves in a plane or 3-D space.

Color.   A 3-vector can represent a color, with its entries giving the Red, Green, and Blue (RGB) intensity values (often between 0 and 1). The vector $(0,0,0)$ represents black, the vector $(0,1,0)$ represents a bright pure green color, and the vector $(1,0.5,0.5)$ represents a shade of pink. This is illustrated in figure 1.2.

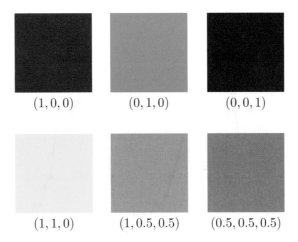

$(1,0,0)$ $\qquad$ $(0,1,0)$ $\qquad$ $(0,0,1)$

$(1,1,0)$ $\qquad$ $(1,0.5,0.5)$ $\qquad$ $(0.5,0.5,0.5)$

Figure 1.2 Six colors and their RGB vectors.

Quantities. An n-vector q can represent the amounts or quantities of n different resources or products held (or produced, or required) by an entity such as a company. Negative entries mean an amount of the resource owed to another party (or consumed, or to be disposed of). For example, a *bill of materials* is a vector that gives the amounts of n resources required to create a product or carry out a task.

Portfolio. An n-vector s can represent a stock portfolio or investment in n different assets, with s_i giving the number of shares of asset i held. The vector $(100, 50, 20)$ represents a portfolio consisting of 100 shares of asset 1, 50 shares of asset 2, and 20 shares of asset 3. Short positions (*i.e.*, shares that you owe another party) are represented by negative entries in a portfolio vector. The entries of the portfolio vector can also be given in dollar values, or fractions of the total dollar amount invested.

Values across a population. An n-vector can give the values of some quantity across a population of individuals or entities. For example, an n-vector b can give the blood pressure of a collection of n patients, with b_i the blood pressure of patient i, for $i = 1, \ldots, n$.

Proportions. A vector w can be used to give fractions or proportions out of n choices, outcomes, or options, with w_i the fraction with choice or outcome i. In this case the entries are nonnegative and add up to one. Such vectors can also be interpreted as the recipes for a mixture of n items, an allocation across n entities, or as probability values in a probability space with n outcomes. For example, a uniform mixture of 4 outcomes is represented as the 4-vector $(1/4, 1/4, 1/4, 1/4)$.

Time series. An n-vector can represent a *time series* or *signal*, that is, the value of some quantity at different times. (The entries in a vector that represents a time series are sometimes called *samples*, especially when the quantity is something

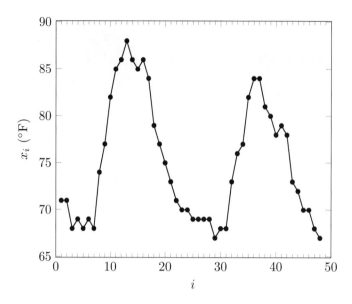

Figure 1.3 Hourly temperature in downtown Los Angeles on August 5 and 6, 2015 (starting at 12:47AM, ending at 11:47PM).

measured.) An audio (sound) signal can be represented as a vector whose entries give the value of acoustic pressure at equally spaced times (typically 48000 or 44100 per second). A vector might give the hourly rainfall (or temperature, or barometric pressure) at some location, over some time period. When a vector represents a time series, it is natural to plot x_i versus i with lines connecting consecutive time series values. (These lines carry no information; they are added only to make the plot easier to understand visually.) An example is shown in figure 1.3, where the 48-vector x gives the hourly temperature in downtown Los Angeles over two days.

Daily return. A vector can represent the daily return of a stock, *i.e.*, its fractional increase (or decrease if negative) in value over the day. For example the return time series vector $(-0.022, +0.014, +0.004)$ means the stock price went down 2.2% on the first day, then up 1.4% the next day, and up again 0.4% on the third day. In this example, the samples are not uniformly spaced in time; the index refers to trading days, and does not include weekends or market holidays. A vector can represent the daily (or quarterly, hourly, or minute-by-minute) value of any other quantity of interest for an asset, such as price or volume.

Cash flow. A cash flow into and out of an entity (say, a company) can be represented by a vector, with positive entries representing payments to the entity, and negative entries representing payments by the entity. For example, with entries giving cash flow each quarter, the vector $(1000, -10, -10, -10, -1010)$ represents a one year loan of $1000, with 1% interest only payments made each quarter, and the principal and last interest payment at the end.

0.65 0.05 0.20

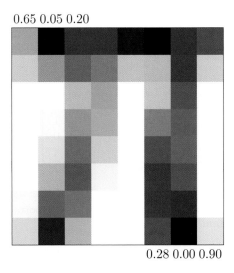

0.28 0.00 0.90

Figure 1.4 8×8 image and the grayscale levels at six pixels.

Images.   A monochrome (black and white) image is an array of $M \times N$ pixels (square patches with uniform grayscale level) with M rows and N columns. Each of the MN pixels has a grayscale or intensity value, with 0 corresponding to black and 1 corresponding to bright white. (Other ranges are also used.) An image can be represented by a vector of length MN, with the elements giving grayscale levels at the pixel locations, typically ordered column-wise or row-wise.

Figure 1.4 shows a simple example, an 8×8 image. (This is a very low resolution; typical values of M and N are in the hundreds or thousands.) With the vector entries arranged row-wise, the associated 64-vector is

$$x = (0.65, 0.05, 0.20, \ldots, 0.28, 0.00, 0.90).$$

A color $M \times N$ pixel image is described by a vector of length $3MN$, with the entries giving the R, G, and B values for each pixel, in some agreed-upon order.

Video.   A monochrome video, *i.e.*, a sequence of length K of images with $M \times N$ pixels, can be represented by a vector of length KMN (again, in some particular order).

Word count and histogram.   A vector of length n can represent the number of times each word in a dictionary of n words appears in a document. For example, $(25, 2, 0)$ means that the first dictionary word appears 25 times, the second one twice, and the third one not at all. (Typical dictionaries used for document word counts have many more than 3 elements.) A small example is shown in figure 1.5. A variation is to have the entries of the vector give the *histogram* of word frequencies in the document, so that, *e.g.*, $x_5 = 0.003$ means that 0.3% of all the words in the document are the fifth word in the dictionary.

It is common practice to count variations of a word (say, the same word stem with different endings) as the same word; for example, 'rain', 'rains', 'raining', and

Word count vectors are used in computer based document analysis. Each entry of the word count vector is the number of times the associated dictionary word appears in the document.

$$
\begin{array}{ll}
\text{word} \\
\text{in} \\
\text{number} \\
\text{horse} \\
\text{the} \\
\text{document}
\end{array}
\qquad
\begin{bmatrix}
3 \\
2 \\
1 \\
0 \\
4 \\
2
\end{bmatrix}
$$

Figure 1.5 A snippet of text (top), the dictionary (bottom left), and word count vector (bottom right).

'rained' might all be counted as 'rain'. Reducing each word to its stem is called *stemming*. It is also common practice to exclude words that are too common (such as 'a' or 'the'), which are referred to as *stop words*, as well as words that are extremely rare.

Customer purchases. An n-vector p can be used to record a particular customer's purchases from some business over some period of time, with p_i the quantity of item i the customer has purchased, for $i = 1, \ldots, n$. (Unless n is small, we would expect many of these entries to be zero, meaning the customer has not purchased those items.) In one variation, p_i represents the total dollar value of item i the customer has purchased.

Occurrence or subsets. An n-vector o can be used to record whether or not each of n different events has occurred, with $o_i = 0$ meaning that event i did not occur, and $o_i = 1$ meaning that it did occur. Such a vector encodes a subset of a collection of n objects, with $o_i = 1$ meaning that object i is contained in the subset, and $o_i = 0$ meaning that object i is not in the subset. Each entry of the vector a is either 0 or 1; such vectors are called *Boolean*, after the mathematician George Boole, a pioneer in the study of logic.

Features or attributes. In many applications a vector collects together n different quantities that pertain to a single thing or object. The quantities can be measurements, or quantities that can be measured or derived from the object. Such a vector is sometimes called a *feature vector*, and its entries are called the *features* or *attributes*. For example, a 6-vector f could give the age, height, weight, blood pressure, temperature, and gender of a patient admitted to a hospital. (The last entry of the vector could be encoded as $f_6 = 0$ for male, $f_6 = 1$ for female.) In this example, the quantities represented by the entries of the vector are quite different, with different physical units.

Vector entry labels.   In applications such as the ones described above, each entry of a vector has a meaning, such as the count of a specific word in a document, the number of shares of a specific stock held in a portfolio, or the rainfall in a specific hour. It is common to keep a separate list of labels or tags that explain or annotate the meaning of the vector entries. As an example, we might associate the portfolio vector $(100, 50, 20)$ with the list of ticker symbols (AAPL, INTC, AMZN), so we know that assets 1, 2, and 3 are Apple, Intel, and Amazon. In some applications, such as an image, the meaning or ordering of the entries follow known conventions or standards.

1.2   Vector addition

Two vectors *of the same size* can be added together by adding the corresponding elements, to form another vector of the same size, called the *sum* of the vectors. Vector addition is denoted by the symbol $+$. (Thus the symbol $+$ is overloaded to mean scalar addition when scalars appear on its left- and right-hand sides, and vector addition when vectors appear on its left- and right-hand sides.) For example,

$$\begin{bmatrix} 0 \\ 7 \\ 3 \end{bmatrix} + \begin{bmatrix} 1 \\ 2 \\ 0 \end{bmatrix} = \begin{bmatrix} 1 \\ 9 \\ 3 \end{bmatrix}.$$

Vector subtraction is similar. As an example,

$$\begin{bmatrix} 1 \\ 9 \end{bmatrix} - \begin{bmatrix} 1 \\ 1 \end{bmatrix} = \begin{bmatrix} 0 \\ 8 \end{bmatrix}.$$

The result of vector subtraction is called the *difference* of the two vectors.

Properties.   Several properties of vector addition are easily verified. For any vectors a, b, and c of the same size we have the following.

- Vector addition is *commutative*: $a + b = b + a$.

- Vector addition is *associative*: $(a + b) + c = a + (b + c)$. We can therefore write both as $a + b + c$.

- $a + 0 = 0 + a = a$. Adding the zero vector to a vector has no effect. (This is an example where the size of the zero vector follows from the context: It must be the same as the size of a.)

- $a - a = 0$. Subtracting a vector from itself yields the zero vector. (Here too the size of 0 is the size of a.)

To show that these properties hold, we argue using the definition of vector addition and vector equality. As an example, let us show that for any n-vectors a and b, we have $a + b = b + a$. The ith entry of $a + b$ is, by the definition of vector

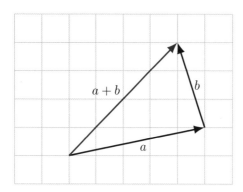 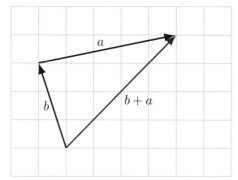

Figure 1.6 *Left.* The lower blue arrow shows the displacement a; the displacement b, shown as the shorter blue arrow, starts from the head of the displacement a and ends at the sum displacement $a + b$, shown as the red arrow. *Right.* The displacement $b + a$.

addition, $a_i + b_i$. The ith entry of $b + a$ is $b_i + a_i$. For any two numbers we have $a_i + b_i = b_i + a_i$, so the ith entries of the vectors $a + b$ and $b + a$ are the same. This is true for all of the entries, so by the definition of vector equality, we have $a + b = b + a$.

Verifying identities like the ones above, and many others we will encounter later, can be tedious. But it is important to understand that the various properties we will list can be derived using elementary arguments like the one above. We recommend that the reader select a few of the properties we will see, and attempt to derive them, just to see that it can be done. (Deriving all of them is overkill.)

Examples.

- *Displacements.* When vectors a and b represent displacements, the sum $a + b$ is the net displacement found by first displacing by a, then displacing by b, as shown in figure 1.6. Note that we arrive at the same vector if we first displace by b and then a. If the vector p represents a position and the vector a represents a displacement, then $p + a$ is the position of the point p, displaced by a, as shown in figure 1.7.

- *Displacements between two points.* If the vectors p and q represent the positions of two points in 2-D or 3-D space, then $p - q$ is the displacement vector from q to p, as illustrated in figure 1.8.

- *Word counts.* If a and b are word count vectors (using the same dictionary) for two documents, the sum $a + b$ is the word count vector of a new document created by combining the original two (in either order). The word count difference vector $a - b$ gives the number of times more each word appears in the first document than the second.

- *Bill of materials.* Suppose $q_1, \ldots, q_N$ are n-vectors that give the quantities of n different resources required to accomplish N tasks. Then the sum n-vector $q_1 + \cdots + q_N$ gives the bill of materials for completing all N tasks.

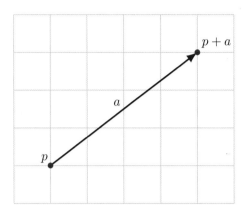

Figure 1.7 The vector $p + a$ is the position of the point represented by p displaced by the displacement represented by a.

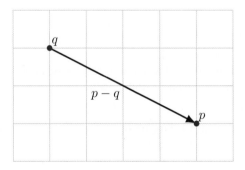

Figure 1.8 The vector $p - q$ represents the displacement from the point represented by q to the point represented by p.

- *Market clearing.* Suppose the n-vector q_i represents the quantities of n goods or resources produced (when positive) or consumed (when negative) by agent i, for $i = 1, \ldots, N$, so $(q_5)_4 = -3.2$ means that agent 5 consumes 3.2 units of resource 4. The sum $s = q_1 + \cdots + q_N$ is the n-vector of total net surplus of the resources (or shortfall, when the entries are negative). When $s = 0$, we have a closed market, which means that the total quantity of each resource produced by the agents balances the total quantity consumed. In other words, the n resources are *exchanged* among the agents. In this case we say that the *market clears* (with the resource vectors $q_1, \ldots, q_N$).

- *Audio addition.* When a and b are vectors representing audio signals over the same period of time, the sum $a + b$ is an audio signal that is perceived as containing both audio signals combined into one. If a represents a recording of a voice, and b a recording of music (of the same length), the audio signal $a + b$ will be perceived as containing both the voice recording and, simultaneously, the music.

- *Feature differences.* If f and g are n-vectors that give n feature values for two items, the difference vector $d = f - g$ gives the difference in feature values for the two objects. For example, $d_7 = 0$ means that the two objects have the same value for feature 7; $d_3 = 1.67$ means that the first object's third feature value exceeds the second object's third feature value by 1.67.

- *Time series.* If a and b represent time series of the same quantity, such as daily profit at two different stores, then $a + b$ represents a time series which is the total daily profit at the two stores. An example (with monthly rainfall) is shown in figure 1.9.

- *Portfolio trading.* Suppose s is an n-vector giving the number of shares of n assets in a portfolio, and b is an n-vector giving the number of shares of the assets that we buy (when b_i is positive) or sell (when b_i is negative). After the asset purchases and sales, our portfolio is given by $s + b$, the sum of the original portfolio vector and the purchase vector b, which is also called the *trade vector* or *trade list*. (The same interpretation works when the portfolio and trade vectors are given in dollar value.)

Addition notation in computer languages.   Some computer languages for manipulating vectors define the sum of a vector and a scalar as the vector obtained by adding the scalar to each element of the vector. This is not standard mathematical notation, however, so we will not use it. Even more confusing, in some computer languages the plus symbol is used to denote concatenation of arrays, which means putting one array after another, as in $(1, 2) + (3, 4, 5) = (1, 2, 3, 4, 5)$. While this notation might give a valid expression in some computer languages, it is not standard mathematical notation, and we will not use it in this book. In general, it is very important to distinguish between mathematical notation for vectors (which we use) and the syntax of specific computer languages or software packages for manipulating vectors.

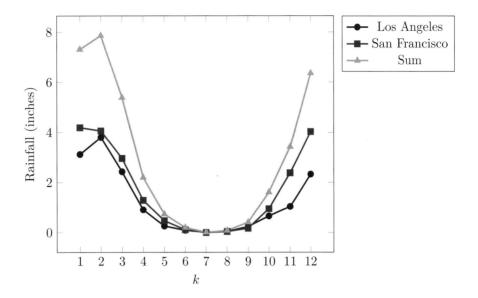

Figure 1.9 Average monthly rainfall in inches measured in downtown Los Angeles and San Francisco International Airport, and their sum. Averages are 30-year averages (1981–2010).

1.3   Scalar-vector multiplication

Another operation is *scalar multiplication* or *scalar-vector multiplication*, in which a vector is multiplied by a scalar (*i.e.*, number), which is done by multiplying every element of the vector by the scalar. Scalar multiplication is denoted by juxtaposition, typically with the scalar on the left, as in

$$(-2) \begin{bmatrix} 1 \\ 9 \\ 6 \end{bmatrix} = \begin{bmatrix} -2 \\ -18 \\ -12 \end{bmatrix}.$$

Scalar-vector multiplication can also be written with the scalar on the right, as in

$$\begin{bmatrix} 1 \\ 9 \\ 6 \end{bmatrix} (1.5) = \begin{bmatrix} 1.5 \\ 13.5 \\ 9 \end{bmatrix}.$$

The meaning is the same: It is the vector obtained by multiplying each element by the scalar. A similar notation is $a/2$, where a is a vector, meaning $(1/2)a$. The scalar-vector product $(-1)a$ is written simply as $-a$. Note that $0\,a = 0$ (where the left-hand zero is the scalar zero, and the right-hand zero is a vector zero of the same size as a).

Properties.   By definition, we have $\alpha a = a\alpha$, for any scalar α and any vector a. This is called the *commutative property* of scalar-vector multiplication; it means that scalar-vector multiplication can be written in either order.

Scalar multiplication obeys several other laws that are easy to figure out from the definition. For example, it satisfies the associative property: If a is a vector and β and γ are scalars, we have

$$(\beta\gamma)a = \beta(\gamma a).$$

On the left-hand side we see scalar-scalar multiplication $(\beta\gamma)$ and scalar-vector multiplication; on the right-hand side we see two scalar-vector products. As a consequence, we can write the vector above as $\beta\gamma a$, since it does not matter whether we interpret this as $\beta(\gamma a)$ or $(\beta\gamma)a$.

The associative property holds also when we denote scalar-vector multiplication with the scalar on the right. For example, we have $\beta(\gamma a) = (\beta a)\gamma$, and consequently we can write both as $\beta a\gamma$. As a convention, however, this vector is normally written as $\beta\gamma a$ or as $(\beta\gamma)a$.

If a is a vector and β, γ are scalars, then

$$(\beta + \gamma)a = \beta a + \gamma a.$$

(This is the left-distributive property of scalar-vector multiplication.) Scalar multiplication, like ordinary multiplication, has higher precedence in equations than vector addition, so the right-hand side here, $\beta a + \gamma a$, means $(\beta a) + (\gamma a)$. It is useful to identify the symbols appearing in this formula above. The $+$ symbol on the left is addition of scalars, while the $+$ symbol on the right denotes vector addition. When scalar multiplication is written with the scalar on the right, we have the right-distributive property:

$$a(\beta + \gamma) = a\beta + a\gamma.$$

Scalar-vector multiplication also satisfies another version of the right-distributive property:

$$\beta(a + b) = \beta a + \beta b$$

for any scalar β and any n-vectors a and b. In this equation, both of the $+$ symbols refer to the addition of n-vectors.

Examples.

- *Displacements.* When a vector a represents a displacement, and $\beta > 0$, βa is a displacement in the same direction of a, with its magnitude scaled by β. When $\beta < 0$, βa represents a displacement in the opposite direction of a, with magnitude scaled by $|\beta|$. This is illustrated in figure 1.10.

- *Materials requirements.* Suppose the n-vector q is the bill of materials for producing one unit of some product, *i.e.*, q_i is the amount of raw material required to produce one unit of product. To produce α units of the product will then require raw materials given by αq. (Here we assume that $\alpha \geq 0$.)

- *Audio scaling.* If a is a vector representing an audio signal, the scalar-vector product βa is perceived as the same audio signal, but changed in volume (loudness) by the factor $|\beta|$. For example, when $\beta = 1/2$ (or $\beta = -1/2$), βa is perceived as the same audio signal, but quieter.

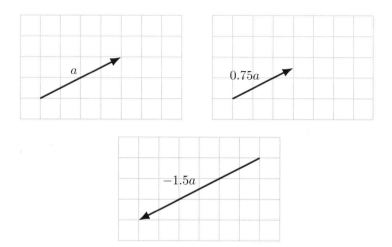

Figure 1.10 The vector $0.75a$ represents the displacement in the direction of the displacement a, with magnitude scaled by 0.75; $(-1.5)a$ represents the displacement in the opposite direction, with magnitude scaled by 1.5.

Linear combinations.   If $a_1, \ldots, a_m$ are n-vectors, and $\beta_1, \ldots, \beta_m$ are scalars, the n-vector
$$\beta_1 a_1 + \cdots + \beta_m a_m$$
is called a *linear combination* of the vectors $a_1, \ldots, a_n$. The scalars $\beta_1, \ldots, \beta_m$ are called the *coefficients* of the linear combination.

Linear combination of unit vectors.   We can write any n-vector b as a linear combination of the standard unit vectors, as
$$b = b_1 e_1 + \cdots + b_n e_n. \tag{1.1}$$
In this equation b_i is the ith entry of b (*i.e.*, a scalar), and e_i is the ith unit vector. In the linear combination of $e_1, \ldots, e_n$ given in (1.1), the coefficients are the entries of the vector b. A specific example is
$$\begin{bmatrix} -1 \\ 3 \\ 5 \end{bmatrix} = (-1) \begin{bmatrix} 1 \\ 0 \\ 0 \end{bmatrix} + 3 \begin{bmatrix} 0 \\ 1 \\ 0 \end{bmatrix} + 5 \begin{bmatrix} 0 \\ 0 \\ 1 \end{bmatrix}.$$

Special linear combinations.   Some linear combinations of the vectors $a_1, \ldots, a_m$ have special names. For example, the linear combination with $\beta_1 = \cdots = \beta_m = 1$, given by $a_1 + \cdots + a_m$, is the *sum* of the vectors, and the linear combination with $\beta_1 = \cdots = \beta_m = 1/m$, given by $(1/m)(a_1 + \cdots + a_m)$, is the *average* of the vectors. When the coefficients sum to one, *i.e.*, $\beta_1 + \cdots + \beta_m = 1$, the linear combination is called an *affine combination*. When the coefficients in an affine combination are nonnegative, it is called a *convex combination*, a *mixture*, or a *weighted average*. The coefficients in an affine or convex combination are sometimes given as percentages, which add up to 100%.

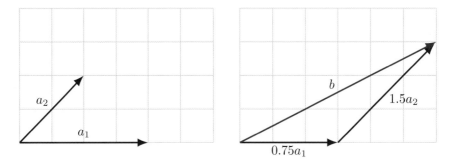

Figure 1.11 *Left.* Two 2-vectors a_1 and a_2. *Right.* The linear combination $b = 0.75a_1 + 1.5a_2$.

Examples.

- *Displacements.* When the vectors represent displacements, a linear combination is the sum of the scaled displacements. This is illustrated in figure 1.11.

- *Audio mixing.* When $a_1, \ldots, a_m$ are vectors representing audio signals (over the same period of time, for example, simultaneously recorded), they are called *tracks.* The linear combination $\beta_1 a_1 + \cdots + \beta_m a_m$ is perceived as a mixture (also called a *mix*) of the audio tracks, with relative loudness given by $|\beta_1|, \ldots, |\beta_m|$. A producer in a studio, or a sound engineer at a live show, chooses values of $\beta_1, \ldots, \beta_m$ to give a good balance between the different instruments, vocals, and drums.

- *Cash flow replication.* Suppose that $c_1, \ldots, c_m$ are vectors that represent cash flows, such as particular types of loans or investments. The linear combination $f = \beta_1 c_1 + \cdots + \beta_m c_m$ represents another cash flow. We say that the cash flow f has been *replicated* by the (linear combination of the) original cash flows $c_1, \ldots, c_m$. As an example, $c_1 = (1, -1.1, 0)$ represents a \$1 loan from period 1 to period 2 with 10% interest, and $c_2 = (0, 1, -1.1)$ represents a \$1 loan from period 2 to period 3 with 10% interest. The linear combination

$$d = c_1 + 1.1c_2 = (1, 0, -1.21)$$

  represents a two period loan of \$1 in period 1, with compounded 10% interest. Here we have replicated a two period loan from two one period loans.

- *Line and segment.* When a and b are different n-vectors, the affine combination $c = (1 - \theta)a + \theta b$, where θ is a scalar, describes a point on the *line* passing through a and b. When $0 \le \theta \le 1$, c is a convex combination of a and b, and is said to lie on the *segment* between a and b. For $n = 2$ and $n = 3$, with the vectors representing coordinates of 2-D or 3-D points, this agrees with the usual geometric notion of line and segment. But we can also talk about the line passing through two vectors of dimension 100. This is illustrated in figure 1.12.

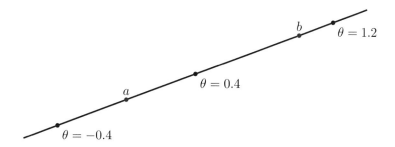

Figure 1.12 The affine combination $(1 - \theta)a + \theta b$ for different values of θ. These points are on the line passing through a and b; for θ between 0 and 1, the points are on the line segment between a and b.

1.4 Inner product

The (standard) *inner product* (also called *dot product*) of two n-vectors is defined as the scalar

$$a^T b = a_1 b_1 + a_2 b_2 + \cdots + a_n b_n,$$

the sum of the products of corresponding entries. (The origin of the superscript 'T' in the inner product notation $a^T b$ will be explained in chapter 6.) Some other notations for the inner product (that we will not use in this book) are $\langle a, b \rangle$, $\langle a | b \rangle$, (a, b), and $a \cdot b$. (In the notation used in this book, (a, b) denotes a stacked vector of length $2n$.) As you might guess, there is also a vector *outer product*, which we will encounter later, in §10.1. As a specific example of the inner product, we have

$$\begin{bmatrix} -1 \\ 2 \\ 2 \end{bmatrix}^T \begin{bmatrix} 1 \\ 0 \\ -3 \end{bmatrix} = (-1)(1) + (2)(0) + (2)(-3) = -7.$$

When $n = 1$, the inner product reduces to the usual product of two numbers.

Properties. The inner product satisfies some simple properties that are easily verified from the definition. If a, b, and c are vectors of the same size, and γ is a scalar, we have the following.

- *Commutativity.* $a^T b = b^T a$. The order of the two vector arguments in the inner product does not matter.

- *Associativity with scalar multiplication.* $(\gamma a)^T b = \gamma(a^T b)$, so we can write both as $\gamma a^T b$.

- *Distributivity with vector addition.* $(a+b)^T c = a^T c + b^T c$. The inner product can be distributed across vector addition.

These can be combined to obtain other identities, such as $a^T(\gamma b) = \gamma(a^T b)$, or $a^T(b + \gamma c) = a^T b + \gamma a^T c$. As another useful example, we have, for any vectors a, b, c, d of the same size,

$$(a + b)^T (c + d) = a^T c + a^T d + b^T c + b^T d.$$

This formula expresses an inner product on the left-hand side as a sum of four inner products on the right-hand side, and is analogous to expanding a product of sums in algebra. Note that on the left-hand side, the two addition symbols refer to vector addition, whereas on the right-hand side, the three addition symbols refer to scalar (number) addition.

General examples.

- *Unit vector.* $e_i^T a = a_i$. The inner product of a vector with the ith standard unit vector gives (or 'picks out') the ith element a.

- *Sum.* $\mathbf{1}^T a = a_1 + \cdots + a_n$. The inner product of a vector with the vector of ones gives the sum of the elements of the vector.

- *Average.* $(\mathbf{1}/n)^T a = (a_1 + \cdots + a_n)/n$. The inner product of an n-vector with the vector $\mathbf{1}/n$ gives the average or mean of the elements of the vector. The average of the entries of a vector is denoted by $\mathbf{avg}(x)$. The Greek letter μ is a traditional symbol used to denote the average or mean.

- *Sum of squares.* $a^T a = a_1^2 + \cdots + a_n^2$. The inner product of a vector with itself gives the sum of the squares of the elements of the vector.

- *Selective sum.* Let b be a vector all of whose entries are either 0 or 1. Then $b^T a$ is the sum of the elements in a for which $b_i = 1$.

Block vectors. If the vectors a and b are block vectors, and the corresponding blocks have the same sizes (in which case we say they *conform*), then we have

$$
a^T b = \begin{bmatrix} a_1 \\ \vdots \\ a_k \end{bmatrix}^T \begin{bmatrix} b_1 \\ \vdots \\ b_k \end{bmatrix} = a_1^T b_1 + \cdots + a_k^T b_k.
$$

The inner product of block vectors is the sum of the inner products of the blocks.

Applications. The inner product is useful in many applications, a few of which we list here.

- *Co-occurrence.* If a and b are n-vectors that describe occurrence, *i.e.*, each of their elements is either 0 or 1, then $a^T b$ gives the total number of indices for which a_i and b_i are both one, that is, the total number of co-occurrences. If we interpret the vectors a and b as describing subsets of n objects, then $a^T b$ gives the number of objects in the intersection of the two subsets. This is illustrated in figure 1.13, for two subsets A and B of 7 objects, labeled $1, \ldots, 7$, with corresponding occurrence vectors

$$
a = (0, 1, 1, 1, 1, 1, 1), \qquad b = (1, 0, 1, 0, 1, 0, 0).
$$

Here we have $a^T b = 2$, which is the number of objects in both A and B (*i.e.*, objects 3 and 5).

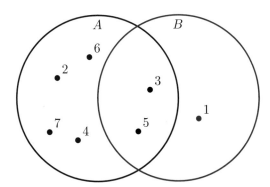

Figure 1.13 Two sets A and B, containing seven objects.

- *Weights, features, and score.* When the vector f represents a set of features of an object, and w is a vector of the same size (often called a *weight vector*), the inner product $w^T f$ is the sum of the feature values, scaled (or weighted) by the weights, and is sometimes called a *score*. For example, if the features are associated with a loan applicant (*e.g.*, age, income, ...), we might interpret $s = w^T f$ as a credit score. In this example we can interpret w_i as the weight given to feature i in forming the score.

- *Price-quantity.* If p represents a vector of prices of n goods, and q is a vector of quantities of the n goods (say, the bill of materials for a product), then their inner product $p^T q$ is the total cost of the goods given by the vector q.

- *Speed-time.* A vehicle travels over n segments with constant speed in each segment. Suppose the n-vector s gives the speed in the segments, and the n-vector t gives the times taken to traverse the segments. Then $s^T t$ is the total distance traveled.

- *Probability and expected values.* Suppose the n-vector p has nonnegative entries that sum to one, so it describes a set of proportions among n items, or a set of probabilities of n outcomes, one of which must occur. Suppose f is another n-vector, where we interpret f_i as the value of some quantity if outcome i occurs. Then $f^T p$ gives the expected value or mean of the quantity, under the probabilities (or fractions) given by p.

- *Polynomial evaluation.* Suppose the n-vector c represents the coefficients of a polynomial p of degree $n-1$ or less:

$$p(x) = c_1 + c_2 x + \cdots + c_{n-1} x^{n-2} + c_n x^{n-1}.$$

Let t be a number, and let $z = (1, t, t^2, \ldots, t^{n-1})$ be the n-vector of powers of t. Then $c^T z = p(t)$, the value of the polynomial p at the point t. So the inner product of a polynomial coefficient vector and vector of powers of a number evaluates the polynomial at the number.

- *Discounted total.* Let c be an n-vector representing a cash flow, with c_i the cash received (when $c_i > 0$) in period i. Let d be the n-vector defined as

$$d = (1, 1/(1+r), \ldots, 1/(1+r)^{n-1}),$$

where $r \geq 0$ is an interest rate. Then

$$d^T c = c_1 + c_2/(1+r) + \cdots + c_n/(1+r)^{n-1}$$

is the discounted total of the cash flow, *i.e.*, its *net present value* (NPV), with interest rate r.

- *Portfolio value.* Suppose s is an n-vector representing the holdings in shares of a portfolio of n different assets, with negative values meaning short positions. If p is an n-vector giving the prices of the assets, then $p^T s$ is the total (or net) value of the portfolio.

- *Portfolio return.* Suppose r is the vector of (fractional) returns of n assets over some time period, *i.e.*, the asset relative price changes

$$r_i = \frac{p_i^{\text{final}} - p_i^{\text{initial}}}{p_i^{\text{initial}}}, \quad i = 1, \ldots, n,$$

where p_i^{initial} and p_i^{final} are the (positive) prices of asset i at the beginning and end of the investment period. If h is an n-vector giving our portfolio, with h_i denoting the dollar value of asset i held, then the inner product $r^T h$ is the total return of the portfolio, in dollars, over the period. If w represents the fractional (dollar) holdings of our portfolio, then $r^T w$ gives the total return of the portfolio. For example, if $r^T w = 0.09$, then our portfolio return is 9%. If we had invested $10000 initially, we would have earned $900.

- *Document sentiment analysis.* Suppose the n-vector x represents the histogram of word occurrences in a document, from a dictionary of n words. Each word in the dictionary is assigned to one of three sentiment categories: *Positive*, *Negative*, and *Neutral*. The list of positive words might include 'nice' and 'superb'; the list of negative words might include 'bad' and 'terrible'. Neutral words are those that are neither positive nor negative. We encode the word categories as an n-vector w, with $w_i = 1$ if word i is positive, with $w_i = -1$ if word i is negative, and $w_i = 0$ if word i is neutral. The number $w^T x$ gives a (crude) measure of the sentiment of the document.

1.5   Complexity of vector computations

Computer representation of numbers and vectors. Real numbers are stored in computers using *floating point format*, which represents a real number using a block of 64 *bits* (0s and 1s), or 8 *bytes* (groups of 8 bits). Each of the 2^{64} possible sequences of bits corresponds to a specific real number. The floating point numbers span a very wide range of values, and are very closely spaced, so numbers that arise in applications can be approximated as floating point numbers to an accuracy of around 10 digits, which is good enough for almost all practical applications. Integers are stored in a more compact format, and are represented exactly.

Vectors are stored as arrays of floating point numbers (or integers, when the entries are all integers). Storing an n-vector requires $8n$ bytes to store. Current memory and storage devices, with capacities measured in many gigabytes (10^9 bytes), can easily store vectors with dimensions in the millions or billions. Sparse vectors are stored in a more efficient way that keeps track of indices and values of the nonzero entries.

Floating point operations.   When computers carry out addition, subtraction, multiplication, division, or other arithmetic operations on numbers represented in floating point format, the result is rounded to the nearest floating point number. These operations are called *floating point operations*. The very small error in the computed result is called (floating point) *round-off error*. In most applications, these very small errors have no practical effect. Floating point round-off errors, and methods to mitigate their effect, are studied in a field called *numerical analysis*. In this book we will not consider floating point round-off error, but you should be aware that it exists. For example, when a computer evaluates the left-hand and right-hand sides of a mathematical identity, you should not be surprised if the two numbers are not equal. They should, however, be very close.

Flop counts and complexity.   So far we have seen only a few vector operations, like scalar multiplication, vector addition, and the inner product. How quickly these operations can be carried out by a computer depends very much on the computer hardware and software, and the size of the vector.

A very rough estimate of the time required to carry out some computation, such as an inner product, can be found by counting the total number of floating point operations, or FLOPs. This term is in such common use that the acronym is now written in lower case letters, as flops, and the speed with which a computer can carry out flops is expressed in Gflop/s (gigaflops per second, *i.e.*, billions of flops per second). Typical current values are in the range of 1–10 Gflop/s, but this can vary by several orders of magnitude. The actual time it takes a computer to carry out some computation depends on many other factors beyond the total number of flops required, so time estimates based on counting flops are very crude, and are not meant to be more accurate than a factor of ten or so. For this reason, gross approximations (such as ignoring a factor of 2) can be used when counting the flops required in a computation.

The *complexity* of an operation is the number of flops required to carry it out, as a function of the size or sizes of the input to the operation. Usually the complexity is highly simplified, dropping terms that are small or negligible (compared to other terms) when the sizes of the inputs are large. In theoretical computer science, the term 'complexity' is used in a different way, to mean the number of flops of the best method to carry out the computation, *i.e.*, the one that requires the fewest flops. In this book, we use the term complexity to mean the number of flops required by a specific method.

Complexity of vector operations. Scalar-vector multiplication ax, where x is an n-vector, requires n multiplications, *i.e.*, ax_i for $i = 1, \ldots, n$. Vector addition $x + y$ of two n-vectors takes n additions, *i.e.*, $x_i + y_i$ for $i = 1, \ldots, n$. Computing the inner product $x^T y = x_1 y_1 + \cdots + x_n y_n$ of two n-vectors takes $2n - 1$ flops, n scalar multiplications and $n - 1$ scalar additions. So scalar multiplication, vector addition, and the inner product of n-vectors require n, n, and $2n - 1$ flops, respectively. We only need an estimate, so we simplify the last to $2n$ flops, and say that the *complexity* of scalar multiplication, vector addition, and the inner product of n-vectors is n, n, and $2n$ flops, respectively. We can guess that a 1 Gflop/s computer can compute the inner product of two vectors of size one million in around one thousandth of a second, but we should not be surprised if the actual time differs by a factor of 10 from this value.

The *order* of the computation is obtained by ignoring any constant that multiplies a power of the dimension. So we say that the three vector operations scalar multiplication, vector addition, and inner product have order n. Ignoring the factor of 2 dropped in the actual complexity of the inner product is reasonable, since we do not expect flop counts to predict the running time with an accuracy better than a factor of 2. The order is useful in understanding how the time to execute the computation will scale when the size of the operands changes. An order n computation should take around 10 times longer to carry out its computation on an input that is 10 times bigger.

Complexity of sparse vector operations. If x is sparse, then computing ax requires $\mathbf{nnz}(x)$ flops. If x and y are sparse, computing $x + y$ requires no more than $\min\{\mathbf{nnz}(x), \mathbf{nnz}(y)\}$ flops (since no arithmetic operations are required to compute $(x + y)_i$ when either x_i or y_i is zero). If the sparsity patterns of x and y do not overlap (intersect), then zero flops are needed to compute $x + y$. The inner product calculation is similar: computing $x^T y$ requires no more than $2 \min\{\mathbf{nnz}(x), \mathbf{nnz}(y)\}$ flops. When the sparsity patterns of x and y do not overlap, computing $x^T y$ requires zero flops, since $x^T y = 0$ in this case.

Exercises

1.1 *Vector equations.* Determine whether each of the equations below is true, false, or contains bad notation (and therefore does not make sense).

(a) $\begin{bmatrix} 1 \\ 2 \\ 1 \end{bmatrix} = (1, 2, 1)$.

(b) $\begin{bmatrix} 1 \\ 2 \\ 1 \end{bmatrix} = \begin{bmatrix} 1, & 2, & 1 \end{bmatrix}$.

(c) $(1, (2, 1)) = ((1, 2), 1)$.

1.2 *Vector notation.* Which of the following expressions uses correct notation? When the expression does make sense, give its length. In the following, a and b are 10-vectors, and c is a 20-vector.

(a) $a + b - c_{3:12}$.

(b) $(a, b, c_{3:13})$.

(c) $2a + c$.

(d) $(a, 1) + (c_1, b)$.

(e) $((a, b), a)$.

(f) $\begin{bmatrix} a & b \end{bmatrix} + 4c$.

(g) $\begin{bmatrix} a \\ b \end{bmatrix} + 4c$.

1.3 *Overloading.* Which of the following expressions uses correct notation? If the notation is correct, is it also unambiguous? Assume that a is a 10-vector and b is a 20-vector.

(a) $b = (0, a)$.

(b) $a = (0, b)$.

(c) $b = (0, a, 0)$.

(d) $a = 0 = b$.

1.4 *Periodic energy usage.* The 168-vector w gives the hourly electricity consumption of a manufacturing plant, starting on Sunday midnight to 1AM, over one week, in MWh (megawatt-hours). The consumption pattern is the same each day, *i.e.*, it is 24-periodic, which means that $w_{t+24} = w_t$ for $t = 1, \ldots, 144$. Let d be the 24-vector that gives the energy consumption over one day, starting at midnight.

(a) Use vector notation to express w in terms of d.

(b) Use vector notation to express d in terms of w.

1.5 *Interpreting sparsity.* Suppose the n-vector x is sparse, *i.e.*, has only a few nonzero entries. Give a short sentence or two explaining what this means in each of the following contexts.

(a) x represents the daily cash flow of some business over n days.

(b) x represents the annual dollar value purchases by a customer of n products or services.

(c) x represents a portfolio, say, the dollar value holdings of n stocks.

(d) x represents a bill of materials for a project, *i.e.*, the amounts of n materials needed.

(e) x represents a monochrome image, *i.e.*, the brightness values of n pixels.

(f) x is the daily rainfall in a location over one year.

1.6 *Vector of differences.* Suppose x is an n-vector. The associated vector of differences is the $(n-1)$-vector d given by $d = (x_2 - x_1, x_3 - x_2, \ldots, x_n - x_{n-1})$. Express d in terms of x using vector operations (*e.g.*, slicing notation, sum, difference, linear combinations, inner product). The difference vector has a simple interpretation when x represents a time series. For example, if x gives the daily value of some quantity, d gives the day-to-day changes in the quantity.

1.7 *Transforming between two encodings for Boolean vectors.* A Boolean n-vector is one for which all entries are either 0 or 1. Such vectors are used to encode whether each of n conditions holds, with $a_i = 1$ meaning that condition i holds. Another common encoding of the same information uses the two values -1 and $+1$ for the entries. For example the Boolean vector $(0, 1, 1, 0)$ would be written using this alternative encoding as $(-1, +1, +1, -1)$. Suppose that x is a Boolean vector with entries that are 0 or 1, and y is a vector encoding the same information using the values -1 and $+1$. Express y in terms of x using vector notation. Also, express x in terms of y using vector notation.

1.8 *Profit and sales vectors.* A company sells n different products or items. The n-vector p gives the profit, in dollars per unit, for each of the n items. (The entries of p are typically positive, but a few items might have negative entries. These items are called *loss leaders*, and are used to increase customer engagement in the hope that the customer will make other, profitable purchases.) The n-vector s gives the total sales of each of the items, over some period (such as a month), *i.e.*, s_i is the total number of units of item i sold. (These are also typically nonnegative, but negative entries can be used to reflect items that were purchased in a previous time period and returned in this one.) Express the total profit in terms of p and s using vector notation.

1.9 *Symptoms vector.* A 20-vector s records whether each of 20 different symptoms is present in a medical patient, with $s_i = 1$ meaning the patient has the symptom and $s_i = 0$ meaning she does not. Express the following using vector notation.

(a) The total number of symptoms the patient has.

(b) The patient exhibits five out of the first ten symptoms.

1.10 *Total score from course record.* The record for each student in a class is given as a 10-vector r, where $r_1, \ldots, r_8$ are the grades for the 8 homework assignments, each on a 0–10 scale, r_9 is the midterm exam grade on a 0–120 scale, and r_{10} is final exam score on a 0–160 scale. The student's total course score s, on a 0–100 scale, is based 25% on the homework, 35% on the midterm exam, and 40% on the final exam. Express s in the form $s = w^T r$. (That is, determine the 10-vector w.) You can give the coefficients of w to 4 digits after the decimal point.

1.11 *Word count and word count histogram vectors.* Suppose the n-vector w is the word count vector associated with a document and a dictionary of n words. For simplicity we will assume that all words in the document appear in the dictionary.

(a) What is $\mathbf{1}^T w$?

(b) What does $w_{282} = 0$ mean?

(c) Let h be the n-vector that gives the histogram of the word counts, *i.e.*, h_i is the fraction of the words in the document that are word i. Use vector notation to express h in terms of w. (You can assume that the document contains at least one word.)

1.12 *Total cash value.* An international company holds cash in five currencies: USD (US dollar), RMB (Chinese yuan), EUR (euro), GBP (British pound), and JPY (Japanese yen), in amounts given by the 5-vector c. For example, c_2 gives the number of RMB held. Negative entries in c represent liabilities or amounts owed. Express the total (net) value of the cash in USD, using vector notation. Be sure to give the size and define the entries of any vectors that you introduce in your solution. Your solution can refer to currency exchange rates.

1.13 *Average age in a population.* Suppose the 100-vector x represents the distribution of ages in some population of people, with x_i being the number of $i-1$ year olds, for $i = 1, \ldots, 100$. (You can assume that $x \neq 0$, and that there is no one in the population over age 99.) Find expressions, using vector notation, for the following quantities.

(a) The total number of people in the population.

(b) The total number of people in the population age 65 and over.

(c) The average age of the population. (You can use ordinary division of numbers in your expression.)

1.14 *Industry or sector exposure.* Consider a set of n assets or stocks that we invest in. Let f be an n-vector that encodes whether each asset is in some specific industry or sector, *e.g.*, pharmaceuticals or consumer electronics. Specifically, we take $f_i = 1$ if asset i is in the sector, and $f_i = 0$ if it is not. Let the n-vector h denote a portfolio, with h_i the dollar value held in asset i (with negative meaning a short position). The inner product $f^T h$ is called the (dollar value) *exposure* of our portfolio to the sector. It gives the net dollar value of the portfolio that is invested in assets from the sector. A portfolio h is called *neutral* (to a sector or industry) if $f^T h = 0$.

A portfolio h is called *long only* if each entry is nonnegative, *i.e.*, $h_i \geq 0$ for each i. This means the portfolio does not include any short positions.

What does it mean if a long-only portfolio is neutral to a sector, say, pharmaceuticals? Your answer should be in simple English, but you should back up your conclusion with an argument.

1.15 *Cheapest supplier.* You must buy n raw materials in quantities given by the n-vector q, where q_i is the amount of raw material i that you must buy. A set of K potential suppliers offer the raw materials at prices given by the n-vectors $p_1, \ldots, p_K$. (Note that p_k is an n-vector; $(p_k)_i$ is the price that supplier k charges per unit of raw material i.) We will assume that all quantities and prices are positive.

If you must choose just one supplier, how would you do it? Your answer should use vector notation.

A (highly paid) consultant tells you that you might do better (*i.e.*, get a better total cost) by splitting your order into two, by choosing two suppliers and ordering $(1/2)q$ (*i.e.*, half the quantities) from each of the two. He argues that having a diversity of suppliers is better. Is he right? If so, explain how to find the two suppliers you would use to fill half the order.

1.16 *Inner product of nonnegative vectors.* A vector is called *nonnegative* if all its entries are nonnegative.

(a) Explain why the inner product of two nonnegative vectors is nonnegative.

(b) Suppose the inner product of two nonnegative vectors is zero. What can you say about them? Your answer should be in terms of their respective sparsity patterns, *i.e.*, which entries are zero and nonzero.

1.17 *Linear combinations of cash flows.* We consider cash flow vectors over T time periods, with a positive entry meaning a payment received, and negative meaning a payment made. A (unit) *single period loan*, at time period t, is the T-vector l_t that corresponds to a payment received of \$1 in period t and a payment made of \$$(1 + r)$ in period $t + 1$, with all other payments zero. Here $r > 0$ is the interest rate (over one period).

Let c be a \$1 $T - 1$ period loan, starting at period 1. This means that \$1 is received in period 1, \$$(1 + r)^{T-1}$ is paid in period T, and all other payments (*i.e.*, $c_2, \ldots, c_{T-1}$) are zero. Express c as a linear combination of single period loans.

1.18 *Linear combinations of linear combinations.* Suppose that each of the vectors $b_1, \ldots, b_k$ is a linear combination of the vectors $a_1, \ldots, a_m$, and c is a linear combination of $b_1, \ldots, b_k$. Then c is a linear combination of $a_1, \ldots, a_m$. Show this for the case with $m = k = 2$. (Showing it in general is not much more difficult, but the notation gets more complicated.)

1.19 *Auto-regressive model.* Suppose that $z_1, z_2, \ldots$ is a time series, with the number z_t giving the value in period or time t. For example z_t could be the gross sales at a particular store on day t. An *auto-regressive* (AR) model is used to predict z_{t+1} from the previous M values, $z_t, z_{t-1}, \ldots, z_{t-M+1}$:

$$\hat{z}_{t+1} = (z_t, z_{t-1}, \ldots, z_{t-M+1})^T \beta, \quad t = M, M+1, \ldots.$$

Here $\hat{z}_{t+1}$ denotes the AR model's prediction of z_{t+1}, M is the memory length of the AR model, and the M-vector β is the AR model coefficient vector. For this problem we will assume that the time period is daily, and $M = 10$. Thus, the AR model predicts tomorrow's value, given the values over the last 10 days.

For each of the following cases, give a short interpretation or description of the AR model in English, without referring to mathematical concepts like vectors, inner product, and so on. You can use words like 'yesterday' or 'today'.

(a) $\beta \approx e_1$.

(b) $\beta \approx 2e_1 - e_2$.

(c) $\beta \approx e_6$.

(d) $\beta \approx 0.5e_1 + 0.5e_2$.

1.20 How many bytes does it take to store 100 vectors of length 10^5? How many flops does it take to form a linear combination of them (with 100 nonzero coefficients)? About how long would this take on a computer capable of carrying out 1 Gflop/s?

Chapter 2

Linear functions

In this chapter we introduce linear and affine functions, and describe some common settings where they arise, including regression models.

2.1 Linear functions

Function notation. The notation $f : \mathbf{R}^n \to \mathbf{R}$ means that f is a *function* that maps real n-vectors to real numbers, *i.e.*, it is a scalar-valued function of n-vectors. If x is an n-vector, then $f(x)$, which is a scalar, denotes the *value* of the function f at x. (In the notation $f(x)$, x is referred to as the *argument* of the function.) We can also interpret f as a function of n scalar arguments, the entries of the vector argument, in which case we write $f(x)$ as

$$f(x) = f(x_1, x_2, \ldots, x_n).$$

Here we refer to $x_1, \ldots, x_n$ as the arguments of f. We sometimes say that f is real-valued, or scalar-valued, to emphasize that $f(x)$ is a real number or scalar.

To describe a function $f : \mathbf{R}^n \to \mathbf{R}$, we have to specify what its value is for any possible argument $x \in \mathbf{R}^n$. For example, we can define a function $f : \mathbf{R}^4 \to \mathbf{R}$ by

$$f(x) = x_1 + x_2 - x_4^2$$

for any 4-vector x. In words, we might describe f as the sum of the first two elements of its argument, minus the square of the last entry of the argument. (This particular function does not depend on the third element of its argument.)

Sometimes we introduce a function without formally assigning a symbol for it, by directly giving a formula for its value in terms of its arguments, or describing how to find its value from its arguments. An example is the *sum function*, whose value is $x_1 + \cdots + x_n$. We can give a name to the value of the function, as in $y = x_1 + \cdots + x_n$, and say that y is a function of x, in this case, the sum of its entries.

Many functions are not given by formulas or equations. As an example, suppose $f : \mathbf{R}^3 \to \mathbf{R}$ is the function that gives the lift (vertical upward force) on a particular airplane, as a function of the 3-vector x, where x_1 is the angle of attack of the airplane (*i.e.*, the angle between the airplane body and its direction of motion), x_2 is its air speed, and x_3 is the air density.

The inner product function.   Suppose a is an n-vector. We can define a scalar-valued function f of n-vectors, given by

$$f(x) = a^T x = a_1 x_1 + a_2 x_2 + \cdots + a_n x_n \tag{2.1}$$

for any n-vector x. This function gives the inner product of its n-vector argument x with some (fixed) n-vector a. We can also think of f as forming a weighted sum of the elements of x; the elements of a give the weights used in forming the weighted sum.

Superposition and linearity.   The inner product function f defined in (2.1) satisfies the property

$$
\begin{aligned}
f(\alpha x + \beta y) &= a^T(\alpha x + \beta y) \\
&= a^T(\alpha x) + a^T(\beta y) \\
&= \alpha(a^T x) + \beta(a^T y) \\
&= \alpha f(x) + \beta f(y)
\end{aligned}
$$

for all n-vectors x, y, and all scalars α, β. This property is called *superposition*. A function that satisfies the superposition property is called *linear*. We have just shown that the inner product with a fixed vector is a linear function.

The superposition equality

$$f(\alpha x + \beta y) = \alpha f(x) + \beta f(y) \tag{2.2}$$

looks deceptively simple; it is easy to read it as just a re-arrangement of the parentheses and the order of a few terms. But in fact it says a lot. On the left-hand side, the term $\alpha x + \beta y$ involves *scalar-vector* multiplication and *vector addition*. On the right-hand side, $\alpha f(x) + \beta f(y)$ involves ordinary *scalar multiplication* and *scalar addition*.

If a function f is linear, superposition extends to linear combinations of any number of vectors, and not just linear combinations of two vectors: We have

$$f(\alpha_1 x_1 + \cdots + \alpha_k x_k) = \alpha_1 f(x_1) + \cdots + \alpha_k f(x_k),$$

for any n vectors $x_1, \ldots, x_k$, and any scalars $\alpha_1, \ldots, \alpha_k$. (This more general k-term form of superposition reduces to the two-term form given above when $k = 2$.) To see this, we note that

$$
\begin{aligned}
f(\alpha_1 x_1 + \cdots + \alpha_k x_k) &= \alpha_1 f(x_1) + f(\alpha_2 x_2 + \cdots + \alpha_k x_k) \\
&= \alpha_1 f(x_1) + \alpha_2 f(x_2) + f(\alpha_3 x_3 + \cdots + \alpha_k x_k) \\
&\ \ \vdots \\
&= \alpha_1 f(x_1) + \cdots + \alpha_k f(x_k).
\end{aligned}
$$

In the first line here, we apply (two-term) superposition to the argument

$$\alpha_1 x_1 + (1)(\alpha_2 x_2 + \cdots + \alpha_k x_k),$$

and in the other lines we apply this recursively.

The superposition equality (2.2) is sometimes broken down into two properties, one involving the scalar-vector product and one involving vector addition in the argument. A function $f : \mathbf{R}^n \to \mathbf{R}$ is linear if it satisfies the following two properties.

- *Homogeneity.* For any n-vector x and any scalar α, $f(\alpha x) = \alpha f(x)$.

- *Additivity.* For any n-vectors x and y, $f(x + y) = f(x) + f(y)$.

Homogeneity states that scaling the (vector) argument is the same as scaling the function value; additivity says that adding (vector) arguments is the same as adding the function values.

Inner product representation of a linear function.   We saw above that a function defined as the inner product of its argument with some fixed vector is linear. The converse is also true: If a function is linear, then it can be expressed as the inner product of its argument with some fixed vector.

Suppose f is a scalar-valued function of n-vectors, and is linear, *i.e.*, (2.2) holds for all n-vectors x, y, and all scalars α, β. Then there is an n-vector a such that $f(x) = a^T x$ for all x. We call $a^T x$ the *inner product representation* of f.

To see this, we use the identity (1.1) to express an arbitrary n-vector x as $x = x_1 e_1 + \cdots + x_n e_n$. If f is linear, then by multi-term superposition we have

$$
\begin{aligned}
f(x) &= f(x_1 e_1 + \cdots + x_n e_n) \\
     &= x_1 f(e_1) + \cdots + x_n f(e_n) \\
     &= a^T x,
\end{aligned}
$$

with $a = (f(e_1), f(e_2), \ldots, f(e_n))$. The formula just derived,

$$f(x) = x_1 f(e_1) + x_2 f(e_2) + \cdots + x_n f(e_n) \tag{2.3}$$

which holds for any linear scalar-valued function f, has several interesting implications. Suppose, for example, that the linear function f is given as a subroutine (or a physical system) that computes (or results in the output) $f(x)$ when we give the argument (or input) x. Once we have found $f(e_1)$, ..., $f(e_n)$, by n calls to the subroutine (or n experiments), we can predict (or simulate) what $f(x)$ will be, for *any* vector x, using the formula (2.3).

The representation of a linear function f as $f(x) = a^T x$ is *unique*, which means that there is only one vector a for which $f(x) = a^T x$ holds for all x. To see this, suppose that we have $f(x) = a^T x$ for all x, and also $f(x) = b^T x$ for all x. Taking $x = e_i$, we have $f(e_i) = a^T e_i = a_i$, using the formula $f(x) = a^T x$. Using the formula $f(x) = b^T x$, we have $f(e_i) = b^T e_i = b_i$. These two numbers must be the same, so we have $a_i = b_i$. Repeating this argument for $i = 1, \ldots, n$, we conclude that the corresponding elements in a and b are the same, so $a = b$.

Examples.

- *Average.* The *mean* or *average* value of an n-vector is defined as

$$f(x) = (x_1 + x_2 + \cdots + x_n)/n,$$

  and is denoted $\mathbf{avg}(x)$ (and sometimes $\bar{x}$). The average of a vector is a linear function. It can be expressed as $\mathbf{avg}(x) = a^T x$ with

$$a = (1/n, \ldots, 1/n) = \mathbf{1}/n.$$

- *Maximum.* The maximum element of an n-vector x, $f(x) = \max\{x_1, \ldots, x_n\}$, is not a linear function (except when $n = 1$). We can show this by a counterexample for $n = 2$. Take $x = (1, -1)$, $y = (-1, 1)$, $\alpha = 1/2$, $\beta = 1/2$. Then

$$f(\alpha x + \beta y) = 0 \neq \alpha f(x) + \beta f(y) = 1.$$

Affine functions.   A linear function plus a constant is called an *affine* function. A function $f : \mathbf{R}^n \to \mathbf{R}$ is affine if and only if it can be expressed as $f(x) = a^T x + b$ for some n-vector a and scalar b, which is sometimes called the *offset*. For example, the function on 3-vectors defined by

$$f(x) = 2.3 - 2x_1 + 1.3x_2 - x_3,$$

is affine, with $b = 2.3$, $a = (-2, 1.3, -1)$.

Any affine scalar-valued function satisfies the following variation on the superposition property:

$$f(\alpha x + \beta y) = \alpha f(x) + \beta f(y),$$

for all n-vectors x, y, and all scalars α, β that satisfy $\alpha + \beta = 1$. For linear functions, superposition holds for *any* coefficients α and β; for affine functions, it holds *when the coefficients sum to one* (*i.e.*, when the argument is an affine combination).

To see that the restricted superposition property holds for an affine function $f(x) = a^T x + b$, we note that, for any vectors x, y and scalars α and β that satisfy $\alpha + \beta = 1$,

$$\begin{aligned}
f(\alpha x + \beta y) &= a^T(\alpha x + \beta y) + b \\
&= \alpha a^T x + \beta a^T y + (\alpha + \beta)b \\
&= \alpha(a^T x + b) + \beta(a^T y + b) \\
&= \alpha f(x) + \beta f(y).
\end{aligned}$$

(In the second line we use $\alpha + \beta = 1$.)

This restricted superposition property for affine functions is useful in showing that a function f is *not* affine: We find vectors x, y, and numbers α and β with $\alpha + \beta = 1$, and verify that $f(\alpha x + \beta y) \neq \alpha f(x) + \beta f(y)$. This shows that f cannot be affine. As an example, we verified above that superposition does not hold for the maximum function (with $n > 1$); the coefficients in our counterexample are $\alpha = \beta = 1/2$, which sum to one, which allows us to conclude that the maximum function is not affine.

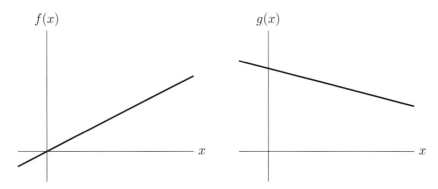

Figure 2.1 *Left.* The function f is linear. *Right.* The function g is affine, but not linear.

The converse is also true: Any scalar-valued function that satisfies the restricted superposition property is affine. An analog of the formula (2.3) is

$$f(x) = f(0) + x_1 \left(f(e_1) - f(0) \right) + \cdots + x_n \left(f(e_n) - f(0) \right), \quad (2.4)$$

which holds when f is affine, and x is any n-vector. (See exercise 2.7.) This formula shows that for an affine function, once we know the $n+1$ numbers $f(0)$, $f(e_1)$, ..., $f(e_n)$, we can predict (or reconstruct or evaluate) $f(x)$ for any n-vector x. It also shows how the vector a and constant b in the representation $f(x) = a^T x + b$ can be found from the function f: $a_i = f(e_i) - f(0)$, and $b = f(0)$.

In some contexts affine functions are called linear. For example, when x is a scalar, the function f defined as $f(x) = \alpha x + \beta$ is sometimes referred to as a linear function of x, perhaps because its graph is a line. But when $\beta \neq 0$, f is not a linear function of x, in the standard mathematical sense; it *is* an affine function of x. In this book we will distinguish between linear and affine functions. Two simple examples are shown in figure 2.1.

A civil engineering example. Many scalar-valued functions that arise in science and engineering are well approximated by linear or affine functions. As a typical example, consider a steel structure like a bridge, and let w be an n-vector that gives the weight of the load on the bridge in n specific locations, in metric tons. These loads will cause the bridge to deform (move and change shape) slightly. Let s denote the distance that a specific point on the bridge sags, in millimeters, due to the load w. This is shown in figure 2.2. For weights the bridge is designed to handle, the sag is very well approximated as a linear function $s = f(x)$. This function can be expressed as an inner product, $s = c^T w$, for some n-vector c. From the equation $s = c_1 w_1 + \cdots + c_n w_n$, we see that $c_1 w_1$ is the amount of the sag that is due to the weight w_1, and similarly for the other weights. The coefficients c_i, which have units of mm/ton, are called *compliances*, and give the sensitivity of the sag with respect to loads applied at the n locations.

The vector c can be computed by (numerically) solving a partial differential equation, given the detailed design of the bridge and the mechanical properties of

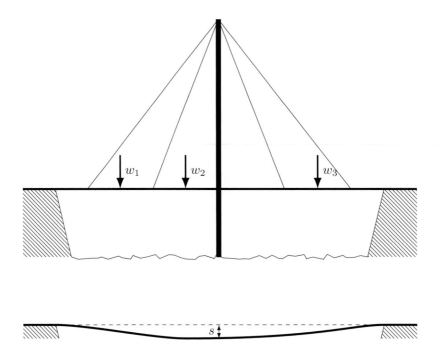

Figure 2.2 A bridge with weights w_1, w_2, w_3 applied in 3 locations. These weights cause the bridge to sag in the middle, by an amount s. (The sag is exaggerated in this diagram.)

w_1	w_2	w_3	Measured sag	Predicted sag
1	0	0	0.12	—
0	1	0	0.31	—
0	0	1	0.26	—
0.5	1.1	0.3	0.481	0.479
1.5	0.8	1.2	0.736	0.740

Table 2.1 Loadings on a bridge (first three columns), the associated measured sag at a certain point (fourth column), and the predicted sag using the linear model constructed from the first three experiments (fifth column).

the steel used to construct it. This is always done during the design of a bridge. The vector c can also be *measured* once the bridge is built, using the formula (2.3). We apply the load $w = e_1$, which means that we place a one ton load at the first load position on the bridge, with no load at the other positions. We can then measure the sag, which is c_1. We repeat this experiment, moving the one ton load to positions $2, 3, \ldots, n$, which gives us the coefficients $c_2, \ldots, c_n$. At this point we have the vector c, so we can now *predict* what the sag will be with any other loading. To check our measurements (and linearity of the sag function) we might measure the sag under other more complicated loadings, and in each case compare our prediction (*i.e.*, $c^T w$) with the actual measured sag.

Table 2.1 shows what the results of these experiments might look like, with each row representing an experiment (*i.e.*, placing the loads and measuring the sag). In the last two rows we compare the measured sag and the predicted sag, using the linear function with coefficients found in the first three experiments.

2.2 Taylor approximation

In many applications, scalar-valued functions of n variables, or relations between n variables and a scalar one, can be *approximated* as linear or affine functions. In these cases we sometimes refer to the linear or affine function relating the variables and the scalar variable as a *model*, to remind us that the relation is only an approximation, and not exact.

Differential calculus gives us an organized way to find an approximate affine model. Suppose that $f : \mathbf{R}^n \to \mathbf{R}$ is differentiable, which means that its partial derivatives exist (see §C.1). Let z be an n-vector. The (first-order) *Taylor approximation* of f near (or at) the point z is the function $\hat{f}(x)$ of x defined as

$$\hat{f}(x) = f(z) + \frac{\partial f}{\partial x_1}(z)(x_1 - z_1) + \cdots + \frac{\partial f}{\partial x_n}(z)(x_n - z_n),$$

where $\frac{\partial f}{\partial x_i}(z)$ denotes the partial derivative of f with respect to its ith argument, evaluated at the n-vector z. The hat appearing over f on the left-hand side is a common notational hint that it is an approximation of the function f. (The approximation is named after the mathematician Brook Taylor.)

The first-order Taylor approximation $\hat{f}(x)$ is a very good approximation of $f(x)$ when all x_i are near the associated z_i. Sometimes $\hat{f}$ is written with a second vector argument, as $\hat{f}(x; z)$, to show the point z at which the approximation is developed. The first term in the Taylor approximation is a constant; the other terms can be interpreted as the contributions to the (approximate) change in the function value (from $f(z)$) due to the changes in the components of x (from z).

Evidently $\hat{f}$ is an affine function of x. (It is sometimes called the *linear approximation* of f near z, even though it is in general affine, and not linear.) It can be written compactly using inner product notation as

$$\hat{f}(x) = f(z) + \nabla f(z)^T (x - z), \tag{2.5}$$

where $\nabla f(z)$ is an n-vector, the *gradient of* f (at the point z),

$$\nabla f(z) = \begin{bmatrix} \frac{\partial f}{\partial x_1}(z) \\ \vdots \\ \frac{\partial f}{\partial x_n}(z) \end{bmatrix}. \tag{2.6}$$

The first term in the Taylor approximation (2.5) is the constant $f(z)$, the value of the function when $x = z$. The second term is the inner product of the gradient of f at z and the *deviation* or *perturbation* of x from z, i.e., $x - z$.

We can express the first-order Taylor approximation as a linear function plus a constant,

$$\hat{f}(x) = \nabla f(z)^T x + (f(z) - \nabla f(z)^T z),$$

but the form (2.5) is perhaps easier to interpret.

The first-order Taylor approximation gives us an organized way to construct an affine approximation of a function $f : \mathbf{R}^n \to \mathbf{R}$, near a given point z, when there is a formula or equation that describes f, and it is differentiable. A simple example, for $n = 1$, is shown in figure 2.3. Over the full x-axis scale shown, the Taylor approximation $\hat{f}$ does not give a good approximation of the function f. But for x near z, the Taylor approximation is very good.

Example. Consider the function $f : \mathbf{R}^2 \to \mathbf{R}$ given by $f(x) = x_1 + \exp(x_2 - x_1)$, which is not linear or affine. To find the Taylor approximation $\hat{f}$ near the point $z = (1, 2)$, we take partial derivatives to obtain

$$\nabla f(z) = \begin{bmatrix} 1 - \exp(z_2 - z_1) \\ \exp(z_2 - z_1) \end{bmatrix},$$

which evaluates to $(-1.7183, 2.7183)$ at $z = (1, 2)$. The Taylor approximation at $z = (1, 2)$ is then

$$\begin{aligned} \hat{f}(x) &= 3.7183 + (-1.7183, 2.7183)^T (x - (1, 2)) \\ &= 3.7183 - 1.7183(x_1 - 1) + 2.7183(x_2 - 2). \end{aligned}$$

Table 2.2 shows $f(x)$ and $\hat{f}(x)$, and the approximation error $|\hat{f}(x) - f(x)|$, for some values of x relatively near z. We can see that $\hat{f}$ is indeed a very good approximation of f, especially when x is near z.

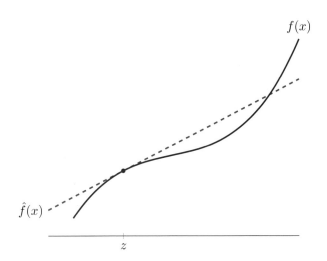

Figure 2.3 A function f of one variable, and the first-order Taylor approximation $\hat{f}(x) = f(z) + f'(z)(x - z)$ at z.

x	$f(x)$	$\hat{f}(x)$	$|\hat{f}(x) - f(x)|$
$(1.00, 2.00)$	3.7183	3.7183	0.0000
$(0.96, 1.98)$	3.7332	3.7326	0.0005
$(1.10, 2.11)$	3.8456	3.8455	0.0001
$(0.85, 2.05)$	4.1701	4.1119	0.0582
$(1.25, 2.41)$	4.4399	4.4032	0.0367

Table 2.2 Some values of x (first column), the function value $f(x)$ (second column), the Taylor approximation $\hat{f}(x)$ (third column), and the error (fourth column).

2.3   Regression model

In this section we describe a very commonly used affine function, especially when the n-vector x represents a feature vector. The affine function of x given by

$$\hat{y} = x^T \beta + v, \qquad (2.7)$$

where β is an n-vector and v is a scalar, is called a *regression model*. In this context, the entries of x are called the *regressors*, and $\hat{y}$ is called the *prediction*, since the regression model is typically an approximation or prediction of some true value y, which is called the *dependent variable*, *outcome*, or *label*.

The vector β is called the *weight vector* or *coefficient vector*, and the scalar v is called the *offset* or *intercept* in the regression model. Together, β and v are called the *parameters* in the regression model. (We will see in chapter 13 how the parameters in a regression model can be estimated or guessed, based on some past or known observations of the feature vector x and the associated outcome y.) The symbol $\hat{y}$ is used in the regression model to emphasize that it is an *estimate* or *prediction* of some outcome y.

The entries in the weight vector have a simple interpretation: β_i is the amount by which $\hat{y}$ increases (if $\beta_i > 0$) when feature i increases by one (with all other features the same). If β_i is small, the prediction $\hat{y}$ doesn't depend too strongly on feature i. The offset v is the value of $\hat{y}$ when all features have the value 0.

The regression model is very interpretable when all of the features are Boolean, *i.e.*, have values that are either 0 or 1, which occurs when the features represent which of two outcomes holds. As a simple example consider a regression model for the lifespan of a person in some group, with $x_1 = 0$ if the person is female ($x_1 = 1$ if male), $x_2 = 1$ if the person has type II diabetes, and $x_3 = 1$ if the person smokes cigarettes. In this case, v is the regression model estimate for the lifespan of a female nondiabetic nonsmoker; β_1 is the increase in estimated lifespan if the person is male, β_2 is the increase in estimated lifespan if the person is diabetic, and β_3 is the increase in estimated lifespan if the person smokes cigarettes. (In a model that fits real data, all three of these coefficients would be negative, meaning that they decrease the regression model estimate of lifespan.)

Simplified regression model notation.   Vector stacking can be used to lump the weights and offset in the regression model (2.7) into a single parameter vector, which simplifies the regression model notation a bit. We create a new regressor vector $\tilde{x}$, with $n + 1$ entries, as $\tilde{x} = (1, x)$. We can think of $\tilde{x}$ as a new feature vector, consisting of all n original features, and one new feature added ($\tilde{x}_1$) at the beginning, which always has the value one. We define the parameter vector $\tilde{\beta} = (v, \beta)$, so the regression model (2.7) has the simple inner product form

$$\hat{y} = x^T \beta + v = \begin{bmatrix} 1 \\ x \end{bmatrix}^T \begin{bmatrix} v \\ \beta \end{bmatrix} = \tilde{x}^T \tilde{\beta}. \qquad (2.8)$$

Often we omit the tildes, and simply write this as $\hat{y} = x^T \beta$, where we assume that the first feature in x is the constant 1. A feature that always has the value 1 is not particularly informative or interesting, but it does simplify the notation in a regression model.

House	x_1 (area)	x_2 (beds)	y (price)	$\hat{y}$ (prediction)
1	0.846	1	115.00	161.37
2	1.324	2	234.50	213.61
3	1.150	3	198.00	168.88
4	3.037	4	528.00	430.67
5	3.984	5	572.50	552.66

Table 2.3 Five houses with associated feature vectors shown in the second and third columns. The fourth and fifth column give the actual price, and the price predicted by the regression model.

House price regression model.   As a simple example of a regression model, suppose that y is the selling price of a house in some neighborhood, over some time period, and the 2-vector x contains attributes of the house:

- x_1 is the house area (in 1000 square feet),

- x_2 is the number of bedrooms.

If y represents the selling price of the house, in thousands of dollars, the regression model

$$\hat{y} = x^T \beta + v = \beta_1 x_1 + \beta_2 x_2 + v$$

predicts the price in terms of the attributes or features. This regression model is not meant to describe an exact relationship between the house attributes and its selling price; it is a model or approximation. Indeed, we would expect such a model to give, at best, only a crude approximation of selling price.

As a specific numerical example, consider the regression model parameters

$$\beta = (148.73, -18.85), \qquad v = 54.40. \tag{2.9}$$

These parameter values were found using the methods we will see in chapter 13, based on records of sales for 774 houses in the Sacramento area. Table 2.3 shows the feature vectors x for five houses that sold during the period, the actual sale price y, and the predicted price $\hat{y}$ from the regression model above. Figure 2.4 shows the predicted and actual sale prices for 774 houses, including the five houses in the table, on a scatter plot, with actual price on the horizontal axis and predicted price on the vertical axis.

We can see that this particular regression model gives reasonable, but not very accurate, predictions of the actual sale price. (Regression models for house prices that are used in practice use many more than two regressors, and are much more accurate.)

The model parameters in (2.9) are readily interpreted. The parameter $\beta_1 = 148.73$ is the amount the regression model price prediction increases (in thousands of dollars) when the house area increases by 1000 square feet (with the same number of bedrooms). The parameter $\beta_2 = -18.85$ is the price prediction increase with the addition of one bedroom, with the total house area held constant, in units of

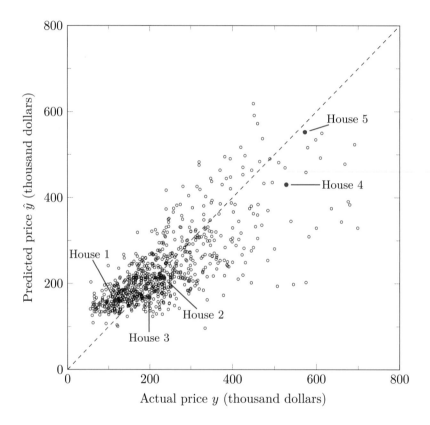

Figure 2.4 Scatter plot of actual and predicted sale prices for 774 houses sold in Sacramento during a five-day period.

thousands of dollars per bedroom. It might seem strange that β_2 is negative, since one imagines that adding a bedroom to a house would *increase* its sale price, not decrease it. To understand why β_2 might be negative, we note that it gives the change in predicted price when we add a bedroom, without adding any additional area to the house. If we remodel a house by adding a bedroom that *also* adds more than around 127 square feet to the house area, the regression model (2.9) *does* predict an increase in house sale price. The offset $v = 54.40$ is the predicted price for a house with no area and no bedrooms, which we might interpret as the model's prediction of the value of the lot. But this regression model is crude enough that these interpretations are dubious.

Exercises

2.1 *Linear or not?* Determine whether each of the following scalar-valued functions of n-vectors is linear. If it is a linear function, give its inner product representation, *i.e.*, an n-vector a for which $f(x) = a^T x$ for all x. If it is not linear, give specific x, y, α, and β for which superposition fails, *i.e.*,

$$f(\alpha x + \beta y) \neq \alpha f(x) + \beta f(y).$$

(a) The spread of values of the vector, defined as $f(x) = \max_k x_k - \min_k x_k$.

(b) The difference of the last element and the first, $f(x) = x_n - x_1$.

(c) The median of an n-vector, where we will assume $n = 2k + 1$ is odd. The median of the vector x is defined as the $(k + 1)$st largest number among the entries of x. For example, the median of $(-7.1, 3.2, -1.5)$ is -1.5.

(d) The average of the entries with odd indices, minus the average of the entries with even indices. You can assume that $n = 2k$ is even.

(e) Vector extrapolation, defined as $x_n + (x_n - x_{n-1})$, for $n \geq 2$. (This is a simple prediction of what x_{n+1} would be, based on a straight line drawn through x_n and x_{n-1}.)

2.2 *Processor powers and temperature.* The temperature T of an electronic device containing three processors is an affine function of the power dissipated by the three processors, $P = (P_1, P_2, P_3)$. When all three processors are idling, we have $P = (10, 10, 10)$, which results in a temperature $T = 30$. When the first processor operates at full power and the other two are idling, we have $P = (100, 10, 10)$, and the temperature rises to $T = 60$. When the second processor operates at full power and the other two are idling, we have $P = (10, 100, 10)$ and $T = 70$. When the third processor operates at full power and the other two are idling, we have $P = (10, 10, 100)$ and $T = 65$. Now suppose that all three processors are operated at the same power P^{same}. How large can P^{same} be, if we require that $T \leq 85$? *Hint.* From the given data, find the 3-vector a and number b for which $T = a^T P + b$.

2.3 *Motion of a mass in response to applied force.* A unit mass moves on a straight line (in one dimension). The position of the mass at time t (in seconds) is denoted by $s(t)$, and its derivatives (the velocity and acceleration) by $s'(t)$ and $s''(t)$. The position as a function of time can be determined from Newton's second law

$$s''(t) = F(t),$$

where $F(t)$ is the force applied at time t, and the initial conditions $s(0)$, $s'(0)$. We assume $F(t)$ is piecewise-constant, and is kept constant in intervals of one second. The sequence of forces $F(t)$, for $0 \leq t < 10$, can then be represented by a 10-vector f, with

$$F(t) = f_k, \quad k - 1 \leq t < k.$$

Derive expressions for the final velocity $s'(10)$ and final position $s(10)$. Show that $s(10)$ and $s'(10)$ are affine functions of x, and give 10-vectors a, c and constants b, d for which

$$s'(10) = a^T f + b, \qquad s(10) = c^T f + d.$$

This means that the mapping from the applied force sequence to the final position and velocity is affine.

Hint. You can use

$$s'(t) = s'(0) + \int_0^t F(\tau) \, d\tau, \qquad s(t) = s(0) + \int_0^t s'(\tau) \, d\tau.$$

You will find that the mass velocity $s'(t)$ is piecewise-linear.

2.4 *Linear function?* The function $\phi : \mathbf{R}^3 \to \mathbf{R}$ satisfies

$$\phi(1,1,0) = -1, \qquad \phi(-1,1,1) = 1, \qquad \phi(1,-1,-1) = 1.$$

Choose one of the following, and justify your choice: ϕ must be linear; ϕ could be linear; ϕ cannot be linear.

2.5 *Affine function.* Suppose $\psi : \mathbf{R}^2 \to \mathbf{R}$ is an affine function, with $\psi(1,0) = 1$, $\psi(1,-2) = 2$.

(a) What can you say about $\psi(1,-1)$? Either give the value of $\psi(1,-1)$, or state that it cannot be determined.

(b) What can you say about $\psi(2,-2)$? Either give the value of $\psi(2,-2)$, or state that it cannot be determined.

Justify your answers.

2.6 *Questionnaire scoring.* A questionnaire in a magazine has 30 questions, broken into two sets of 15 questions. Someone taking the questionnaire answers each question with 'Rarely', 'Sometimes', or 'Often'. The answers are recorded as a 30-vector a, with $a_i = 1, 2, 3$ if question i is answered Rarely, Sometimes, or Often, respectively. The total score on a completed questionnaire is found by adding up 1 point for every question answered Sometimes and 2 points for every question answered Often on questions 1–15, and by adding 2 points and 4 points for those responses on questions 16–30. (Nothing is added to the score for Rarely responses.) Express the total score s in the form of an affine function $s = w^T a + v$, where w is a 30-vector and v is a scalar (number).

2.7 *General formula for affine functions.* Verify that formula (2.4) holds for any affine function $f : \mathbf{R}^n \to \mathbf{R}$. You can use the fact that $f(x) = a^T x + b$ for some n-vector a and scalar b.

2.8 *Integral and derivative of polynomial.* Suppose the n-vector c gives the coefficients of a polynomial $p(x) = c_1 + c_2 x + \cdots + c_n x^{n-1}$.

(a) Let α and β be numbers with $\alpha < \beta$. Find an n-vector a for which

$$a^T c = \int_\alpha^\beta p(x)\, dx$$

always holds. This means that the integral of a polynomial over an interval is a linear function of its coefficients.

(b) Let α be a number. Find an n-vector b for which

$$b^T c = p'(\alpha).$$

This means that the derivative of the polynomial at a given point is a linear function of its coefficients.

2.9 *Taylor approximation.* Consider the function $f : \mathbf{R}^2 \to \mathbf{R}$ given by $f(x_1, x_2) = x_1 x_2$. Find the Taylor approximation $\hat{f}$ at the point $z = (1,1)$. Compare $f(x)$ and $\hat{f}(x)$ for the following values of x:

$$x = (1,1), \quad x = (1.05, 0.95), \quad x = (0.85, 1.25), \quad x = (-1, 2).$$

Make a brief comment about the accuracy of the Taylor approximation in each case.

2.10 *Regression model.* Consider the regression model $\hat{y} = x^T \beta + v$, where $\hat{y}$ is the predicted response, x is an 8-vector of features, β is an 8-vector of coefficients, and v is the offset term. Determine whether each of the following statements is true or false.

(a) If $\beta_3 > 0$ and $x_3 > 0$, then $\hat{y} \geq 0$.

(b) If $\beta_2 = 0$ then the prediction $\hat{y}$ does not depend on the second feature x_2.

(c) If $\beta_6 = -0.8$, then increasing x_6 (keeping all other x_is the same) will decrease $\hat{y}$.

2.11 *Sparse regression weight vector.* Suppose that x is an n-vector that gives n features for some object, and the scalar y is some outcome associated with the object. What does it mean if a regression model $\hat{y} = x^T \beta + v$ uses a sparse weight vector β? Give your answer in English, referring to $\hat{y}$ as our prediction of the outcome.

2.12 *Price change to maximize profit.* A business sells n products, and is considering changing the price of *one* of the products to increase its total profits. A business analyst develops a regression model that (reasonably accurately) predicts the total profit when the product prices are changed, given by $\hat{P} = \beta^T x + P$, where the n-vector x denotes the fractional change in the product prices, $x_i = (p_i^{\text{new}} - p_i)/p_i$. Here P is the profit with the current prices, $\hat{P}$ is the predicted profit with the changed prices, p_i is the current (positive) price of product i, and p_i^{new} is the new price of product i.

(a) What does it mean if $\beta_3 < 0$? (And yes, this can occur.)

(b) Suppose that you are given permission to change the price of *one* product, by up to 1%, to increase total profit. Which product would you choose, and would you increase or decrease the price? By how much?

(c) Repeat part (b) assuming you are allowed to change the price of two products, each by up to 1%.

Chapter 3

Norm and distance

In this chapter we focus on the norm of a vector, a measure of its magnitude, and on related concepts like distance, angle, standard deviation, and correlation.

3.1 Norm

The *Euclidean norm* of an n-vector x (named after the Greek mathematician Euclid), denoted $\|x\|$, is the squareroot of the sum of the squares of its elements,

$$\|x\| = \sqrt{x_1^2 + x_2^2 + \cdots + x_n^2}.$$

The Euclidean norm can also be expressed as the squareroot of the inner product of the vector with itself, *i.e.*, $\|x\| = \sqrt{x^T x}$.

The Euclidean norm is sometimes written with a subscript 2, as $\|x\|_2$. (The subscript 2 indicates that the entries of x are raised to the second power.) Other less widely used terms for the Euclidean norm of a vector are the *magnitude*, or *length*, of a vector. (The term *length* should be avoided, since it is also often used to refer to the dimension of the vector.) We use the same notation for the norms of vectors of different dimensions.

As simple examples, we have

$$\left\| \begin{bmatrix} 2 \\ -1 \\ 2 \end{bmatrix} \right\| = \sqrt{9} = 3, \qquad \left\| \begin{bmatrix} 0 \\ -1 \end{bmatrix} \right\| = 1.$$

When x is a scalar, *i.e.*, a 1-vector, the Euclidean norm is the same as the absolute value of x. Indeed, the Euclidean norm can be considered a generalization or extension of the absolute value or magnitude, that applies to vectors. The double bar notation is meant to suggest this. Like the absolute value of a number, the norm of a vector is a (numerical) measure of its magnitude. We say a vector is *small* if its norm is a small number, and we say it is *large* if its norm is a large number. (The numerical values of the norm that qualify for small or large depend on the particular application and context.)

Properties of norm. Some important properties of the Euclidean norm are given below. Here x and y are vectors of the same size, and β is a scalar.

- *Nonnegative homogeneity.* $\|\beta x\| = |\beta|\|x\|$. Multiplying a vector by a scalar multiplies the norm by the absolute value of the scalar.

- *Triangle inequality.* $\|x+y\| \le \|x\|+\|y\|$. The Euclidean norm of a sum of two vectors is no more than the sum of their norms. (The name of this property will be explained later.) Another name for this inequality is *subadditivity*.

- *Nonnegativity.* $\|x\| \ge 0$.

- *Definiteness.* $\|x\| = 0$ only if $x = 0$.

The last two properties together, which state that the norm is always nonnegative, and zero only when the vector is zero, are called *positive definiteness*. The first, third, and fourth properties are easy to show directly from the definition of the norm. As an example, let's verify the definiteness property. If $\|x\| = 0$, then we also have $\|x\|^2 = 0$, which means that $x_1^2 + \cdots + x_n^2 = 0$. This is a sum of n nonnegative numbers, which is zero. We can conclude that each of the n numbers is zero, since if any of them were nonzero the sum would be positive. So we conclude that $x_i^2 = 0$ for $i = 1, \ldots, n$, and therefore $x_i = 0$ for $i = 1, \ldots, n$; and thus, $x = 0$. Establishing the second property, the triangle inequality, is not as easy; we will give a derivation on page 57.

General norms. Any real-valued function of an n-vector that satisfies the four properties listed above is called a (general) norm. But in this book we will only use the Euclidean norm, so from now on, we refer to the Euclidean norm as the norm. (See exercise 3.5, which describes some other useful norms.)

Root-mean-square value. The norm is related to the *root-mean-square* (RMS) value of an n-vector x, defined as

$$\mathbf{rms}(x) = \sqrt{\frac{x_1^2 + \cdots + x_n^2}{n}} = \frac{\|x\|}{\sqrt{n}}.$$

The argument of the squareroot in the middle expression is called the *mean square* value of x, denoted $\mathbf{ms}(x)$, and the RMS value is the squareroot of the mean square value. The RMS value of a vector x is useful when comparing norms of vectors with different dimensions; the RMS value tells us what a 'typical' value of $|x_i|$ is. For example, the norm of $\mathbf{1}$, the n-vector of all ones, is $\sqrt{n}$, but its RMS value is 1, independent of n. More generally, if all the entries of a vector are the same, say, α, then the RMS value of the vector is $|\alpha|$.

Norm of a sum. A useful formula for the norm of the sum of two vectors x and y is

$$\|x + y\| = \sqrt{\|x\|^2 + 2x^T y + \|y\|^2}. \tag{3.1}$$

To derive this formula, we start with the square of the norm of $x+y$ and use various properties of the inner product:

$$
\begin{aligned}
\|x + y\|^2 &= (x + y)^T (x + y) \\
&= x^T x + x^T y + y^T x + y^T y \\
&= \|x\|^2 + 2x^T y + \|y\|^2.
\end{aligned}
$$

Taking the squareroot of both sides yields the formula (3.1) above. In the first line, we use the definition of the norm. In the second line, we expand the inner product. In the fourth line we use the definition of the norm, and the fact that $x^T y = y^T x$. Some other identities relating norms, sums, and inner products of vectors are explored in exercise 3.4.

Norm of block vectors.   The norm-squared of a stacked vector is the sum of the norm-squared values of its subvectors. For example, with $d = (a, b, c)$ (where a, b, and c are vectors), we have

$$
\|d\|^2 = d^T d = a^T a + b^T b + c^T c = \|a\|^2 + \|b\|^2 + \|c\|^2.
$$

This idea is often used in reverse, to express the sum of the norm-squared values of some vectors as the norm-square value of a block vector formed from them.

We can write the equality above in terms of norms as

$$
\|(a, b, c)\| = \sqrt{\|a\|^2 + \|b\|^2 + \|c\|^2} = \|(\|a\|, \|b\|, \|c\|)\|.
$$

In words: The norm of a stacked vector is the norm of the vector formed from the norms of the subvectors. The right-hand side of the equation above should be carefully read. The outer norm symbols enclose a 3-vector, with (scalar) entries $\|a\|$, $\|b\|$, and $\|c\|$.

Chebyshev inequality.   Suppose that x is an n-vector, and that k of its entries satisfy $|x_i| \geq a$, where $a > 0$. Then k of its entries satisfy $x_i^2 \geq a^2$. It follows that

$$
\|x\|^2 = x_1^2 + \cdots + x_n^2 \geq ka^2,
$$

since k of the numbers in the sum are at least a^2, and the other $n - k$ numbers are nonnegative. We can conclude that $k \leq \|x\|^2/a^2$, which is called the *Chebyshev inequality*, after the mathematician Pafnuty Chebyshev. When $\|x\|^2/a^2 \geq n$, the inequality tells us nothing, since we always have $k \leq n$. In other cases it limits the number of entries in a vector that can be large. For $a > \|x\|$, the inequality is $k \leq \|x\|^2/a^2 < 1$, so we conclude that $k = 0$ (since k is an integer). In other words, no entry of a vector can be larger in magnitude than the norm of the vector.

The Chebyshev inequality is easier to interpret in terms of the RMS value of a vector. We can write it as

$$
\frac{k}{n} \leq \left(\frac{\mathbf{rms}(x)}{a} \right)^2, \tag{3.2}
$$

where k is, as above, the number of entries of x with absolute value at least a. The left-hand side is the fraction of entries of the vector that are at least a in absolute

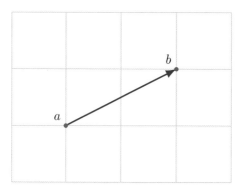

Figure 3.1 The norm of the displacement $b - a$ is the distance between the points with coordinates a and b.

value. The right-hand side is the inverse square of the ratio of a to $\mathbf{rms}(x)$. It says, for example, that no more than $1/25 = 4\%$ of the entries of a vector can exceed its RMS value by more than a factor of 5. The Chebyshev inequality partially justifies the idea that the RMS value of a vector gives an idea of the size of a typical entry: It states that not too many of the entries of a vector can be much bigger (in absolute value) than its RMS value. (A converse statement can also be made: At least one entry of a vector has absolute value as large as the RMS value of the vector; see exercise 3.8.)

3.2   Distance

Euclidean distance.   We can use the norm to define the *Euclidean distance* between two vectors a and b as the norm of their difference:

$$\mathbf{dist}(a, b) = \|a - b\|.$$

For one, two, and three dimensions, this distance is exactly the usual distance between points with coordinates a and b, as illustrated in figure 3.1. But the Euclidean distance is defined for vectors of any dimension; we can refer to the distance between two vectors of dimension 100. Since we only use the Euclidean norm in this book, we will refer to the Euclidean distance between vectors as, simply, the distance between the vectors. If a and b are n-vectors, we refer to the RMS value of the difference, $\|a - b\|/\sqrt{n}$, as the *RMS deviation* between the two vectors.

When the distance between two n-vectors x and y is small, we say they are 'close' or 'nearby', and when the distance $\|x - y\|$ is large, we say they are 'far'. The particular numerical values of $\|x - y\|$ that correspond to 'close' or 'far' depend on the particular application.

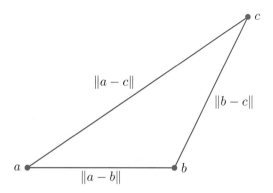

Figure 3.2 Triangle inequality.

As an example, consider the 4-vectors

$$u = \begin{bmatrix} 1.8 \\ 2.0 \\ -3.7 \\ 4.7 \end{bmatrix}, \qquad v = \begin{bmatrix} 0.6 \\ 2.1 \\ 1.9 \\ -1.4 \end{bmatrix}, \qquad w = \begin{bmatrix} 2.0 \\ 1.9 \\ -4.0 \\ 4.6 \end{bmatrix}.$$

The distances between pairs of them are

$$\|u - v\| = 8.368, \qquad \|u - w\| = 0.387, \qquad \|v - w\| = 8.533,$$

so we can say that u is much nearer (or closer) to w than it is to v. We can also say that w is much nearer to u than it is to v.

Triangle inequality.   We can now explain where the triangle inequality gets its name. Consider a triangle in two or three dimensions, whose vertices have coordinates a, b, and c. The lengths of the sides are the distances between the vertices,

$$\mathbf{dist}(a, b) = \|a - b\|, \qquad \mathbf{dist}(b, c) = \|b - c\|, \qquad \mathbf{dist}(a, c) = \|a - c\|.$$

Geometric intuition tells us that the length of any side of a triangle cannot exceed the sum of the lengths of the other two sides. For example, we have

$$\|a - c\| \le \|a - b\| + \|b - c\|. \tag{3.3}$$

This follows from the triangle inequality, since

$$\|a - c\| = \|(a - b) + (b - c)\| \le \|a - b\| + \|b - c\|.$$

This is illustrated in figure 3.2.

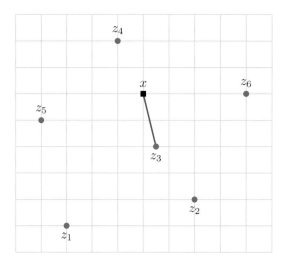

Figure 3.3 A point x, shown as a square, and six other points $z_1, \ldots, z_6$. The point z_3 is the nearest neighbor of x among the points $z_1, \ldots, z_6$.

Examples.

- *Feature distance.* If x and y represent vectors of n features of two objects, the quantity $\|x - y\|$ is called the *feature distance*, and gives a measure of how different the objects are (in terms of their feature values). Suppose for example the feature vectors are associated with patients in a hospital, with entries such as weight, age, presence of chest pain, difficulty breathing, and the results of tests. We can use feature vector distance to say that one patient case is near another one (at least in terms of their feature vectors).

- *RMS prediction error.* Suppose that the n-vector y represents a time series of some quantity, for example, hourly temperature at some location, and $\hat{y}$ is another n-vector that represents an estimate or prediction of the time series y, based on other information. The difference $y - \hat{y}$ is called the *prediction error*, and its RMS value $\mathbf{rms}(y - \hat{y})$ is called the *RMS prediction error*. If this value is small (say, compared to $\mathbf{rms}(y)$) the prediction is good.

- *Nearest neighbor.* Suppose $z_1, \ldots, z_m$ is a collection of m n-vectors, and that x is another n-vector. We say that z_j is the *nearest neighbor* of x (among $z_1, \ldots, z_m$) if
$$\|x - z_j\| \leq \|x - z_i\|, \quad i = 1, \ldots, m.$$
In words: z_j is the closest vector to x among the vectors $z_1, \ldots, z_m$. This is illustrated in figure 3.3. The idea of nearest neighbor, and generalizations such as the k-nearest neighbors, are used in many applications.

- *Document dissimilarity.* Suppose n-vectors x and y represent the histograms of word occurrences for two documents. Then $\|x - y\|$ represents a measure of the dissimilarity of the two documents. We might expect the dissimilarity

	Veterans Day	Memorial Day	Academy Awards	Golden Globe Awards	Super Bowl
Veterans Day	0	0.095	0.130	0.153	0.170
Memorial Day	0.095	0	0.122	0.147	0.164
Academy A.	0.130	0.122	0	0.108	0.164
Golden Globe A.	0.153	0.147	0.108	0	0.181
Super Bowl	0.170	0.164	0.164	0.181	0

Table 3.1 Pairwise word count histogram distances between five Wikipedia articles.

to be smaller when the two documents have the same genre, topic, or author; we would expect it to be larger when they are on different topics, or have different authors. As an example we form the word count histograms for the 5 Wikipedia articles with titles 'Veterans Day', 'Memorial Day', 'Academy Awards', 'Golden Globe Awards', and 'Super Bowl', using a dictionary of 4423 words. (More detail is given in §4.4.) The pairwise distances between the word count histograms are shown in table 3.1. We can see that pairs of related articles have smaller word count histogram distances than less related pairs of articles.

Units for heterogeneous vector entries.   The square of the distance between two n-vectors x and y is given by

$$\|x - y\|^2 = (x_1 - y_1)^2 + \cdots + (x_n - y_n)^2,$$

the sum of the squares of the differences between their respective entries. Roughly speaking, the entries in the vectors all have equal status in determining the distance between them. For example, if x_2 and y_2 differ by one, the contribution to the square of the distance between them is the same as the contribution when x_3 and y_3 differ by one. This makes sense when the entries of the vectors x and y represent the same type of quantity, using the same units (say, at different times or locations), for example meters or dollars. For example if x and y are word count histograms, their entries are all word occurrence frequencies, and it makes sense to say they are close when their distance is small.

When the entries of a vector represent different types of quantities, for example when the vector entries represent different types of features associated with an object, we must be careful about choosing the units used to represent the numerical values of the entries. If we want the different entries to have approximately equal status in determining distance, their numerical values should be approximately of the same magnitude. For this reason units for different entries in vectors are often chosen in such a way that their typical numerical values are similar in magnitude, so that the different entries play similar roles in determining distance.

As an example suppose that the 2-vectors x, y, and z are the feature vectors for three houses that were sold, as in the example described on page 39. The first entry of each vector gives the house area and the second entry gives the number of bedrooms. These are very different types of features, since the first one is a physical area, and the second one is a count, $i.e.$, an integer. In the example on page 39, we chose the unit used to represent the first feature, area, to be thousands of square feet. With this choice of unit used to represent house area, the numerical values of both of these features range from around 1 to 5; their values have roughly the same magnitude. When we determine the distance between feature vectors associated with two houses, the difference in the area (in thousands of square feet), and the difference in the number of bedrooms, play equal roles.

For example, consider three houses with feature vectors

$$x = (1.6, 2), \quad y = (1.5, 2), \quad z = (1.6, 4).$$

The first two are 'close' or 'similar' since $\|x - y\| = 0.1$ is small (compared to the norms of x and y, which are around 2.5). This matches our intuition that the first two houses are similar, since they both have two bedrooms and are close in area. The third house would be considered 'far' or 'different' from the first two houses, and rightly so since it has four bedrooms instead of two.

To appreciate the significance of our choice of units in this example, suppose we had chosen instead to represent house area directly in square feet, and not thousands of square feet. The three houses above would then be represented by feature vectors

$$\tilde{x} = (1600, 2), \quad \tilde{y} = (1500, 2), \quad \tilde{z} = (1600, 4).$$

The distance between the first and third houses is now 2, which is very small compared to the norms of the vectors (which are around 1600). The distance between the first and second houses is much larger. It seems strange to consider a two-bedroom house and a four-bedroom house as 'very close', while two houses with the same number of bedrooms and similar areas are much more dissimilar. The reason is simple: With our choice of square feet as the unit to measure house area, distances are very strongly influenced by differences in area, with number of bedrooms playing a much smaller (relative) role.

3.3   Standard deviation

For any vector x, the vector $\tilde{x} = x - \mathbf{avg}(x)\mathbf{1}$ is called the associated *de-meaned* vector, obtained by subtracting from each entry of x the mean value of the entries. (This is not standard notation; $i.e.$, $\tilde{x}$ is not generally used to denote the de-meaned vector.) The mean value of the entries of $\tilde{x}$ is zero, $i.e.$, $\mathbf{avg}(\tilde{x}) = 0$. This explains why $\tilde{x}$ is called the de-meaned version of x; it is x with its mean removed. The de-meaned vector is useful for understanding how the entries of a vector deviate from their mean value. It is zero if all the entries in the original vector x are the same.

The *standard deviation* of an n-vector x is defined as the RMS value of the de-meaned vector $x - \mathbf{avg}(x)\mathbf{1}$, $i.e.$,

$$\mathbf{std}(x) = \sqrt{\frac{(x_1 - \mathbf{avg}(x))^2 + \cdots + (x_n - \mathbf{avg}(x))^2}{n}}.$$

This is the same as the RMS deviation between a vector x and the vector all of whose entries are $\mathbf{avg}(x)$. It can be written using the inner product and norm as

$$\mathbf{std}(x) = \frac{\|x - (\mathbf{1}^T x/n)\mathbf{1}\|}{\sqrt{n}}. \tag{3.4}$$

The standard deviation of a vector x tells us the typical amount by which its entries deviate from their average value. The standard deviation of a vector is zero only when all its entries are equal. The standard deviation of a vector is small when the entries of the vector are nearly the same.

As a simple example consider the vector $x = (1, -2, 3, 2)$. Its mean or average value is $\mathbf{avg}(x) = 1$, so the de-meaned vector is $\tilde{x} = (0, -3, 2, 1)$. Its standard deviation is $\mathbf{std}(x) = 1.872$. We interpret this number as a 'typical' value by which the entries differ from the mean of the entries. These numbers are 0, 3, 2, and 1, so 1.872 is reasonable.

We should warn the reader that another slightly different definition of the standard deviation of a vector is widely used, in which the denominator $\sqrt{n}$ in (3.4) is replaced with $\sqrt{n-1}$ (for $n \geq 2$). In this book we will only use the definition (3.4).

In some applications the Greek letter σ (sigma) is traditionally used to denote standard deviation, while the mean is denoted μ (mu). In this notation we have, for an n-vector x,

$$\mu = \mathbf{1}^T x/n, \qquad \sigma = \|x - \mu\mathbf{1}\|/\sqrt{n}.$$

We will use the symbols $\mathbf{avg}(x)$ and $\mathbf{std}(x)$, switching to μ and σ only with explanation, when describing an application that traditionally uses these symbols.

Average, RMS value, and standard deviation.   The average, RMS value, and standard deviation of a vector are related by the formula

$$\mathbf{rms}(x)^2 = \mathbf{avg}(x)^2 + \mathbf{std}(x)^2. \tag{3.5}$$

This formula makes sense: $\mathbf{rms}(x)^2$ is the mean square value of the entries of x, which can be expressed as the square of the mean value, plus the mean square fluctuation of the entries of x around their mean value. We can derive this formula from our vector notation formula for $\mathbf{std}(x)$ given above. We have

$$
\begin{aligned}
\mathbf{std}(x)^2 &= (1/n)\|x - (\mathbf{1}^T x/n)\mathbf{1}\|^2 \\
&= (1/n)(x^T x - 2x^T(\mathbf{1}^T x/n)\mathbf{1} + ((\mathbf{1}^T x/n)\mathbf{1})^T((\mathbf{1}^T x/n)\mathbf{1})) \\
&= (1/n)(x^T x - (2/n)(\mathbf{1}^T x)^2 + n(\mathbf{1}^T x/n)^2) \\
&= (1/n)x^T x - (\mathbf{1}^T x/n)^2 \\
&= \mathbf{rms}(x)^2 - \mathbf{avg}(x)^2,
\end{aligned}
$$

which can be re-arranged to obtain the identity (3.5) above. This derivation uses many of the properties for norms and inner products, and should be read carefully to understand every step. In the second line, we expand the norm-square of the sum of two vectors. In the third line, we use the commutative property of scalar-vector multiplication, moving scalars such as $(\mathbf{1}^T x/n)$ to the front of each term, and also the fact that $\mathbf{1}^T \mathbf{1} = n$.

Examples.

- *Mean return and risk.* Suppose that an n-vector represents a time series of return on an investment, expressed as a percentage, in n time periods over some interval of time. Its average gives the mean return over the whole interval, often shortened to its *return*. Its standard deviation is a measure of how variable the return is, from period to period, over the time interval, *i.e.*, how much it typically varies from its mean, and is often called the (per period) *risk* of the investment. Multiple investments can be compared by plotting them on a *risk-return plot*, which gives the mean and standard deviation of the returns of each of the investments over some interval. A desirable return history vector has high mean return and low risk; this means that the returns in the different periods are consistently high. Figure 3.4 shows an example.

- *Temperature or rainfall.* Suppose that an n-vector is a time series of the daily average temperature at a particular location, over a one year period. Its average gives the average temperature at that location (over the year) and its standard deviation is a measure of how much the temperature varied from its average value. We would expect the average temperature to be high and the standard deviation to be low in a tropical location, and the opposite for a location with high latitude.

Chebyshev inequality for standard deviation. The Chebyshev inequality (3.2) can be transcribed to an inequality expressed in terms of the mean and standard deviation: If k is the number of entries of x that satisfy $|x_i - \mathbf{avg}(x)| \geq a$, then $k/n \leq (\mathbf{std}(x)/a)^2$. (This inequality is only interesting for $a > \mathbf{std}(x)$.) For example, at most $1/9 = 11.1\%$ of the entries of a vector can deviate from the mean value $\mathbf{avg}(x)$ by 3 standard deviations or more. Another way to state this is: The fraction of entries of x within α standard deviations of $\mathbf{avg}(x)$ is at least $1 - 1/\alpha^2$ (for $\alpha > 1$).

As an example, consider a time series of return on an investment, with a mean return of 8%, and a risk (standard deviation) 3%. By the Chebyshev inequality, the fraction of periods with a loss (*i.e.*, $x_i \leq 0$) is no more than $(3/8)^2 = 14.1\%$. (In fact, the fraction of periods when the return is either a loss, $x_i \leq 0$, or very good, $x_i \geq 16\%$, is together no more than 14.1%.)

Properties of standard deviation.

- *Adding a constant.* For any vector x and any number a, we have $\mathbf{std}(x+a\mathbf{1}) = \mathbf{std}(x)$. Adding a constant to every entry of a vector does not change its standard deviation.

- *Multiplying by a scalar.* For any vector x and any number a, we have $\mathbf{std}(ax) = |a|\,\mathbf{std}(x)$. Multiplying a vector by a scalar multiplies the standard deviation by the absolute value of the scalar.

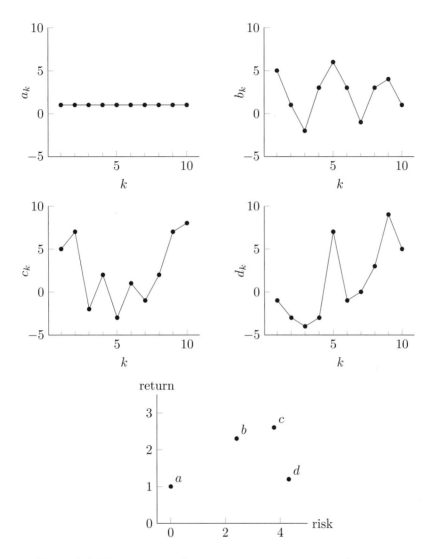

Figure 3.4 The vectors a, b, c, d represent time series of returns on investments over 10 periods. The bottom plot shows the investments in a risk-return plane, with return defined as the average value and risk as the standard deviation of the corresponding vector.

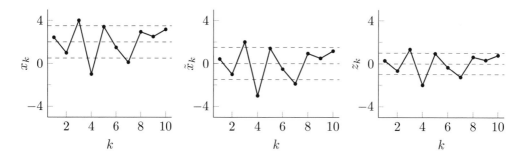

Figure 3.5 A 10-vector x, the de-meaned vector $\tilde{x} = x - \mathbf{avg}(x)\mathbf{1}$, and the standardized vector $z = (1/\mathbf{std}(x))\tilde{x}$. The horizontal dashed lines indicate the mean and the standard deviation of each vector. The middle line is the mean; the distance between the middle line and the other two is the standard deviation.

Standardization.   For any vector x, we refer to $\tilde{x} = x - \mathbf{avg}(x)\mathbf{1}$ as the de-meaned version of x, since it has average or mean value zero. If we then divide by the RMS value of $\tilde{x}$ (which is the standard deviation of x), we obtain the vector

$$z = \frac{1}{\mathbf{std}(x)}(x - \mathbf{avg}(x)\mathbf{1}).$$

This vector is called the *standardized* version of x. It has mean zero, and standard deviation one. Its entries are sometimes called the *z-scores* associated with the original entries of x. For example, $z_4 = 1.4$ means that x_4 is 1.4 standard deviations above the mean of the entries of x. Figure 3.5 shows an example.

The standardized values for a vector give a simple way to interpret the original values in the vectors. For example, if an n-vector x gives the values of some medical test of n patients admitted to a hospital, the standardized values or z-scores tell us how high or low, compared to the population, that patient's value is. A value $z_6 = -3.2$, for example, means that patient 6 has a very low value of the measurement; whereas $z_{22} = 0.3$ says that patient 22's value is quite close to the average value.

3.4   Angle

Cauchy–Schwarz inequality.   An important inequality that relates norms and inner products is the *Cauchy–Schwarz inequality*:

$$|a^T b| \leq \|a\| \, \|b\|$$

for any n-vectors a and b. Written out in terms of the entries, this is

$$|a_1 b_1 + \cdots + a_n b_n| \leq \left(a_1^2 + \cdots + a_n^2\right)^{1/2} \left(b_1^2 + \cdots + b_n^2\right)^{1/2},$$

which looks more intimidating. This inequality is attributed to the mathematician Augustin-Louis Cauchy; Hermann Schwarz gave the derivation given below.

The Cauchy–Schwarz inequality can be shown as follows. The inequality clearly holds if $a = 0$ or $b = 0$ (in this case, both sides of the inequality are zero). So we suppose now that $a \neq 0$, $b \neq 0$, and define $\alpha = \|a\|$, $\beta = \|b\|$. We observe that

$$
\begin{aligned}
0 &\leq \|\beta a - \alpha b\|^2 \\
&= \|\beta a\|^2 - 2(\beta a)^T(\alpha b) + \|\alpha b\|^2 \\
&= \beta^2 \|a\|^2 - 2\beta\alpha(a^T b) + \alpha^2 \|b\|^2 \\
&= \|b\|^2 \|a\|^2 - 2\|b\|\|a\|(a^T b) + \|a\|^2 \|b\|^2 \\
&= 2\|a\|^2 \|b\|^2 - 2\|a\|\,\|b\|(a^T b).
\end{aligned}
$$

Dividing by $2\|a\|\,\|b\|$ yields $a^T b \leq \|a\|\,\|b\|$. Applying this inequality to $-a$ and b we obtain $-a^T b \leq \|a\|\,\|b\|$. Putting these two inequalities together we get the Cauchy–Schwarz inequality, $|a^T b| \leq \|a\|\,\|b\|$.

This argument also reveals the conditions on a and b under which they satisfy the Cauchy–Schwarz inequality with equality. This occurs only if $\|\beta a - \alpha b\| = 0$, i.e., $\beta a = \alpha b$. This means that each vector is a scalar multiple of the other (in the case when they are nonzero). This statement remains true when either a or b is zero. So the Cauchy–Schwarz inequality holds with equality when one of the vectors is a multiple of the other; in all other cases, it holds with strict inequality.

Verification of triangle inequality. We can use the Cauchy–Schwarz inequality to verify the triangle inequality. Let a and b be any vectors. Then

$$
\begin{aligned}
\|a + b\|^2 &= \|a\|^2 + 2a^T b + \|b\|^2 \\
&\leq \|a\|^2 + 2\|a\|\|b\| + \|b\|^2 \\
&= (\|a\| + \|b\|)^2,
\end{aligned}
$$

where we used the Cauchy–Schwarz inequality in the second line. Taking the squareroot we get the triangle inequality, $\|a + b\| \leq \|a\| + \|b\|$.

Angle between vectors. The *angle* between two nonzero vectors a, b is defined as

$$
\theta = \arccos\left(\frac{a^T b}{\|a\|\,\|b\|}\right)
$$

where arccos denotes the inverse cosine, normalized to lie in the interval $[0, \pi]$. In other words, we define θ as the unique number between 0 and π that satisfies

$$
a^T b = \|a\|\,\|b\|\cos\theta.
$$

The angle between a and b is written as $\angle(a, b)$, and is sometimes expressed in degrees. (The default angle unit is *radians*; $360°$ is 2π radians.) For example, $\angle(a, b) = 60°$ means $\angle(a, b) = \pi/3$, i.e., $a^T b = (1/2)\|a\|\|b\|$.

The angle coincides with the usual notion of angle between vectors, when they have dimension two or three, and they are thought of as displacements from a common point. For example, the angle between the vectors $a = (1, 2, -1)$ and $b = (2, 0, -3)$ is

$$\arccos\left(\frac{5}{\sqrt{6}\sqrt{13}}\right) = \arccos(0.5661) = 0.9690 = 55.52°$$

(to 4 digits). But the definition of angle is more general; we can refer to the angle between two vectors with dimension 100.

The angle is a symmetric function of a and b: We have $\angle(a, b) = \angle(b, a)$. The angle is not affected by scaling each of the vectors by a positive scalar: We have, for any vectors a and b, and any positive numbers α and β,

$$\angle(\alpha a, \beta b) = \angle(a, b).$$

Acute and obtuse angles. Angles are classified according to the sign of $a^T b$. Suppose a and b are nonzero vectors of the same size.

- If the angle is $\pi/2 = 90°$, *i.e.*, $a^T b = 0$, the vectors are said to be *orthogonal*. We write $a \perp b$ if a and b are orthogonal. (By convention, we also say that a zero vector is orthogonal to any vector.)

- If the angle is zero, which means $a^T b = \|a\|\|b\|$, the vectors are *aligned*. Each vector is a positive multiple of the other.

- If the angle is $\pi = 180°$, which means $a^T b = -\|a\|\|b\|$, the vectors are *anti-aligned*. Each vector is a negative multiple of the other.

- If $\angle(a, b) < \pi/2 = 90°$, the vectors are said to make an *acute angle*. This is the same as $a^T b > 0$, *i.e.*, the vectors have positive inner product.

- If $\angle(a, b) > \pi/2 = 90°$, the vectors are said to make an *obtuse angle*. This is the same as $a^T b < 0$, *i.e.*, the vectors have negative inner product.

These definitions are illustrated in figure 3.6.

Examples.

- *Spherical distance.* Suppose a and b are 3-vectors that represent two points that lie on a sphere of radius R (for example, locations on earth). The spherical distance between them, measured along the sphere, is given by $R\angle(a, b)$. This is illustrated in figure 3.7.

- *Document similarity via angles.* If n-vectors x and y represent the word counts for two documents, their angle $\angle(x, y)$ can be used as a measure of document dissimilarity. (When using angle to measure document dissimilarity, either word counts or histograms can be used; they produce the same result.) As an example, table 3.2 gives the angles in degrees between the word histograms in the example at the end of §3.2.

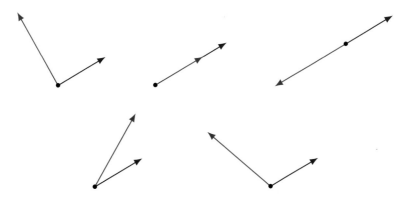

Figure 3.6 *Top row.* Examples of orthogonal, aligned, and anti-aligned vectors. *Bottom row.* Vectors that make an obtuse and an acute angle.

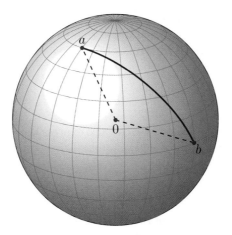

Figure 3.7 Two points a and b on a sphere with radius R and center at the origin. The spherical distance between the points is equal to $R\angle(a, b)$.

	Veterans Day	Memorial Day	Academy Awards	Golden Globe Awards	Super Bowl
Veterans Day	0	60.6	85.7	87.0	87.7
Memorial Day	60.6	0	85.6	87.5	87.5
Academy A.	85.7	85.6	0	58.7	85.7
Golden Globe A.	87.0	87.5	58.7	0	86.0
Super Bowl	87.7	87.5	86.1	86.0	0

Table 3.2 Pairwise angles (in degrees) between word histograms of five Wikipedia articles.

Norm of sum via angles.   For vectors x and y we have

$$\|x + y\|^2 = \|x\|^2 + 2x^T y + \|y\|^2 = \|x\|^2 + 2\|x\|\|y\| \cos\theta + \|y\|^2, \tag{3.6}$$

where $\theta = \angle(x, y)$. (The first equality comes from (3.1).) From this we can make several observations.

- If x and y are aligned ($\theta = 0$), we have $\|x + y\| = \|x\| + \|y\|$. Thus, their norms add.

- If x and y are orthogonal ($\theta = 90°$), we have $\|x + y\|^2 = \|x\|^2 + \|y\|^2$. In this case the norm-squared values add, and we have $\|x + y\| = \sqrt{\|x\|^2 + \|y\|^2}$. This formula is sometimes called the *Pythagorean theorem*, after the Greek mathematician Pythagoras of Samos.

Correlation coefficient.   Suppose a and b are n-vectors, with associated de-meaned vectors

$$\tilde{a} = a - \mathbf{avg}(a)\mathbf{1}, \qquad \tilde{b} = b - \mathbf{avg}(b)\mathbf{1}.$$

Assuming these de-meaned vectors are not zero (which occurs when the original vectors have all equal entries), we define their *correlation coefficient* as

$$\rho = \frac{\tilde{a}^T \tilde{b}}{\|\tilde{a}\|\,\|\tilde{b}\|}. \tag{3.7}$$

Thus, $\rho = \cos\theta$, where $\theta = \angle(\tilde{a}, \tilde{b})$. We can also express the correlation coefficient in terms of the vectors u and v obtained by standardizing a and b. With $u = \tilde{a}/\mathbf{std}(a)$ and $v = \tilde{b}/\mathbf{std}(b)$, we have

$$\rho = u^T v/n. \tag{3.8}$$

(We use $\|u\| = \|v\| = \sqrt{n}$.)

   This is a symmetric function of the vectors: The correlation coefficient between a and b is the same as the correlation coefficient between b and a. The Cauchy–Schwarz inequality tells us that the correlation coefficient ranges between -1 and $+1$. For this reason, the correlation coefficient is sometimes expressed as a percentage. For example, $\rho = 30\%$ means $\rho = 0.3$. When $\rho = 0$, we say the vectors are *uncorrelated*. (By convention, we say that a vector with all entries equal is uncorrelated with any vector.)

   The correlation coefficient tells us how the entries in the two vectors vary together. High correlation (say, $\rho = 0.8$) means that entries of a and b are typically above their mean for many of the same entries. The extreme case $\rho = 1$ occurs only if the vectors $\tilde{a}$ and $\tilde{b}$ are aligned, which means that each is a positive multiple of the other, and the other extreme case $\rho = -1$ occurs only when $\tilde{a}$ and $\tilde{b}$ are negative multiples of each other. This idea is illustrated in figure 3.8, which shows the entries of two vectors, as well as a scatter plot of them, for cases with correlation near 1, near -1, and near 0.

   The correlation coefficient is often used when the vectors represent time series, such as the returns on two investments over some time interval, or the rainfall in two locations over some time interval. If they are highly correlated (say, $\rho > 0.8$),

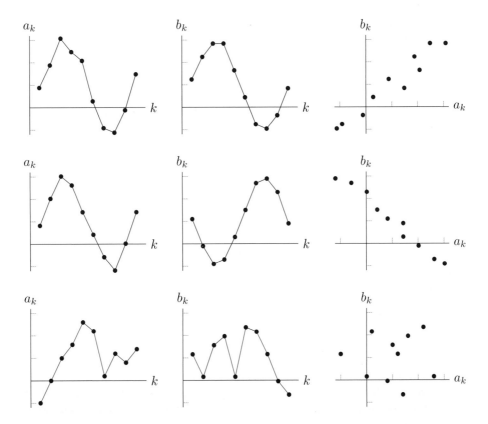

Figure 3.8 Three pairs of vectors a, b of length 10, with correlation coefficients 0.968 (top), -0.988 (middle), and 0.004 (bottom).

the two time series are typically above their mean values at the same times. For example, we would expect the rainfall time series at two nearby locations to be highly correlated. As another example, we might expect the returns of two similar companies, in the same business area, to be highly correlated.

Standard deviation of sum.   We can derive a formula for the standard deviation of a sum from (3.6):

$$\mathbf{std}(a+b) = \sqrt{\mathbf{std}(a)^2 + 2\rho\,\mathbf{std}(a)\,\mathbf{std}(b) + \mathbf{std}(b)^2}. \tag{3.9}$$

To derive this from (3.6) we let $\tilde{a}$ and $\tilde{b}$ denote the de-meaned versions of a and b. Then $\tilde{a} + \tilde{b}$ is the de-meaned version of $a + b$, and $\mathbf{std}(a+b)^2 = \|\tilde{a}+\tilde{b}\|^2/n$. Now using (3.6) and $\rho = \cos \angle(\tilde{a}, \tilde{b})$, we get

$$\begin{aligned} n\,\mathbf{std}(a+b)^2 &= \|\tilde{a}+\tilde{b}\|^2 \\ &= \|\tilde{a}\|^2 + 2\rho\|\tilde{a}\|\|\tilde{b}\| + \|\tilde{b}\|^2 \\ &= n\,\mathbf{std}(a)^2 + 2\rho n\,\mathbf{std}(a)\,\mathbf{std}(b) + n\,\mathbf{std}(b)^2. \end{aligned}$$

Dividing by n and taking the squareroot yields the formula above.

If $\rho = 1$, the standard deviation of the sum of vectors is the sum of their standard deviations, *i.e.*,

$$\mathbf{std}(a+b) = \mathbf{std}(a) + \mathbf{std}(b).$$

As ρ decreases, the standard deviation of the sum decreases. When $\rho = 0$, *i.e.*, a and b are uncorrelated, the standard deviation of the sum $a + b$ is

$$\mathbf{std}(a+b) = \sqrt{\mathbf{std}(a)^2 + \mathbf{std}(b)^2},$$

which is smaller than $\mathbf{std}(a) + \mathbf{std}(b)$ (unless one of them is zero). When $\rho = -1$, the standard deviation of the sum is as small as it can be,

$$\mathbf{std}(a+b) = |\,\mathbf{std}(a) - \mathbf{std}(b)|.$$

Hedging investments.   Suppose that vectors a and b are time series of returns for two assets with the same return (average) μ and risk (standard deviation) σ, and correlation coefficient ρ. (These are the traditional symbols used.) The vector $c = (a+b)/2$ is the time series of returns for an investment with 50% in each of the assets. This blended investment has the same return as the original assets, since

$$\mathbf{avg}(c) = \mathbf{avg}((a+b)/2) = (\mathbf{avg}(a) + \mathbf{avg}(b))/2 = \mu.$$

The risk (standard deviation) of this blended investment is

$$\mathbf{std}(c) = \sqrt{2\sigma^2 + 2\rho\sigma^2}/2 = \sigma\sqrt{(1+\rho)/2},$$

using (3.9). From this we see that the risk of the blended investment is never more than the risk of the original assets, and is smaller when the correlation of the original asset returns is smaller. When the returns are uncorrelated, the risk is a factor $1/\sqrt{2} = 0.707$ smaller than the risk of the original assets. If the asset returns are strongly negatively correlated (*i.e.*, ρ is near -1), the risk of the blended investment is much smaller than the risk of the original assets. Investing in two assets with uncorrelated, or negatively correlated, returns is called *hedging* (which is short for 'hedging your bets'). Hedging reduces risk.

Units for heterogeneous vector entries.   When the entries of vectors represent different types of quantities, the choice of units used to represent each entry affects the angle, standard deviation, and correlation between a pair of vectors. The discussion on page 51, about how the choice of units can affect distances between pairs of vectors, therefore applies to these quantities as well. The general rule of thumb is to choose units for different entries so the typical vector entries have similar sizes or ranges of values.

3.5   Complexity

Computing the norm of an n-vector requires n multiplications (to square each entry), $n - 1$ additions (to add the squares), and one squareroot. Even though computing the squareroot typically takes more time than computing the product or sum of two numbers, it is counted as just one flop. So computing the norm takes $2n$ flops. The cost of computing the RMS value of an n-vector is the same, since we can ignore the two flops involved in division by $\sqrt{n}$. Computing the distance between two vectors costs $3n$ flops, and computing the angle between them costs $6n$ flops. All of these operations have order n.

De-meaning an n-vector requires $2n$ flops (n for forming the average and another n flops for subtracting the average from each entry). The standard deviation is the RMS value of the de-meaned vector, and this calculation takes $4n$ flops ($2n$ for computing the de-meaned vector and $2n$ for computing its RMS value). Equation (3.5) suggests a slightly more efficient method with a complexity of $3n$ flops: first compute the average (n flops) and RMS value ($2n$ flops), and then find the standard deviation as $\mathbf{std}(x) = (\mathbf{rms}(x)^2 - \mathbf{avg}(x)^2)^{1/2}$. Standardizing an n-vector costs $5n$ flops. The correlation coefficient between two vectors costs $10n$ flops to compute. These operations also have order n.

As a slightly more involved computation, suppose that we wish to determine the nearest neighbor among a collection of k n-vectors $z_1, \ldots, z_k$ to another n-vector x. (This will come up in the next chapter.) The simple approach is to compute the distances $\|x - z_i\|$ for $i = 1, \ldots, k$, and then find the minimum of these. (Sometimes a comparison of two numbers is also counted as a flop.) The cost of this is $3kn$ flops to compute the distances, and $k - 1$ comparisons to find the minimum. The latter term can be ignored, so the flop count is $3kn$. The order of finding the nearest neighbor in a collection of k n-vectors is kn.

Exercises

3.1 *Distance between Boolean vectors.* Suppose that x and y are Boolean n-vectors, which means that each of their entries is either 0 or 1. What is their distance $\|x - y\|$?

3.2 *RMS value and average of block vectors.* Let x be a block vector with two vector elements, $x = (a, b)$, where a and b are vectors of size n and m, respectively.

(a) Express $\mathbf{rms}(x)$ in terms of $\mathbf{rms}(a)$, $\mathbf{rms}(b)$, m, and n.

(b) Express $\mathbf{avg}(x)$ in terms of $\mathbf{avg}(a)$, $\mathbf{avg}(b)$, m, and n.

3.3 *Reverse triangle inequality.* Suppose a and b are vectors of the same size. The triangle inequality states that $\|a + b\| \le \|a\| + \|b\|$. Show that we also have $\|a + b\| \ge \|a\| - \|b\|$. *Hints.* Draw a picture to get the idea. To show the inequality, apply the triangle inequality to $(a + b) + (-b)$.

3.4 *Norm identities.* Verify that the following identities hold for any two vectors a and b of the same size.

(a) $(a + b)^T (a - b) = \|a\|^2 - \|b\|^2$.

(b) $\|a + b\|^2 + \|a - b\|^2 = 2(\|a\|^2 + \|b\|^2)$. This is called the *parallelogram law.*

3.5 *General norms.* Any real-valued function f that satisfies the four properties given on page 46 (nonnegative homogeneity, triangle inequality, nonnegativity, and definiteness) is called a *vector norm*, and is usually written as $f(x) = \|x\|_{\mathrm{mn}}$, where the subscript is some kind of identifier or mnemonic to identify it. The most commonly used norm is the one we use in this book, the Euclidean norm, which is sometimes written with the subscript 2, as $\|x\|_2$. Two other common vector norms for n-vectors are the 1-*norm* $\|x\|_1$ and the ∞-*norm* $\|x\|_\infty$, defined as

$$\|x\|_1 = |x_1| + \cdots + |x_n|, \qquad \|x\|_\infty = \max\{|x_1|, \ldots, |x_n|\}.$$

These norms are the sum and the maximum of the absolute values of the entries in the vector, respectively. The 1-norm and the ∞-norm arise in some recent and advanced applications, but we will not encounter them in this book.

Verify that the 1-norm and the ∞-norm satisfy the four norm properties listed on page 46.

3.6 *Taylor approximation of norm.* Find a general formula for the Taylor approximation of the function $f(x) = \|x\|$ near a given nonzero vector z. You can express the approximation in the form $\hat{f}(x) = a^T(x - z) + b$.

3.7 *Chebyshev inequality.* Suppose x is a 100-vector with $\mathbf{rms}(x) = 1$. What is the maximum number of entries of x that can satisfy $|x_i| \ge 3$? If your answer is k, explain why no such vector can have $k + 1$ entries with absolute values at least 3, and give an example of a specific 100-vector that has RMS value 1, with k of its entries larger than 3 in absolute value.

3.8 *Converse Chebyshev inequality.* Show that at least one entry of a vector has absolute value at least as large as the RMS value of the vector.

3.9 *Difference of squared distances.* Determine whether the difference of the squared distances to two fixed vectors c and d, defined as

$$f(x) = \|x - c\|^2 - \|x - d\|^2,$$

is linear, affine, or neither. If it is linear, give its inner product representation, *i.e.*, an n-vector a for which $f(x) = a^T x$ for all x. If it is affine, give a and b for which $f(x) = a^T x + b$ holds for all x. If it is neither linear nor affine, give specific x, y, α, and β for which superposition fails, *i.e.*,

$$f(\alpha x + \beta y) \ne \alpha f(x) + \beta f(y).$$

(Provided $\alpha + \beta = 1$, this shows the function is neither linear nor affine.)

3.10 *Nearest neighbor document.* Consider the 5 Wikipedia pages in table 3.1 on page 51. What is the nearest neighbor of (the word count histogram vector of) 'Veterans Day' among the others? Does the answer make sense?

3.11 *Neighboring electronic health records.* Let $x_1, \ldots, x_N$ be n-vectors that contain n features extracted from a set of N electronic health records (EHRs), for a population of N patients. (The features might involve patient attributes and current and past symptoms, diagnoses, test results, hospitalizations, procedures, and medications.) Briefly describe in words a practical use for identifying the 10 nearest neighbors of a given EHR (as measured by their associated feature vectors), among the other EHRs.

3.12 *Nearest point to a line.* Let a and b be different n-vectors. The line passing through a and b is given by the set of vectors of the form $(1 - \theta)a + \theta b$, where θ is a scalar that determines the particular point on the line. (See page 18.)

Let x be any n-vector. Find a formula for the point p on the line that is closest to x. The point p is called the *projection* of x onto the line. Show that $(p - x) \perp (a - b)$, and draw a simple picture illustrating this in 2-D. *Hint.* Work with the square of the distance between a point on the line and x, i.e., $\|(1 - \theta)a + \theta b - x\|^2$. Expand this, and minimize over θ.

3.13 *Nearest nonnegative vector.* Let x be an n-vector and define y as the nonnegative vector (*i.e.*, the vector with nonnegative entries) closest to x. Give an expression for the elements of y. Show that the vector $z = y - x$ is also nonnegative and that $z^T y = 0$.

3.14 *Nearest unit vector.* What is the nearest neighbor of the n-vector x among the unit vectors $e_1, \ldots, e_n$?

3.15 *Average, RMS value, and standard deviation.* Use the formula (3.5) to show that for any vector x, the following two inequalities hold:

$$|\mathbf{avg}(x)| \leq \mathbf{rms}(x), \qquad \mathbf{std}(x) \leq \mathbf{rms}(x).$$

Is it possible to have equality in these inequalities? If $|\mathbf{avg}(x)| = \mathbf{rms}(x)$ is possible, give the conditions on x under which it holds. Repeat for $\mathbf{std}(x) = \mathbf{rms}(x)$.

3.16 *Effect of scaling and offset on average and standard deviation.* Suppose x is an n-vector and α and β are scalars.

   (a) Show that $\mathbf{avg}(\alpha x + \beta \mathbf{1}) = \alpha \, \mathbf{avg}(x) + \beta$.

   (b) Show that $\mathbf{std}(\alpha x + \beta \mathbf{1}) = |\alpha| \, \mathbf{std}(x)$.

3.17 *Average and standard deviation of linear combination.* Let $x_1, \ldots, x_k$ be n-vectors, and $\alpha_1, \ldots, \alpha_k$ be numbers, and consider the linear combination $z = \alpha_1 x_1 + \cdots + \alpha_k x_k$.

   (a) Show that $\mathbf{avg}(z) = \alpha_1 \, \mathbf{avg}(x_1) + \cdots + \alpha_k \, \mathbf{avg}(x_k)$.

   (b) Now suppose the vectors are uncorrelated, which means that for $i \neq j$, x_i and x_j are uncorrelated. Show that $\mathbf{std}(z) = \sqrt{\alpha_1^2 \, \mathbf{std}(x_1)^2 + \cdots + \alpha_k^2 \, \mathbf{std}(x_k)^2}$.

3.18 *Triangle equality.* When does the triangle inequality hold with equality, *i.e.*, what are the conditions on a and b to have $\|a + b\| = \|a\| + \|b\|$?

3.19 *Norm of sum.* Use the formulas (3.1) and (3.6) to show the following:

   (a) $a \perp b$ if and only if $\|a + b\| = \sqrt{\|a\|^2 + \|b\|^2}$.

   (b) Nonzero vectors a and b make an acute angle if and only if $\|a + b\| > \sqrt{\|a\|^2 + \|b\|^2}$.

   (c) Nonzero vectors a and b make an obtuse angle if and only if $\|a + b\| < \sqrt{\|a\|^2 + \|b\|^2}$.

Draw a picture illustrating each case in 2-D.

3.20 *Regression model sensitivity.* Consider the regression model $\hat{y} = x^T \beta + v$, where $\hat{y}$ is the prediction, x is a feature vector, β is a coefficient vector, and v is the offset term. If x and $\tilde{x}$ are feature vectors with corresponding predictions $\hat{y}$ and $\tilde{y}$, show that $|\hat{y} - \tilde{y}| \leq \|\beta\| \|x - \tilde{x}\|$. This means that when $\|\beta\|$ is small, the prediction is not very sensitive to a change in the feature vector.

3.21 *Dirichlet energy of a signal.* Suppose the T-vector x represents a time series or signal. The quantity
$$\mathcal{D}(x) = (x_1 - x_2)^2 + \cdots + (x_{T-1} - x_T)^2,$$
the sum of the differences of adjacent values of the signal, is called the *Dirichlet energy* of the signal, named after the mathematician Peter Gustav Lejeune Dirichlet. The Dirichlet energy is a measure of the roughness or wiggliness of the time series. It is sometimes divided by $T - 1$, to give the mean square difference of adjacent values.

(a) Express $\mathcal{D}(x)$ in vector notation. (You can use vector slicing, vector addition or subtraction, inner product, norm, and angle.)

(b) How small can $\mathcal{D}(x)$ be? What signals x have this minimum value of the Dirichlet energy?

(c) Find a signal x with entries no more than one in absolute value that has the largest possible value of $\mathcal{D}(x)$. Give the value of the Dirichlet energy achieved.

3.22 *Distance from Palo Alto to Beijing.* The surface of the earth is reasonably approximated as a sphere with radius $R = 6367.5$km. A location on the earth's surface is traditionally given by its latitude θ and its longitude λ, which correspond to angular distance from the equator and prime meridian, respectively. The 3-D coordinates of the location are given by

$$\begin{bmatrix} R \sin \lambda \cos \theta \\ R \cos \lambda \cos \theta \\ R \sin \theta \end{bmatrix}.$$

(In this coordinate system $(0, 0, 0)$ is the center of the earth, $R(0, 0, 1)$ is the North pole, and $R(0, 1, 0)$ is the point on the equator on the prime meridian, due south of the Royal Observatory outside London.)

The distance *through the earth* between two locations (3-vectors) a and b is $\|a - b\|$. The distance *along the surface of the earth* between points a and b is $R\angle(a, b)$. Find these two distances between Palo Alto and Beijing, with latitudes and longitudes given below.

City	Latitude θ	Longitude λ
Beijing	$39.914°$	$116.392°$
Palo Alto	$37.429°$	$-122.138°$

3.23 *Angle between two nonnegative vectors.* Let x and y be two nonzero n-vectors with nonnegative entries, *i.e.*, each $x_i \geq 0$ and each $y_i \geq 0$. Show that the angle between x and y lies between 0 and 90°. Draw a picture for the case when $n = 2$, and give a short geometric explanation. When are x and y orthogonal?

3.24 *Distance versus angle nearest neighbor.* Suppose $z_1, \ldots, z_m$ is a collection of n-vectors, and x is another n-vector. The vector z_j is the (distance) nearest neighbor of x (among the given vectors) if
$$\|x - z_j\| \leq \|x - z_i\|, \quad i = 1, \ldots, m,$$
i.e., x has smallest distance to z_j. We say that z_j is the *angle nearest neighbor* of x if
$$\angle(x, z_j) \leq \angle(x, z_i), \quad i = 1, \ldots, m,$$
i.e., x has smallest angle to z_j.

(a) Give a simple specific numerical example where the (distance) nearest neighbor is not the same as the angle nearest neighbor.

(b) Now suppose that the vectors $z_1, \ldots, z_m$ are normalized, which means that $\|z_i\| = 1$, $i = 1, \ldots, m$. Show that in this case the distance nearest neighbor and the angle nearest neighbor are always the same. *Hint.* You can use the fact that arccos is a decreasing function, *i.e.*, for any u and v with $-1 \leq u < v \leq 1$, we have $\arccos(u) > \arccos(v)$.

3.25 *Leveraging.* Consider an asset with return time series over T periods given by the T-vector r. This asset has mean return μ and risk σ, which we assume is positive. We also consider cash as an asset, with return vector $\mu^{\mathrm{rf}}\mathbf{1}$, where μ^{rf} is the cash interest rate per period. Thus, we model cash as an asset with return μ^{rf} and zero risk. (The superscript in μ^{rf} stands for 'risk-free'.) We will create a simple portfolio consisting of the asset and cash. If we invest a fraction θ in the asset, and $1 - \theta$ in cash, our portfolio return is given by the time series

$$p = \theta r + (1 - \theta)\mu^{\mathrm{rf}}\mathbf{1}.$$

We interpret θ as the fraction of our portfolio we hold in the asset. We allow the choices $\theta > 1$, or $\theta < 0$. In the first case we are *borrowing* cash and using the proceeds to buy more of the asset, which is called *leveraging*. In the second case we are *shorting* the asset. When θ is between 0 and 1 we are blending our investment in the asset and cash, which is a form of *hedging*.

(a) Derive a formula for the return and risk of the portfolio, *i.e.*, the mean and standard deviation of p. These should be expressed in terms of μ, σ, μ^{rf}, and θ. Check your formulas for the special cases $\theta = 0$ and $\theta = 1$.

(b) Explain how to choose θ so the portfolio has a given target risk level σ^{tar} (which is positive). If there are multiple values of θ that give the target risk, choose the one that results in the highest portfolio return.

(c) Assume we choose the value of θ as in part (b). When do we use leverage? When do we short the asset? When do we hedge? Your answers should be in English.

3.26 *Time series auto-correlation.* Suppose the T-vector x is a non-constant time series, with x_t the value at time (or period) t. Let $\mu = (\mathbf{1}^T x)/T$ denote its mean value. The *auto-correlation* of x is the function $R(\tau)$, defined for $\tau = 0, 1, \ldots$ as the correlation coefficient of the two vectors $(x, \mu\mathbf{1}_\tau)$ and $(\mu\mathbf{1}_\tau, x)$. (The subscript τ denotes the length of the ones vector.) Both of these vectors also have mean μ. Roughly speaking, $R(\tau)$ tells us how correlated the time series is with a version of itself lagged or shifted by τ periods. (The argument τ is called the lag.)

(a) Explain why $R(0) = 1$, and $R(\tau) = 0$ for $\tau \geq T$.

(b) Let z denote the standardized or z-scored version of x (see page 56). Show that for $\tau = 0, \ldots, T - 1$,

$$R(\tau) = \frac{1}{T} \sum_{t=1}^{T-\tau} z_t z_{t+\tau}.$$

(c) Find the auto-correlation for the time series $x = (+1, -1, +1, -1, \ldots, +1, -1)$. You can assume that T is even.

(d) Suppose x denotes the number of meals served by a restaurant on day τ. It is observed that $R(7)$ is fairly high, and $R(14)$ is also high, but not as high. Give an English explanation of why this might be.

3.27 *Another measure of the spread of the entries of a vector.* The standard deviation is a measure of how much the entries of a vector differ from their mean value. Another measure of how much the entries of an n-vector x differ from each other, called the *mean square difference*, is defined as

$$\mathrm{MSD}(x) = \frac{1}{n^2} \sum_{i,j=1}^{n} (x_i - x_j)^2.$$

(The sum means that you should add up the n^2 terms, as the indices i and j each range from 1 to n.) Show that $\mathrm{MSD}(x) = 2\,\mathbf{std}(x)^2$. *Hint.* First observe that $\mathrm{MSD}(\tilde{x}) = \mathrm{MSD}(x)$, where $\tilde{x} = x - \mathbf{avg}(x)\mathbf{1}$ is the de-meaned vector. Expand the sum and recall that $\sum_{i=1}^{n} \tilde{x}_i = 0$.

3.28 *Weighted norm.* On page 51 we discuss the importance of choosing the units or scaling for the individual entries of vectors, when they represent heterogeneous quantities. Another approach is to use a *weighted norm* of a vector x, defined as

$$\|x\|_w = \sqrt{w_1 x_1^2 + \cdots + w_n x_n^2},$$

where $w_1, \ldots, w_n$ are given positive *weights*, used to assign more or less importance to the different elements of the n-vector x. If all the weights are one, the weighted norm reduces to the usual ('unweighted') norm. It can be shown that the weighted norm is a general norm, *i.e.*, it satisfies the four norm properties listed on page 46. Following the discussion on page 51, one common rule of thumb is to choose the weight w_i as the inverse of the typical value of x_i^2 in the application.

A version of the Cauchy–Schwarz inequality holds for weighted norms: For any n-vector x and y, we have

$$|w_1 x_1 y_1 + \cdots + w_n x_n y_n| \le \|x\|_w \|y\|_w.$$

(The expression inside the absolute value on the left-hand side is sometimes called the weighted inner product of x and y.) Show that this inequality holds. *Hint.* Consider the vectors $\tilde{x} = (x_1 \sqrt{w_1}, \ldots, x_n \sqrt{w_n})$ and $\tilde{y} = (y_1 \sqrt{w_1}, \ldots, y_n \sqrt{w_n})$, and use the (standard) Cauchy–Schwarz inequality.

Chapter 4

Clustering

In this chapter we consider the task of clustering a collection of vectors into groups or clusters of vectors that are close to each other, as measured by the distance between pairs of them. We describe a famous clustering method, called the k-*means algorithm*, and give some typical applications.

The material in this chapter will not be used in the sequel. But the ideas, and the k-means algorithm in particular, are widely used in practical applications, and rely only on the ideas developed in the previous three chapters. So this chapter can be considered an interlude that covers useful material that builds on the ideas developed so far.

4.1 Clustering

Suppose we have N n-vectors, $x_1, \ldots, x_N$. The goal of *clustering* is to group or partition the vectors (if possible) into k groups or clusters, with the vectors in each group close to each other. Clustering is very widely used in many application areas, typically (but not always) when the vectors represent features of objects.

Normally we have k much smaller than N, *i.e.*, there are many more vectors than groups. Typical applications use values of k that range from a handful to a few hundred or more, with values of N that range from hundreds to billions. Part of the task of clustering a collection of vectors is to determine whether or not the vectors can be divided into k groups, with vectors in each group near each other. Of course this depends on k, the number of clusters, and the particular data, *i.e.*, the vectors $x_1, \ldots, x_N$.

Figure 4.1 shows a simple example, with $N = 300$ 2-vectors, shown as small circles. We can easily see that this collection of vectors can be divided into $k = 3$ clusters, shown on the right with the colors representing the different clusters. We could partition these data into other numbers of clusters, but we can see that $k = 3$ is a good value.

This example is not typical in several ways. First, the vectors have dimension $n = 2$. Clustering any set of 2-vectors is easy: We simply scatter plot the values

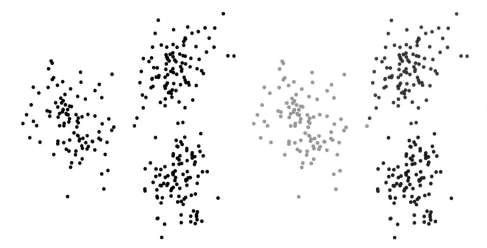

Figure 4.1 300 points in a plane. The points can be clustered in the three groups shown on the right.

and check visually if the data are clustered, and if so, how many clusters there are. In almost all applications n is larger than 2 (and typically, much larger than 2), in which case this simple visual method cannot be used. The second way in which it is not typical is that the points are very well clustered. In most applications, the data are not as cleanly clustered as in this simple example; there are several or even many points that lie in between clusters. Finally, in this example, it is clear that the best choice of k is $k = 3$. In real examples, it can be less clear what the best value of k is. But even when the clustering is not as clean as in this example, and the best value of k is not clear, clustering can be very useful in practice.

Examples.   Before we delve more deeply into the details of clustering and clustering algorithms, we list some common applications where clustering is used.

- *Topic discovery.* Suppose x_i are word histograms associated with N documents. A clustering algorithm partitions the documents into k groups, which typically can be interpreted as groups of documents with the same or similar topics, genre, or author. Since the clustering algorithm runs automatically and without any understanding of what the words in the dictionary mean, this is sometimes called *automatic topic discovery.*

- *Patient clustering.* If x_i are feature vectors associated with N patients admitted to a hospital, a clustering algorithm clusters the patients into k groups of similar patients (at least in terms of their feature vectors).

- *Customer market segmentation.* Suppose the vector x_i gives the quantities (or dollar values) of n items purchased by customer i over some period of time. A clustering algorithm will group the customers into k market segments, which are groups of customers with similar purchasing patterns.

- *ZIP code clustering.* Suppose that x_i is a vector giving n quantities or statistics for the residents of ZIP code i, such as numbers of residents in various age groups, household size, education statistics, and income statistics. (In this example N is around 40000.) A clustering algorithm might be used to cluster the 40000 ZIP codes into, say, $k = 100$ groups of ZIP codes with similar statistics.

- *Student clustering.* Suppose the vector x_i gives the detailed grading record of student i in a course, *i.e.*, her grades on each question in the quizzes, homework assignments, and exams. A clustering algorithm might be used to cluster the students into $k = 10$ groups of students who performed similarly.

- *Survey response clustering.* A group of N people respond to a survey with n questions. Each question contains a statement, such as 'The movie was too long', followed by some ordered options such as

    Strongly Disagree,   Disagree,   Neutral,   Agree,   Strongly Agree.

    (This is called a *Likert scale*, named after the psychologist Rensis Likert.) Suppose the n-vector x_i encodes the selections of respondent i on the n questions, using the numerical coding -2, -1, 0, $+1$, $+2$ for the responses above. A clustering algorithm can be used to cluster the respondents into k groups, each with similar responses to the survey.

- *Weather zones.* For each of N counties we have a 24-vector x_i that gives the average monthly temperature in the first 12 entries and the average monthly rainfall in the last 12 entries. (We can standardize all the temperatures, and all the rainfall data, so they have a typical range between -1 and $+1$.) The vector x_i summarizes the annual weather pattern in county i. A clustering algorithm can be used to cluster the counties into k groups that have similar weather patterns, called *weather zones*. This clustering can be shown on a map, and used to recommend landscape plantings depending on zone.

- *Daily energy use patterns.* The 24-vectors x_i give the average (electric) energy use for N customers over some period (say, a month) for each hour of the day. A clustering algorithm partitions customers into groups, each with similar patterns of daily energy consumption. We might expect a clustering algorithm to 'discover' which customers have a swimming pool, an electric water heater, or solar panels.

- *Financial sectors.* For each of N companies we have an n-vector whose components are financial and business attributes such as total capitalization, quarterly returns and risks, trading volume, profit and loss, or dividends paid. (These quantities would typically be scaled so as to have similar ranges of values.) A clustering algorithm would group companies into *sectors*, *i.e.*, groups of companies with similar attributes.

In each of these examples, it would be quite informative to know that the vectors can be well clustered into, say, $k = 5$ or $k = 37$ groups. This can be used to develop insight into the data. By examining the clusters we can often understand them, and assign labels or descriptions to them.

4.2   A clustering objective

In this section we formalize the idea of clustering, and introduce a natural measure of the quality of a given clustering.

Specifying the cluster assignments.   We specify a clustering of the vectors by saying which cluster or group each vector belongs to. We label the groups $1, \ldots, k$, and specify a clustering or assignment of the N given vectors to groups using an N-vector c, where c_i is the group (number) that the vector x_i is assigned to. As a simple example with $N = 5$ vectors and $k = 3$ groups, $c = (3, 1, 1, 1, 2)$ means that x_1 is assigned to group 3, x_2, x_3, and x_4 are assigned to group 1, and x_5 is assigned to group 2. We will also describe the clustering by the sets of indices for each group. We let G_j be the set of indices corresponding to group j. For our simple example above, we have

$$G_1 = \{2, 3, 4\}, \qquad G_2 = \{5\}, \qquad G_3 = \{1\}.$$

(Here we are using the notation of sets; see appendix A.) Formally, we can express these index sets in terms of the group assignment vector c as

$$G_j = \{i \mid c_i = j\},$$

which means that G_j is the set of all indices i for which $c_i = j$.

Group representatives.   With each of the groups we associate a *group representative n-vector*, which we denote $z_1, \ldots, z_k$. These representatives can be any n-vectors; they do not need to be one of the given vectors. We want each representative to be close to the vectors in its associated group, *i.e.*, we want the quantities

$$\|x_i - z_{c_i}\|$$

to be small. (Note that x_i is in group $j = c_i$, so z_{c_i} is the representative vector associated with data vector x_i.)

A clustering objective.   We can now give a single number that we use to judge a choice of clustering, along with a choice of the group representatives. We define

$$J^{\text{clust}} = \left(\|x_1 - z_{c_1}\|^2 + \cdots + \|x_N - z_{c_N}\|^2 \right) / N, \qquad (4.1)$$

which is the mean square distance from the vectors to their associated representatives. Note that J^{clust} depends on the cluster assignments (*i.e.*, c), as well as the choice of the group representatives $z_1, \ldots, z_k$. The smaller J^{clust} is, the better the clustering. An extreme case is $J^{\text{clust}} = 0$, which means that the distance between every original vector and its assigned representative is zero. This happens only when the original collection of vectors only takes k different values, and each vector is assigned to the representative it is equal to. (This extreme case would probably not occur in practice.)

Our choice of clustering objective J^{clust} makes sense, since it encourages all points to be near their associated representative, but there are other reasonable choices. For example, it is possible to use an objective that encourages more balanced groupings. But we will stick with this basic (and very common) choice of clustering objective.

Optimal and suboptimal clustering. We seek a clustering, *i.e.*, a choice of group assignments $c_1, \ldots, c_N$ and a choice of representatives $z_1, \ldots, z_k$, that minimize the objective J^{clust}. We call such a clustering *optimal*. Unfortunately, for all but the very smallest problems, it is practically impossible to find an optimal clustering. (It can be done in principle, but the amount of computation needed grows extremely rapidly with N.) The good news is that the k-means algorithm described in the next section requires far less computation (and indeed, can be run for problems with N measured in billions), and often finds a very good, if not the absolute best, clustering. (Here, 'very good' means a clustering and choice of representatives that achieves a value of J^{clust} near its smallest possible value.) We say that the clustering choices found by the k-means algorithm are *suboptimal*, which means that they might not give the lowest possible value of J^{clust}.

Even though it is a hard problem to choose the best clustering and the best representatives, it turns out that we *can* find the best clustering, if the representatives are fixed, and we can find the best representatives, if the clustering is fixed. We address these two topics now.

Partitioning the vectors with the representatives fixed. Suppose that the group representatives $z_1, \ldots, z_k$ are fixed, and we seek the group assignments $c_1, \ldots, c_N$ that achieve the smallest possible value of J^{clust}. It turns out that this problem can be solved exactly.

The objective J^{clust} is a sum of N terms. The choice of c_i (*i.e.*, the group to which we assign the vector x_i) only affects the ith term in J^{clust}, which is $(1/N)\|x_i - z_{c_i}\|^2$. We can choose c_i to minimize just this term, since c_i does not affect the other $N - 1$ terms in J^{clust}. How do we choose c_i to minimize this term? This is easy: We simply choose c_i to be the value of j that minimizes $\|x_i - z_j\|$ over j. In other words, we should assign each data vector x_i to its nearest neighbor among the representatives. This choice of assignment is very natural, and easily carried out.

So when the group representatives are fixed, we can readily find the best group assignment (*i.e.*, the one that minimizes J^{clust}), by assigning each vector to its nearest representative. With this choice of group assignment, we have (by the way the assignment is made)

$$\|x_i - z_{c_i}\| = \min_{j=1,\ldots,k} \|x_i - z_j\|,$$

so the value of J^{clust} is given by

$$\left(\min_{j=1,\ldots,k} \|x_1 - z_j\|^2 + \cdots + \min_{j=1,\ldots,k} \|x_N - z_j\|^2 \right) / N.$$

This has a simple interpretation: It is the mean of the squared distance from the data vectors to their closest representative.

Optimizing the group representatives with the assignment fixed.   Now we turn to the problem of choosing the group representatives, with the clustering (group assignments) fixed, in order to minimize our objective J^{clust}. It turns out that this problem also has a simple and natural solution.

We start by re-arranging the sum of N terms into k sums, each associated with one group. We write

$$J^{\text{clust}} = J_1 + \cdots + J_k,$$

where

$$J_j = (1/N) \sum_{i \in G_j} \|x_i - z_j\|^2$$

is the contribution to the objective J^{clust} from the vectors in group j. (The sum here means that we should add up all terms of the form $\|x_i - z_j\|^2$, for any $i \in G_j$, *i.e.*, for any vector x_i in group j; see appendix A.)

The choice of group representative z_j only affects the term J_j; it has no effect on the other terms in J^{clust}. So we can choose each z_j to minimize J_j. Thus we should choose the vector z_j so as to minimize the mean square distance to the vectors in group j. This problem has a very simple solution: We should choose z_j to be the average (or mean or centroid) of the vectors x_i in its group:

$$z_j = (1/|G_j|) \sum_{i \in G_j} x_i,$$

where $|G_j|$ is standard mathematical notation for the number of elements in the set G_j, *i.e.*, the size of group j. (See exercise 4.1.)

So if we fix the group assignments, we minimize J^{clust} by choosing each group representative to be the average or centroid of the vectors assigned to its group. (This is sometimes called the *group centroid* or *cluster centroid*.)

4.3   The k-means algorithm

It might seem that we can now solve the problem of choosing the group assignments and the group representatives to minimize J^{clust}, since we know how to do this when one or the other choice is fixed. But the two choices are circular, *i.e.*, each depends on the other. Instead we rely on a very old idea in computation: We simply *iterate* between the two choices. This means that we repeatedly alternate between updating the group assignments, and then updating the representatives, using the methods developed above. In each step the objective J^{clust} gets better (*i.e.*, goes down) unless the step does not change the choice. Iterating between choosing the group representatives and choosing the group assignments is the celebrated k-means algorithm for clustering a collection of vectors.

The k-means algorithm was first proposed in 1957 by Stuart Lloyd, and independently by Hugo Steinhaus. It is sometimes called the Lloyd algorithm. The name 'k-means' has been used since the 1960s.

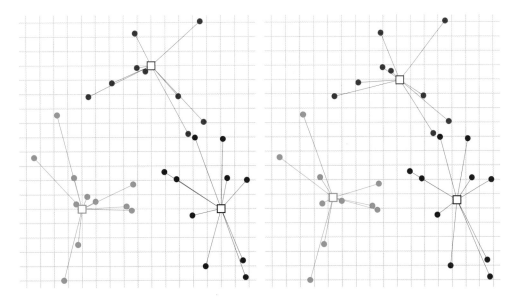

Figure 4.2   *One iteration of the k-means algorithm.* The 30 2-vectors x_i are shown as filled circles, and the 3 group representatives z_j are shown as rectangles. In the left-hand figure the vectors are each assigned to the closest representative. In the right-hand figure, the representatives are replaced by the cluster centroids.

Algorithm 4.1 k-MEANS ALGORITHM

given a list of N vectors $x_1, \ldots, x_N$, and an initial list of k group representative vectors $z_1, \ldots, z_k$

repeat until convergence

1. *Partition the vectors into k groups.* For each vector $i = 1, \ldots, N$, assign x_i to the group associated with the nearest representative.

2. *Update representatives.* For each group $j = 1, \ldots, k$, set z_j to be the mean of the vectors in group j.

One iteration of the k-means algorithm is illustrated in figure 4.2.

Comments and clarifications.

- Ties in step 1 can be broken by assigning x_i to the group associated with one of the closest representatives with the smallest value of j.

- It is possible that in step 1, one or more of the groups can be empty, *i.e.*, contain no vectors. In this case we simply drop this group (and its representative). When this occurs, we end up with a partition of the vectors into fewer than k groups.

- If the group assignments found in step 1 are the same in two successive iterations, the representatives in step 2 will also be the same. It follows that the group assignments and group representatives will never change in future iterations, so we should stop the algorithm. This is what we mean by 'until convergence'. In practice, one often stops the algorithm earlier, as soon as the improvement in J^{clust} in successive iterations becomes very small.

- We start the algorithm with a choice of initial group representatives. One simple method is to pick the representatives randomly from the original vectors; another is to start from a random assignment of the original vectors to k groups, and use the means of the groups as the initial representatives. (There are more sophisticated methods for choosing an initial representatives, but this topic is beyond the scope of this book.)

Convergence. The fact that J^{clust} decreases in each step implies that the k-means algorithm converges in a finite number of steps. However, depending on the initial choice of representatives, the algorithm can, and does, converge to different final partitions, with different objective values.

The k-means algorithm is a *heuristic*, which means it cannot guarantee that the partition it finds minimizes our objective J^{clust}. For this reason it is common to run the k-means algorithm several times, with different initial representatives, and choose the one among them with the smallest final value of J^{clust}. Despite the fact that the k-means algorithm is a heuristic, it is very useful in practical applications, and very widely used.

Figure 4.3 shows a few iterations generated by the k-means algorithm, applied to the example of figure 4.1. We take $k = 3$ and start with randomly chosen group representatives. The final clustering is shown in figure 4.4. Figure 4.5 shows how the clustering objective decreases in each step.

Interpreting the representatives. The representatives $z_1, \ldots, z_k$ associated with a clustering are quite interpretable. Suppose, for example, that voters in some election can be well clustered into 7 groups, on the basis of a data set that includes demographic data and questionnaire or poll data. If the 4th component of our vectors is the age of the voter, then $(z_3)_4 = 37.8$ tells us that the average age of voters in group 3 is 37.8. Insight gained from this data can be used to tune campaign messages, or choose media outlets for campaign advertising.

Another way to interpret the group representatives is to find one or a few of the original data points that are closest to each representive. These can be thought of as archetypes for the group.

Choosing k. It is common to run the k-means algorithm for different values of k, and compare the results. How to choose a value of k among these depends on how the clustering will be used, which we discuss a bit more in §4.5. But some general statements can be made. For example, if the value of J^{clust} with $k = 7$ is quite a bit smaller than the values of J^{clust} for $k = 2, \ldots, 6$, and not much larger than the values of J^{clust} for $k = 8, 9, \ldots$, we could reasonably choose $k = 7$, and conclude that our data (list of vectors) partitions nicely into 7 groups.

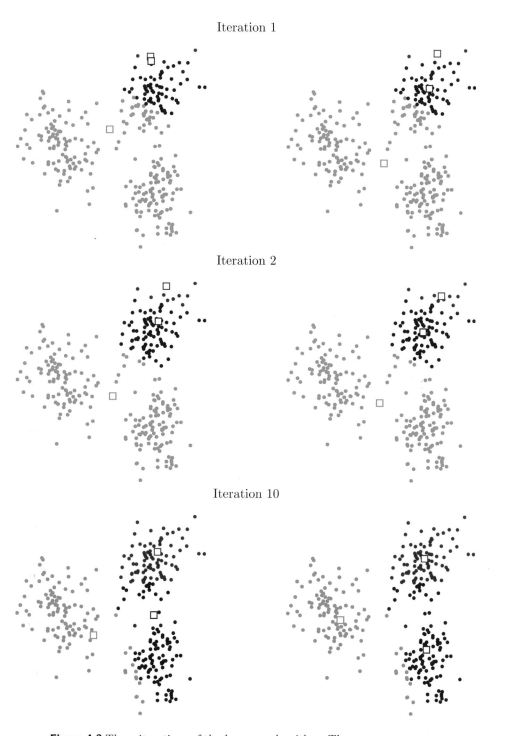

Figure 4.3 Three iterations of the k-means algorithm. The group representatives are shown as squares. In each row, the left-hand plot shows the result of partitioning the vectors in the 3 groups (step 1 of algorithm 4.1). The right-hand plot shows the updated representatives (step 2 of the algorithm).

Figure 4.4 Final clustering.

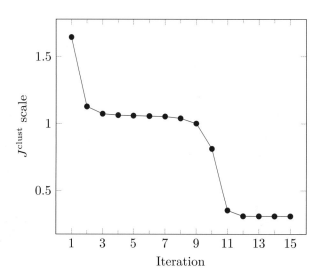

Figure 4.5 The clustering objective J^{clust} after step 1 of each iteration.

Complexity. In step 1 of the k-means algorithm, we find the nearest neighbor to each of N n-vectors, over the list of k centroids. This requires approximately $3Nkn$ flops. In step 2 we average the n-vectors over each of the cluster groups. For a cluster with p vectors, this requires $n(p-1)$ flops, which we approximate as np flops; averaging all clusters requires a total of Nn flops. This is less than the cost of partitioning in step 1. So k-means requires around $(3k+1)Nn$ flops per iteration. Its order is Nkn flops.

Each run of k-means typically takes fewer than a few tens of iterations, and usually k-means is run some modest number of times, like 10. So a very rough guess of the number of flops required to run k-means 10 times (in order to choose the best partition found) is $1000Nkn$ flops.

As an example, suppose we use k-means to partition $N = 100000$ vectors with size $n = 100$ into $k = 10$ groups. On a 1 Gflop/s computer we guess that this will take around 100 seconds. Given the approximations made here (for example, the number of iterations that each run of k-means will take), this is obviously a crude estimate.

4.4   Examples

4.4.1   Image clustering

The MNIST (Mixed National Institute of Standards) database of handwritten digits is a data set containing $N = 60000$ grayscale images of size 28×28, which we represent as n-vectors with $n = 28 \times 28 = 784$. Figure 4.6 shows a few examples from the data set. (The data set is available from Yann LeCun at `yann.lecun.com/exdb/mnist`.)

We use the k-means algorithm to partition these images into $k = 20$ clusters, starting with a random assignment of the vectors to groups, and repeating the experiment 20 times. Figure 4.7 shows the clustering objective versus iteration number for three of the 20 initial assignments, including the two that gave the lowest and the highest final values of the objective.

Figure 4.8 shows the representatives with the lowest final value of the clustering objective. Figure 4.9 shows the set with the highest value. We can see that most of the representatives are recognizable digits, with some reasonable confusion, for example between '4' and '9' or '3' and '8'. This is impressive when you consider that the k-means algorithm knows nothing about digits, handwriting, or even that the 784-vectors represent 28×28 images; it uses only the distances between 784-vectors. One interpretation is that the k-means algorithm has 'discovered' the digits in the data set.

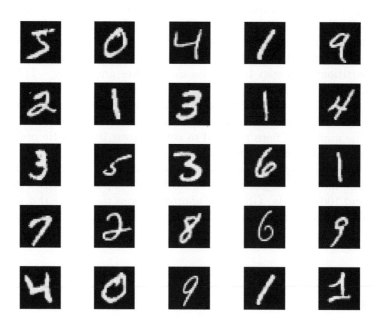

Figure 4.6 25 images of handwritten digits from the MNIST data set. Each image has size 28×28, and can be represented by a 784-vector.

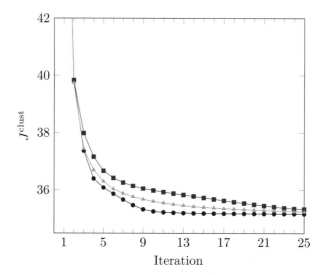

Figure 4.7 Clustering objective J^{clust} after each iteration of the k-means algorithm, for three initial partitions, on digits of the MNIST set.

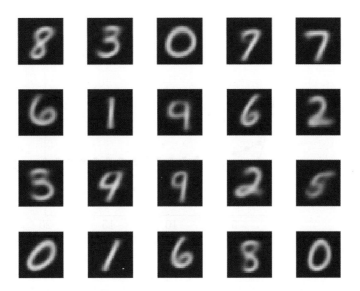

Figure 4.8 Group representatives found by the k-means algorithm applied to the MNIST set.

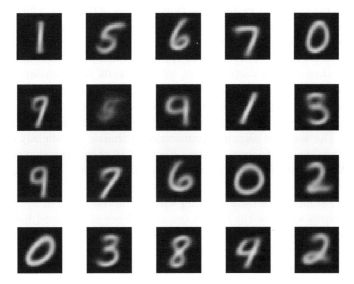

Figure 4.9 Group representatives found by the k-means algorithm applied to the MNIST set, with a different starting point than in figure 4.8.

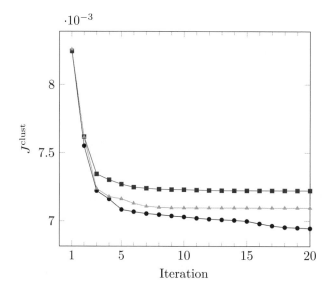

Figure 4.10 Clustering objective J^{clust} after each iteration of the k-means algorithm, for three initial partitions, on Wikipedia word count histograms.

4.4.2   Document topic discovery

We start with a corpus of $N = 500$ Wikipedia articles, compiled from weekly lists of the most popular articles between September 6, 2015, and June 11, 2016. We remove the section titles and reference sections (bibliography, notes, references, further reading), and convert each document to a list of words. The conversion removes numbers and stop words, and applies a stemming algorithm to nouns and verbs. We then form a dictionary of all the words that appear in at least 20 documents. This results in a dictionary of 4423 words. Each document in the corpus is represented by a word histogram vector of length 4423.

We apply the k-means algorithm with $k = 9$, and 20 randomly chosen initial partitions. The k-means algorithm converges to similar but slightly different clusterings of the documents in each case. Figure 4.10 shows the clustering objective versus iteration of the k-means algorithm for three of these, including the one that gave the lowest final value of J^{clust}, which we use below.

Table 4.1 summarizes the clustering with the lowest value of J^{clust}. For each of the nine clusters we show the largest ten coefficients of the word histogram of the cluster representative. Table 4.2 gives the size of each cluster and the titles of the ten articles closest to the cluster representative.

Each of the clusters makes sense, and mostly contains documents on similar topics, or similar themes. The words with largest coefficients in the group representatives also make sense. It is interesting to see that k-means clustering has assigned movies and TV series (mostly) to different clusters (9 and 6). One can also note that clusters 8 and 9 share several top key words but are on separate topics (actors and movies, respectively).

Cluster 1		Cluster 2		Cluster 3	
Word	Coefficient	Word	Coefficient	Word	Coefficient
fight	0.038	holiday	0.012	united	0.004
win	0.022	celebrate	0.009	family	0.003
event	0.019	festival	0.007	party	0.003
champion	0.015	celebration	0.007	president	0.003
fighter	0.015	calendar	0.006	government	0.003
bout	0.015	church	0.006	school	0.003
title	0.013	united	0.005	American	0.003
Ali	0.012	date	0.005	university	0.003
championship	0.011	moon	0.005	city	0.003
bonus	0.010	event	0.005	attack	0.003

Cluster 4		Cluster 5		Cluster 6	
Word	Coefficient	Word	Coefficient	Word	Coefficient
album	0.031	game	0.023	series	0.029
release	0.016	season	0.020	season	0.027
song	0.015	team	0.018	episode	0.013
music	0.014	win	0.017	character	0.011
single	0.011	player	0.014	film	0.008
record	0.010	play	0.013	television	0.007
band	0.009	league	0.010	cast	0.006
perform	0.007	final	0.010	announce	0.006
tour	0.007	score	0.008	release	0.005
chart	0.007	record	0.007	appear	0.005

Cluster 7		Cluster 8		Cluster 9	
Word	Coefficient	Word	Coefficient	Word	Coefficient
match	0.065	film	0.036	film	0.061
win	0.018	star	0.014	million	0.019
championship	0.016	role	0.014	release	0.013
team	0.015	play	0.010	star	0.010
event	0.015	series	0.009	character	0.006
style	0.014	appear	0.008	role	0.006
raw	0.013	award	0.008	movie	0.005
title	0.011	actor	0.007	weekend	0.005
episode	0.010	character	0.006	story	0.005
perform	0.010	release	0.006	gross	0.005

Table 4.1 The 9 cluster representatives. For each representative we show the largest 10 coefficients of the word histogram.

Cluster	Size	Titles
1	21	Floyd Mayweather, Jr; Kimbo Slice; Ronda Rousey; José Aldo; Joe Frazier; Wladimir Klitschko; Saul Álvarez; Gennady Golovkin; Nate Diaz; Conor McGregor.
2	43	Halloween; Guy Fawkes Night; Diwali; Hanukkah; Groundhog Day; Rosh Hashanah; Yom Kippur; Seventh-day Adventist Church; Remembrance Day; Mother's Day.
3	189	Mahatma Gandhi; Sigmund Freud; Carly Fiorina; Frederick Douglass; Marco Rubio; Christopher Columbus; Fidel Castro; Jim Webb; Genie (feral child); Pablo Escobar.
4	46	David Bowie; Kanye West; Celine Dion; Kesha; Ariana Grande; Adele; Gwen Stefani; Anti (album); Dolly Parton; Sia Furler.
5	49	Kobe Bryant; Lamar Odom; Johan Cruyff; Yogi Berra; José Mourinho; Halo 5: Guardians; Tom Brady; Eli Manning; Stephen Curry; Carolina Panthers.
6	39	The X-Files; Game of Thrones; House of Cards (U.S. TV series); Daredevil (TV series); Supergirl (U.S. TV series); American Horror Story; The Flash (2014 TV series); The Man in the High Castle (TV series); Sherlock (TV series); Scream Queens (2015 TV series).
7	16	Wrestlemania 32; Payback (2016); Survivor Series (2015); Royal Rumble (2016); Night of Champions (2015); Fastlane (2016); Extreme Rules (2016); Hell in a Cell (2015); TLC: Tables, Ladders & Chairs (2015); Shane McMahon.
8	58	Ben Affleck; Johnny Depp; Maureen O'Hara; Kate Beckinsale; Leonardo DiCaprio; Keanu Reeves; Charlie Sheen; Kate Winslet; Carrie Fisher; Alan Rickman.
9	39	Star Wars: The Force Awakens; Star Wars Episode I: The Phantom Menace; The Martian (film); The Revenant (2015 film); The Hateful Eight; Spectre (2015 film); The Jungle Book (2016 film); Bajirao Mastani (film); Back to the Future II; Back to the Future.

Table 4.2 Cluster sizes and titles of 10 documents closest to the cluster representatives.

The identification of these separate topics among the documents is impressive, when you consider that the k-means algorithm does not understand the meaning of the words in the documents (and indeed, does not even know the order of the words in each document). It uses only the simple concept of document dissimilarity, as measured by the distance between word count histogram vectors.

4.5 Applications

Clustering, and the k-means algorithm in particular, has many uses and applications. It can be used for exploratory data analysis, to get an idea of what a large collection of vectors 'looks like'. When k is small enough, say less than a few tens, it is common to examine the group representatives, and some of the vectors in the associated groups, to interpret or label the groups. Clustering can also be used for more specific directed tasks, a few of which we describe below.

Classification.   We cluster a large collection of vectors into k groups, and label the groups by hand. We can now assign (or classify) *new* vectors to one of the k groups by choosing the nearest group representative. In our example of the handwritten digits above, this would give us a rudimentary digit classifier, which would automatically guess what a written digit is from its image. In the topic discovery example, we can automatically classify new documents into one of the k topics. (We will see better classification methods in chapter 14.)

Recommendation engine.   Clustering can be used to build a *recommendation engine*, which suggests items that a user or customer might be interested in. Suppose the vectors give the number of times a user has listened to or streamed each song from a library of n songs over some period. These vectors are typically sparse, since each user has listened to only a very small fraction of the music library. Clustering the vectors reveals groups of users with similar musical taste. The group representatives have a nice interpretation: $(z_j)_i$ is the average number of times users in group j listened to song i.

This interpretation allows us to create a set of recommendations for each user. We first identify which cluster j her music listening vector x_i is in. Then we can suggest to her songs that she has not listened to, but others in her group (*i.e.*, those with similar musical taste) have listened to most often. To recommend 5 songs to her, we find the indices l with $(x_i)_l = 0$, with the 5 largest values of $(z_j)_l$.

Guessing missing entries.   Suppose we have a collection of vectors, with some entries of some of the vectors missing or not given. (The symbol '?' or '*' is sometimes used to denote a missing entry in a vector.) For example, suppose the vectors collect attributes of a collection of people, such as age, sex, years of education, income, number of children, and so on. A vector containing the symbol '?' in the age entry means that we do not know that particular person's age. Guessing missing entries of vectors in a collection of vectors is sometimes called *imputing* the missing entries. In our example, we might want to guess the age of the person whose age we do not know.

We can use clustering, and the k-means algorithm in particular, to guess the missing entries. We first carry out k-means clustering on our data, using only those vectors that are complete, *i.e.*, all of their entries are known. Now consider a vector x in our collection that is missing one or more entries. Since some of the entries of x are unknown, we cannot find the distances $\|x - z_j\|$, and therefore we cannot say which group representative is closest to x. Instead we will find the closest group representative to x using only the known entries in x, by finding j that minimizes

$$\sum_{i \in \mathcal{K}} (x_i - (z_j)_i)^2,$$

where $\mathcal{K}$ is the set of indices for the known entries of the vector x. This gives us the closest representative to x, calculated using only its known entries. To guess the missing entries in x, we simply use the corresponding entries in z_j, the nearest representative.

Returning to our example, we would guess the age of the person with a missing age entry by finding the closest representative (ignoring age); then we use the age entry of the representative, which is simply the average age of all the people in that cluster.

Exercises

4.1 *Minimizing mean square distance to a set of vectors.* Let $x_1, \ldots, x_L$ be a collection of n-vectors. In this exercise you will fill in the missing parts of the argument to show that the vector z which minimizes the sum-square distance to the vectors,

$$J(z) = \|x_1 - z\|^2 + \cdots + \|x_L - z\|^2,$$

is the average or centroid of the vectors, $\overline{x} = (1/L)(x_1 + \cdots + x_L)$. (This result is used in one of the steps in the k-means algorithm. But here we have simplified the notation.)

(a) Explain why, for any z, we have

$$J(z) = \sum_{i=1}^{L} \|x_i - \overline{x} - (z - \overline{x})\|^2 = \sum_{i=1}^{L} \left(\|x_i - \overline{x}\|^2 - 2(x_i - \overline{x})^T(z - \overline{x}) \right) + L\|z - \overline{x}\|^2.$$

(b) Explain why $\sum_{i=1}^{L}(x_i - \overline{x})^T(z - \overline{x}) = 0$. *Hint.* Write the left-hand side as

$$\left(\sum_{i=1}^{L}(x_i - \overline{x}) \right)^T (z - \overline{x}),$$

and argue that the left-hand vector is 0.

(c) Combine the results of (a) and (b) to get $J(z) = \sum_{i=1}^{L} \|x_i - \overline{x}\|^2 + L\|z - \overline{x}\|^2$. Explain why for any $z \neq \overline{x}$, we have $J(z) > J(\overline{x})$. This shows that the choice $z = \overline{x}$ minimizes $J(z)$.

4.2 *k-means with nonnegative, proportions, or Boolean vectors.* Suppose that the vectors $x_1, \ldots, x_N$ are clustered using k-means, with group representatives $z_1, \ldots, z_k$.

(a) Suppose the original vectors x_i are nonnegative, *i.e.*, their entries are nonnegative. Explain why the representatives z_j are also nonnegative.

(b) Suppose the original vectors x_i represent proportions, *i.e.*, their entries are nonnegative and sum to one. (This is the case when x_i are word count histograms, for example.) Explain why the representatives z_j also represent proportions, *i.e.*, their entries are nonnegative and sum to one.

(c) Suppose the original vectors x_i are Boolean, *i.e.*, their entries are either 0 or 1. Give an interpretation of $(z_j)_i$, the ith entry of the j group representative.

Hint. Each representative is the average of some of the original vectors.

4.3 *Linear separation in 2-way partitioning.* Clustering a collection of vectors into $k = 2$ groups is called 2-way partitioning, since we are partitioning the vectors into 2 groups, with index sets G_1 and G_2. Suppose we run k-means, with $k = 2$, on the n-vectors $x_1, \ldots, x_N$. Show that there is a nonzero vector w and a scalar v that satisfy

$$w^T x_i + v \geq 0 \text{ for } i \in G_1, \qquad w^T x_i + v \leq 0 \text{ for } i \in G_2.$$

In other words, the affine function $f(x) = w^T x + v$ is greater than or equal to zero on the first group, and less than or equal to zero on the second group. This is called *linear separation* of the two groups (although *affine separation* would be more accurate). *Hint.* Consider the function $\|x - z_1\|^2 - \|x - z_2\|^2$, where z_1 and z_2 are the group representatives.

4.4 *Pre-assigned vectors.* Suppose that some of the vectors $x_1, \ldots, x_N$ are assigned to specific groups. For example, we might insist that x_{27} be assigned to group 5. Suggest a simple modification of the k-means algorithm that respects this requirement. Describe a practical example where this might arise, when each vector represents n features of a medical patient.

Chapter 5

Linear independence

In this chapter we explore the concept of linear independence, which will play an important role in the sequel.

5.1 Linear dependence

A collection or list of n-vectors $a_1, \ldots, a_k$ (with $k \geq 1$) is called *linearly dependent* if

$$\beta_1 a_1 + \cdots + \beta_k a_k = 0$$

holds for some $\beta_1, \ldots, \beta_k$ that are not all zero. In other words, we can form the zero vector as a linear combination of the vectors, with coefficients that are not all zero. Linear dependence of a list of vectors does not depend on the ordering of the vectors in the list.

When a collection of vectors is linearly dependent, at least one of the vectors can be expressed as a linear combination of the other vectors: If $\beta_i \neq 0$ in the equation above (and by definition, this must be true for at least one i), we can move the term $\beta_i a_i$ to the other side of the equation and divide by β_i to get

$$a_i = (-\beta_1/\beta_i)a_1 + \cdots + (-\beta_{i-1}/\beta_i)a_{i-1} + (-\beta_{i+1}/\beta_i)a_{i+1} + \cdots + (-\beta_k/\beta_i)a_k.$$

The converse is also true: If any vector in a collection of vectors is a linear combination of the other vectors, then the collection of vectors is linearly dependent.

Following standard mathematical language usage, we will say "The vectors $a_1, \ldots, a_k$ are linearly dependent" to mean "The list of vectors $a_1, \ldots, a_k$ is linearly dependent". But it must be remembered that linear dependence is an attribute of a *collection* of vectors, and not individual vectors.

Linearly independent vectors. A collection of n-vectors $a_1, \ldots, a_k$ (with $k \geq 1$) is called *linearly independent* if it is not linearly dependent, which means that

$$\beta_1 a_1 + \cdots + \beta_k a_k = 0 \tag{5.1}$$

only holds for $\beta_1 = \cdots = \beta_k = 0$. In other words, the only linear combination of the vectors that equals the zero vector is the linear combination with all coefficients zero.

As with linear dependence, we will say "The vectors $a_1, \ldots, a_k$ are linearly independent" to mean "The list of vectors $a_1, \ldots, a_k$ is linearly independent". But, like linear dependence, linear independence is an attribute of a collection of vectors, and not individual vectors.

It is generally not easy to determine by casual inspection whether or not a list of vectors is linearly dependent or linearly independent. But we will soon see an algorithm that does this.

Examples.

- A list consisting of a single vector is linearly dependent only if the vector is zero. It is linearly independent only if the vector is nonzero.

- Any list of vectors containing the zero vector is linearly dependent.

- A list of two vectors is linearly dependent if and only if one of the vectors is a multiple of the other one. More generally, a list of vectors is linearly dependent if any one of the vectors is a multiple of another one.

- The vectors

$$a_1 = \begin{bmatrix} 0.2 \\ -7.0 \\ 8.6 \end{bmatrix}, \qquad a_2 = \begin{bmatrix} -0.1 \\ 2.0 \\ -1.0 \end{bmatrix}, \qquad a_3 = \begin{bmatrix} 0.0 \\ -1.0 \\ 2.2 \end{bmatrix}$$

  are linearly dependent, since $a_1 + 2a_2 - 3a_3 = 0$. We can express any of these vectors as a linear combination of the other two. For example, we have $a_2 = (-1/2)a_1 + (3/2)a_3$.

- The vectors

$$a_1 = \begin{bmatrix} 1 \\ 0 \\ 0 \end{bmatrix}, \qquad a_2 = \begin{bmatrix} 0 \\ -1 \\ 1 \end{bmatrix}, \qquad a_3 = \begin{bmatrix} -1 \\ 1 \\ 1 \end{bmatrix}$$

  are linearly independent. To see this, suppose $\beta_1 a_1 + \beta_2 a_2 + \beta_3 a_3 = 0$. This means that

$$\beta_1 - \beta_3 = 0, \qquad -\beta_2 + \beta_3 = 0, \qquad \beta_2 + \beta_3 = 0.$$

  Adding the last two equations we find that $2\beta_3 = -0$, so $\beta_3 = 0$. Using this, the first equation is then $\beta_1 = 0$, and the second equation is $\beta_2 = 0$.

- The standard unit n-vectors $e_1, \ldots, e_n$ are linearly independent. To see this, suppose that (5.1) holds. We have

$$0 = \beta_1 e_1 + \cdots + \beta_n e_n = \begin{bmatrix} \beta_1 \\ \vdots \\ \beta_n \end{bmatrix},$$

  so we conclude that $\beta_1 = \cdots = \beta_n = 0$.

Linear combinations of linearly independent vectors. Suppose a vector x is a linear combination of $a_1, \ldots, a_k$,

$$x = \beta_1 a_1 + \cdots + \beta_k a_k.$$

When the vectors $a_1, \ldots, a_k$ are linearly independent, the coefficients that form x are *unique*: If we also have

$$x = \gamma_1 a_1 + \cdots + \gamma_k a_k,$$

then $\beta_i = \gamma_i$ for $i = 1, \ldots, k$. This tells us that, in principle at least, we can find the coefficients that form a vector x as a linear combination of linearly independent vectors.

To see this, we subtract the two equations above to get

$$0 = (\beta_1 - \gamma_1)a_1 + \cdots + (\beta_k - \gamma_k)a_k.$$

Since $a_1, \ldots, a_k$ are linearly independent, we conclude that $\beta_i - \gamma_i$ are all zero.

The converse is also true: If each linear combination of a list of vectors can only be expressed as a linear combination with one set of coefficients, then the list of vectors is linearly independent. This gives a nice interpretation of linear independence: A list of vectors is linearly independent if and only if for any linear combination of them, we can infer or deduce the associated coefficients. (We will see later how to do this.)

Supersets and subsets. If a collection of vectors is linearly dependent, then any superset of it is linearly dependent. In other words: If we add vectors to a linearly dependent collection of vectors, the new collection is also linearly dependent. Any nonempty subset of a linearly independent collection of vectors is linearly independent. In other words: Removing vectors from a collection of vectors preserves linear independence.

5.2   Basis

Independence-dimension inequality. If the n-vectors $a_1, \ldots, a_k$ are linearly independent, then $k \leq n$. In words:

A linearly independent collection of n-vectors can have at most n elements.

Put another way:

Any collection of $n + 1$ or more n-vectors is linearly dependent.

As a very simple example, we can conclude that any three 2-vectors must be linearly dependent. This is illustrated in figure 5.1.

We will prove this fundamental fact below; but first, we describe the concept of basis, which relies on the independence-dimension inequality.

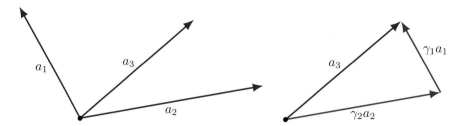

Figure 5.1 *Left.* Three 2-vectors. *Right.* The vector a_3 is a linear combination of a_1 and a_2, which shows that the vectors are linearly dependent.

Basis.   A collection of n linearly independent n-vectors (*i.e.*, a collection of linearly independent vectors of the maximum possible size) is called a *basis*. If the n-vectors $a_1, \ldots, a_n$ are a basis, then any n-vector b can be written as a linear combination of them. To see this, consider the collection of $n+1$ n-vectors $a_1, \ldots, a_n, b$. By the independence-dimension inequality, these vectors are linearly dependent, so there are $\beta_1, \ldots, \beta_{n+1}$, not all zero, that satisfy

$$\beta_1 a_1 + \cdots + \beta_n a_n + \beta_{n+1} b = 0.$$

If $\beta_{n+1} = 0$, then we have

$$\beta_1 a_1 + \cdots + \beta_n a_n = 0,$$

which, since $a_1, \ldots, a_n$ are linearly independent, implies that $\beta_1 = \cdots = \beta_n = 0$. But then all the β_i are zero, a contradiction. So we conclude that $\beta_{n+1} \neq 0$. It follows that

$$b = (-\beta_1/\beta_{n+1})a_1 + \cdots + (-\beta_n/\beta_{n+1})a_n,$$

i.e., b is a linear combination of $a_1, \ldots, a_n$.

Combining this result with the observation above that any linear combination of linearly independent vectors can be expressed in only one way, we conclude:

Any n-vector b can be written in a unique way as a linear combination of a basis $a_1, \ldots, a_n$.

Expansion in a basis.   When we express an n-vector b as a linear combination of a basis $a_1, \ldots, a_n$, we refer to

$$b = \alpha_1 a_1 + \cdots + \alpha_n a_n,$$

as the *expansion of b in the $a_1, \ldots, a_n$ basis*. The numbers $\alpha_1, \ldots, \alpha_n$ are called the *coefficients* of the expansion of b in the basis $a_1, \ldots, a_n$. (We will see later how to find the coefficients in the expansion of a vector in a basis.)

Examples.

- The n standard unit n vectors $e_1, \ldots, e_n$ are a basis. Any n-vector b can be written as the linear combination

$$b = b_1 e_1 + \cdots + b_n e_n.$$

  (This was already observed on page 17.) This expansion is unique, which means that there is no other linear combination of $e_1, \ldots, e_n$ that equals b.

- The vectors

$$a_1 = \begin{bmatrix} 1.2 \\ -2.6 \end{bmatrix}, \qquad a_2 = \begin{bmatrix} -0.3 \\ -3.7 \end{bmatrix}$$

  are a basis. The vector $b = (1, 1)$ can be expressed in only one way as a linear combination of them:

$$b = 0.6513 \, a_1 - 0.7280 \, a_2.$$

  (The coefficients are given here to 4 significant digits. We will see later how these coefficients can be computed.)

Cash flows and single period loans. As a practical example, we consider cash flows over n periods, with positive entries meaning income or cash in and negative entries meaning payments or cash out. We define the single-period loan cash flow vectors as

$$l_i = \begin{bmatrix} 0_{i-1} \\ 1 \\ -(1+r) \\ 0_{n-i-1} \end{bmatrix}, \quad i = 1, \ldots, n-1,$$

where $r \geq 0$ is the per-period interest rate. The cash flow l_i represents a loan of \$1 in period i, which is paid back in period $i+1$ with interest r. (The subscripts on the zero vectors above give their dimensions.) Scaling l_i changes the loan amount; scaling l_i by a negative coefficient converts it into a loan *to* another entity (which is paid back in period $i+1$ with interest).

The vectors $e_1, l_1, \ldots, l_{n-1}$ are a basis. (The first vector e_1 represents income of \$1 in period 1.) To see this, we show that they are linearly independent. Suppose that

$$\beta_1 e_1 + \beta_2 l_1 + \cdots + \beta_n l_{n-1} = 0.$$

We can express this as

$$\begin{bmatrix} \beta_1 + \beta_2 \\ \beta_3 - (1+r)\beta_2 \\ \vdots \\ \beta_n - (1+r)\beta_{n-1} \\ -(1+r)\beta_n \end{bmatrix} = 0.$$

The last entry is $-(1+r)\beta_n = 0$, which implies that $\beta_n = 0$ (since $1 + r > 0$). Using $\beta_n = 0$, the second to last entry becomes $-(1+r)\beta_{n-1} = 0$, so we conclude that $\beta_{n-1} = 0$. Continuing this way we find that $\beta_{n-2}, \ldots, \beta_2$ are all zero. The first entry of the equation above, $\beta_1 + \beta_2 = 0$, then implies $\beta_1 = 0$. We conclude that the vectors $e_1, l_1, \ldots, l_{n-1}$ are linearly independent, and therefore a basis.

This means that any cash flow n-vector c can be expressed as a linear combination of (*i.e.*, replicated by) an initial payment and one period loans:

$$c = \alpha_1 e_1 + \alpha_2 l_1 + \cdots + \alpha_n l_{n-1}.$$

It is possible to work out what the coefficients are (see exercise 5.3). The most interesting one is the first coefficient

$$\alpha_1 = c_1 + \frac{c_2}{1+r} + \cdots + \frac{c_n}{(1+r)^{n-1}},$$

which is exactly the net present value (NPV) of the cash flow, with interest rate r. Thus we see that *any* cash flow can be replicated as an income in period 1 equal to its net present value, plus a linear combination of one-period loans at interest rate r.

Proof of independence-dimension inequality.   The proof is by induction on the dimension n. First consider a linearly independent collection $a_1, \ldots, a_k$ of 1-vectors. We must have $a_1 \neq 0$. This means that every element a_i of the collection can be expressed as a multiple $a_i = (a_i/a_1)a_1$ of the first element a_1. This contradicts linear independence unless $k = 1$.

Next suppose $n \geq 2$ and the independence-dimension inequality holds for dimension $n - 1$. Let $a_1, \ldots, a_k$ be a linearly independent list of n-vectors. We need to show that $k \leq n$. We partition the vectors as

$$a_i = \begin{bmatrix} b_i \\ \alpha_i \end{bmatrix}, \quad i = 1, \ldots, k,$$

where b_i is an $(n-1)$-vector and α_i is a scalar.

First suppose that $\alpha_1 = \cdots = \alpha_k = 0$. Then the vectors $b_1, \ldots, b_k$ are linearly independent: $\sum_{i=1}^{k} \beta_i b_i = 0$ holds if and only if $\sum_{i=1}^{k} \beta_i a_i = 0$, which is only possible for $\beta_1 = \cdots = \beta_k = 0$ because the vectors a_i are linearly independent. The vectors $b_1, \ldots, b_k$ therefore form a linearly independent collection of $(n-1)$-vectors. By the induction hypothesis we have $k \leq n - 1$, so certainly $k \leq n$.

Next suppose that the scalars α_i are not all zero. Assume $\alpha_j \neq 0$. We define a collection of $k - 1$ vectors c_i of length $n - 1$ as follows:

$$c_i = b_i - \frac{\alpha_i}{\alpha_j} b_j, \quad i = 1, \ldots, j-1, \qquad c_i = b_{i+1} - \frac{\alpha_{i+1}}{\alpha_j} b_j, \quad i = j, \ldots, k-1.$$

These $k - 1$ vectors are linearly independent: If $\sum_{i=1}^{k-1} \beta_i c_i = 0$ then

$$\sum_{i=1}^{j-1} \beta_i \begin{bmatrix} b_i \\ \alpha_i \end{bmatrix} + \gamma \begin{bmatrix} b_j \\ \alpha_j \end{bmatrix} + \sum_{i=j+1}^{k} \beta_{i-1} \begin{bmatrix} b_i \\ \alpha_i \end{bmatrix} = 0 \qquad (5.2)$$

with

$$\gamma = -\frac{1}{\alpha_j}\left(\sum_{i=1}^{j-1} \beta_i \alpha_i + \sum_{i=j+1}^{k} \beta_{i-1} \alpha_i \right).$$

Figure 5.2 Orthonormal vectors in a plane.

Since the vectors $a_i = (b_i, \alpha_i)$ are linearly independent, the equality (5.2) only holds when all the coefficients β_i and γ are all zero. This in turns implies that the vectors $c_1, \ldots, c_{k-1}$ are linearly independent. By the induction hypothesis $k - 1 \leq n - 1$, so we have established that $k \leq n$.

5.3 Orthonormal vectors

A collection of vectors $a_1, \ldots, a_k$ is *orthogonal* or *mutually orthogonal* if $a_i \perp a_j$ for any i, j with $i \neq j$, $i, j = 1, \ldots, k$. A collection of vectors $a_1, \ldots, a_k$ is *orthonormal* if it is orthogonal and $\|a_i\| = 1$ for $i = 1, \ldots, k$. (A vector of norm one is called *normalized*; dividing a vector by its norm is called *normalizing* it.) Thus, each vector in an orthonormal collection of vectors is normalized, and two different vectors from the collection are orthogonal. These two conditions can be combined into one statement about the inner products of pairs of vectors in the collection: $a_1, \ldots, a_k$ is orthonormal means that

$$a_i^T a_j = \begin{cases} 1 & i = j \\ 0 & i \neq j. \end{cases}$$

Orthonormality, like linear dependence and independence, is an attribute of a collection of vectors, and not an attribute of vectors individually. By convention, though, we say "The vectors $a_1, \ldots, a_k$ are orthonormal" to mean "The collection of vectors $a_1, \ldots, a_k$ is orthonormal".

Examples. The standard unit n-vectors $e_1, \ldots, e_n$ are orthonormal. As another example, the 3-vectors

$$\begin{bmatrix} 0 \\ 0 \\ -1 \end{bmatrix}, \quad \frac{1}{\sqrt{2}} \begin{bmatrix} 1 \\ 1 \\ 0 \end{bmatrix}, \quad \frac{1}{\sqrt{2}} \begin{bmatrix} 1 \\ -1 \\ 0 \end{bmatrix}, \tag{5.3}$$

are orthonormal. Figure 5.2 shows a set of two orthonormal 2-vectors.

Linear independence of orthonormal vectors. Orthonormal vectors are linearly independent. To see this, suppose $a_1, \ldots, a_k$ are orthonormal, and

$$\beta_1 a_1 + \cdots + \beta_k a_k = 0.$$

Taking the inner product of this equality with a_i yields

$$
\begin{aligned}
0 &= a_i^T(\beta_1 a_1 + \cdots + \beta_k a_k) \\
&= \beta_1(a_i^T a_1) + \cdots + \beta_k(a_i^T a_k) \\
&= \beta_i,
\end{aligned}
$$

since $a_i^T a_j = 0$ for $j \neq i$ and $a_i^T a_i = 1$. Thus, the only linear combination of $a_1, \ldots, a_k$ that is zero is the one with all coefficients zero.

Linear combinations of orthonormal vectors. Suppose a vector x is a linear combination of $a_1, \ldots, a_k$, where $a_1, \ldots, a_k$ are orthonormal,

$$x = \beta_1 a_1 + \cdots + \beta_k a_k.$$

Taking the inner product of the left-hand and right-hand sides of this equation with a_i yields

$$a_i^T x = a_i^T(\beta_1 a_1 + \cdots + \beta_k a_k) = \beta_i,$$

using the same argument as above. So if a vector x is a linear combination of orthonormal vectors, we can easily find the coefficients of the linear combination by taking the inner products with the vectors.

For any x that is a linear combination of orthonormal vectors $a_1, \ldots, a_k$, we have the identity

$$x = (a_1^T x)a_1 + \cdots + (a_k^T x)a_k. \tag{5.4}$$

This identity gives us a simple way to check if an n-vector y is a linear combination of the orthonormal vectors $a_1, \ldots, a_k$. If the identity (5.4) holds for y, $i.e.$,

$$y = (a_1^T y)a_1 + \cdots + (a_k^T y)a_k,$$

then (evidently) y is a linear combination of $a_1, \ldots, a_k$; conversely, if y is a linear combination of $a_1, \ldots, a_k$, the identity (5.4) holds for y.

Orthonormal basis. If the n-vectors $a_1, \ldots, a_n$ are orthonormal, they are linearly independent, and therefore also a basis. In this case they are called an *orthonormal basis*. The three examples above (on page 95) are orthonormal bases.

If $a_1, \ldots, a_n$ is an orthonormal basis, then we have, for any n-vector x, the identity

$$x = (a_1^T x)a_1 + \cdots + (a_n^T x)a_n. \tag{5.5}$$

To see this, we note that since $a_1, \ldots, a_n$ are a basis, x can be expressed as a linear combination of them; hence the identity (5.4) above holds. The equation above is sometimes called the *orthonormal expansion formula*; the right-hand side is called the *expansion of x in the basis $a_1, \ldots, a_n$*. It shows that any n-vector can be expressed as a linear combination of the basis elements, with the coefficients given by taking the inner product of x with the elements of the basis.

As an example, we express the 3-vector $x = (1, 2, 3)$ as a linear combination of the orthonormal basis given in (5.3). The inner products of x with these vectors are

$$\begin{bmatrix} 0 \\ 0 \\ -1 \end{bmatrix}^T x = -3, \qquad \frac{1}{\sqrt{2}} \begin{bmatrix} 1 \\ 1 \\ 0 \end{bmatrix}^T x = \frac{3}{\sqrt{2}}, \qquad \frac{1}{\sqrt{2}} \begin{bmatrix} 1 \\ -1 \\ 0 \end{bmatrix}^T x = \frac{-1}{\sqrt{2}}.$$

It can be verified that the expansion of x in this basis is

$$x = (-3) \begin{bmatrix} 0 \\ 0 \\ -1 \end{bmatrix} + \frac{3}{\sqrt{2}} \left(\frac{1}{\sqrt{2}} \begin{bmatrix} 1 \\ 1 \\ 0 \end{bmatrix} \right) + \frac{-1}{\sqrt{2}} \left(\frac{1}{\sqrt{2}} \begin{bmatrix} 1 \\ -1 \\ 0 \end{bmatrix} \right).$$

5.4 Gram–Schmidt algorithm

In this section we describe an algorithm that can be used to determine if a list of n-vectors $a_1, \ldots, a_k$ is linearly independent. In later chapters we will see that it has many other uses as well. The algorithm is named after the mathematicians Jørgen Pedersen Gram and Erhard Schmidt, although it was already known before their work.

If the vectors are linearly independent, the Gram–Schmidt algorithm produces an orthonormal collection of vectors $q_1, \ldots, q_k$ with the following properties: For each $i = 1, \ldots, k$, a_i is a linear combination of $q_1, \ldots, q_i$, and q_i is a linear combination of $a_1, \ldots, a_i$. If the vectors $a_1, \ldots, a_{j-1}$ are linearly independent, but $a_1, \ldots, a_j$ are linearly dependent, the algorithm detects this and terminates. In other words, the Gram–Schmidt algorithm finds the first vector a_j that is a linear combination of previous vectors $a_1, \ldots, a_{j-1}$.

Algorithm 5.1 Gram–Schmidt algorithm

given n-vectors $a_1, \ldots, a_k$

for $i = 1, \ldots, k$,

    1. *Orthogonalization.* $\tilde{q}_i = a_i - (q_1^T a_i)q_1 - \cdots - (q_{i-1}^T a_i)q_{i-1}$
    2. *Test for linear dependence.* if $\tilde{q}_i = 0$, quit.
    3. *Normalization.* $q_i = \tilde{q}_i/\|\tilde{q}_i\|$

The orthogonalization step, with $i = 1$, reduces to $\tilde{q}_1 = a_1$. If the algorithm does not quit (in step 2), *i.e.*, $\tilde{q}_1, \ldots, \tilde{q}_k$ are all nonzero, we can conclude that the original collection of vectors is linearly independent; if the algorithm does quit early, say, with $\tilde{q}_j = 0$, we can conclude that the original collection of vectors is linearly dependent (and indeed, that a_j is a linear combination of $a_1, \ldots, a_{j-1}$).

Figure 5.3 illustrates the Gram–Schmidt algorithm for two 2-vectors. The top row shows the original vectors; the middle and bottom rows show the first and second iterations of the loop in the Gram–Schmidt algorithm, with the left-hand side showing the orthogonalization step, and the right-hand side showing the normalization step.

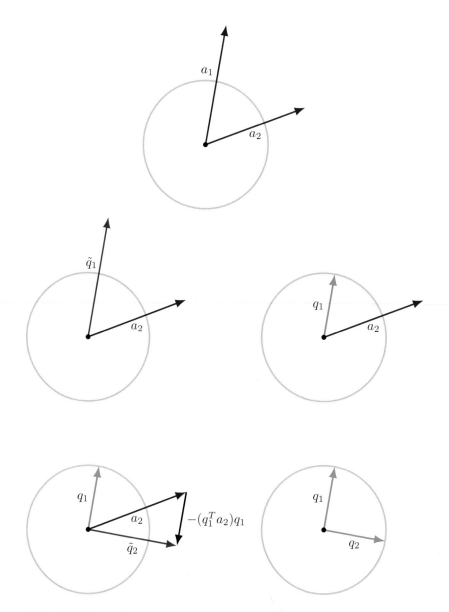

Figure 5.3 Gram–Schmidt algorithm applied to two 2-vectors a_1, a_2. *Top.*
The original vectors a_1 and a_2. The gray circle shows the points with norm
one. *Middle left.* The orthogonalization step in the first iteration yields
$\tilde{q}_1 = a_1$. *Middle right.* The normalization step in the first iteration scales
$\tilde{q}_1$ to have norm one, which yields q_1. *Bottom left.* The orthogonalization
step in the second iteration subtracts a multiple of q_1 to yield the vector
$\tilde{q}_2$, which is orthogonal to q_1. *Bottom right.* The normalization step in the
second iteration scales $\tilde{q}_2$ to have norm one, which yields q_2.

Analysis of Gram–Schmidt algorithm.   Let us show that the following hold, for $i = 1, \ldots, k$, assuming $a_1, \ldots, a_k$ are linearly independent.

1. $\tilde{q}_i \neq 0$, so the linear dependence test in step 2 is not satisfied, and we do not have a divide-by-zero error in step 3.

2. $q_1, \ldots, q_i$ are orthonormal.

3. a_i is a linear combination of $q_1, \ldots, q_i$.

4. q_i is a linear combination of $a_1, \ldots, a_i$.

We show this by induction. For $i = 1$, we have $\tilde{q}_1 = a_1$. Since $a_1, \ldots, a_k$ are linearly independent, we must have $a_1 \neq 0$, and therefore $\tilde{q}_1 \neq 0$, so assertion 1 holds. The single vector q_1 (considered as a list with one element) is evidently orthonormal, since $\|q_1\| = 1$, so assertion 2 holds. We have $a_1 = \|\tilde{q}_1\| q_1$, and $q_1 = (1/\|\tilde{q}_1\|) a_1$, so assertions 3 and 4 hold.

Suppose our assertion holds for some $i-1$, with $i < k$; we will show it holds for i. If $\tilde{q}_i = 0$, then a_i is a linear combination of $q_1, \ldots, q_{i-1}$ (from the first step in the algorithm); but each of these is (by the induction hypothesis) a linear combination of $a_1, \ldots, a_{i-1}$, so it follows that a_i is a linear combination of $a_1, \ldots, a_{i-1}$, which contradicts our assumption that $a_1, \ldots, a_k$ are linearly independent. So assertion 1 holds for i.

Step 3 of the algorithm ensures that $q_1, \ldots, q_i$ are normalized; to show they are orthogonal we will show that $q_i \perp q_j$ for $j = 1, \ldots, i-1$. (Our induction hypothesis tells us that $q_r \perp q_s$ for $r, s < i$.) For any $j = 1, \ldots, i-1$, we have (using step 1)

$$
\begin{aligned}
q_j^T \tilde{q}_i &= q_j^T a_i - (q_1^T a_i)(q_j^T q_1) - \cdots - (q_{i-1}^T a_i)(q_j^T q_{i-1}) \\
&= q_j^T a_i - q_j^T a_i = 0,
\end{aligned}
$$

using $q_j^T q_k = 0$ for $j \neq k$ and $q_j^T q_j = 1$. (This explains why step 1 is called the orthogonalization step: We subtract from a_i a linear combination of $q_1, \ldots, q_{i-1}$ that ensures $q_i \perp \tilde{q}_j$ for $j < i$.) Since $q_i = (1/\|\tilde{q}_i\|) \tilde{q}_i$, we have $q_i^T q_j = 0$ for $j = 1, \ldots, i-1$. So assertion 2 holds for i.

It is immediate that a_i is a linear combination of $q_1, \ldots, q_i$:

$$
\begin{aligned}
a_i &= \tilde{q}_i + (q_1^T a_i)q_1 + \cdots + (q_{i-1}^T a_i)q_{i-1} \\
&= (q_1^T a_i)q_1 + \cdots + (q_{i-1}^T a_i)q_{i-1} + \|\tilde{q}_i\| q_i.
\end{aligned}
$$

From step 1 of the algorithm, we see that $\tilde{q}_i$ is a linear combination of the vectors $a_i, q_1, \ldots, q_{i-1}$. By the induction hypothesis, each of $q_1, \ldots, q_{i-1}$ is a linear combination of $a_1, \ldots, a_{i-1}$, so $\tilde{q}_i$ (and therefore also q_i) is a linear combination of $a_1, \ldots, a_i$. Thus assertions 3 and 4 hold.

Gram–Schmidt completion implies linear independence.   From the properties 1–4 above, we can argue that the original collection of vectors $a_1, \ldots, a_k$ is linearly independent. To see this, suppose that

$$
\beta_1 a_1 + \cdots + \beta_k a_k = 0 \tag{5.6}
$$

holds for some $\beta_1, \ldots, \beta_k$. We will show that $\beta_1 = \cdots = \beta_k = 0$.

We first note that any linear combination of $q_1, \ldots, q_{k-1}$ is orthogonal to any multiple of q_k, since $q_1^T q_k = \cdots = q_{k-1}^T q_k = 0$ (by definition). But each of $a_1, \ldots, a_{k-1}$ is a linear combination of $q_1, \ldots, q_{k-1}$, so we have $q_k^T a_1 = \cdots = q_k^T a_{k-1} = 0$. Taking the inner product of q_k with the left- and right-hand sides of (5.6) we obtain

$$
\begin{aligned}
0 &= q_k^T(\beta_1 a_1 + \cdots + \beta_k a_k) \\
&= \beta_1 q_k^T a_1 + \cdots + \beta_{k-1} q_k^T a_{k-1} + \beta_k q_k^T a_k \\
&= \beta_k \|\tilde{q}_k\|,
\end{aligned}
$$

where we use $q_k^T a_k = \|\tilde{q}_k\|$ in the last line. We conclude that $\beta_k = 0$.

From (5.6) and $\beta_k = 0$ we have

$$
\beta_1 a_1 + \cdots + \beta_{k-1} a_{k-1} = 0.
$$

We now repeat the argument above to conclude that $\beta_{k-1} = 0$. Repeating it k times we conclude that all β_i are zero.

Early termination. Suppose that the Gram–Schmidt algorithm terminates prematurely, in iteration j, because $\tilde{q}_j = 0$. The conclusions 1–4 above hold for $i = 1, \ldots, j-1$, since in those steps $\tilde{q}_i$ is nonzero. Since $\tilde{q}_j = 0$, we have

$$
a_j = (q_1^T a_j)q_1 + \cdots + (q_{j-1}^T a_j)q_{j-1},
$$

which shows that a_j is a linear combination of $q_1, \ldots, q_{j-1}$. But each of these vectors is in turn a linear combination of $a_1, \ldots, a_{j-1}$, by conclusion 3 above. Then a_j is a linear combination of $a_1, \ldots, a_{j-1}$, since it is a linear combination of linear combinations of them (see exercise 1.18). This means that $a_1, \ldots, a_j$ are linearly dependent, which implies that the larger set $a_1, \ldots, a_k$ is linearly dependent.

In summary, the Gram–Schmidt algorithm gives us an explicit method for determining if a list of vectors is linearly dependent or independent.

Example. We define three vectors

$$
a_1 = (-1, 1, -1, 1), \qquad a_2 = (-1, 3, -1, 3), \qquad a_3 = (1, 3, 5, 7).
$$

Applying the Gram–Schmidt algorithm gives the following results.

- $i = 1$. We have $\|\tilde{q}_1\| = 2$, so

$$
q_1 = \frac{1}{\|\tilde{q}_1\|}\tilde{q}_1 = (-1/2, 1/2, -1/2, 1/2),
$$

  which is simply a_1 normalized.

- $i = 2$. We have $q_1^T a_2 = 4$, so

$$
\tilde{q}_2 = a_2 - (q_1^T a_2)q_1 = \begin{bmatrix} -1 \\ 3 \\ -1 \\ 3 \end{bmatrix} - 4 \begin{bmatrix} -1/2 \\ 1/2 \\ -1/2 \\ 1/2 \end{bmatrix} = \begin{bmatrix} 1 \\ 1 \\ 1 \\ 1 \end{bmatrix},
$$

which is indeed orthogonal to q_1 (and a_1). It has norm $\|\tilde{q}_2\| = 2$; normalizing it gives

$$q_2 = \frac{1}{\|\tilde{q}_2\|}\tilde{q}_2 = (1/2, 1/2, 1/2, 1/2).$$

- $i = 3$. We have $q_1^T a_3 = 2$ and $q_2^T a_3 = 8$, so

$$
\begin{aligned}
\tilde{q}_3 &= a_3 - (q_1^T a_3)q_1 - (q_2^T a_3)q_2 \\
&= \begin{bmatrix} 1 \\ 3 \\ 5 \\ 7 \end{bmatrix} - 2\begin{bmatrix} -1/2 \\ 1/2 \\ -1/2 \\ 1/2 \end{bmatrix} - 8\begin{bmatrix} 1/2 \\ 1/2 \\ 1/2 \\ 1/2 \end{bmatrix} \\
&= \begin{bmatrix} -2 \\ -2 \\ 2 \\ 2 \end{bmatrix},
\end{aligned}
$$

which is orthogonal to q_1 and q_2 (and a_1 and a_2). We have $\|\tilde{q}_3\| = 4$, so the normalized vector is

$$q_3 = \frac{1}{\|\tilde{q}_3\|}\tilde{q}_3 = (-1/2, -1/2, 1/2, 1/2).$$

Completion of the Gram–Schmidt algorithm without early termination tells us that the vectors a_1, a_2, a_3 are linearly independent.

Determining if a vector is a linear combination of linearly independent vectors.
Suppose the vectors $a_1, \ldots, a_k$ are linearly independent, and we wish to determine if another vector b is a linear combination of them. (We have already noted on page 91 that if it is a linear combination of them, the coefficients are unique.) The Gram–Schmidt algorithm provides an explicit way to do this. We apply the Gram–Schmidt algorithm to the list of $k + 1$ vectors

$$a_1, \ldots, a_k, b.$$

These vectors are linearly dependent if b is a linear combination of $a_1, \ldots, a_k$; they are linearly independent if b is not a linear combination of $a_1, \ldots, a_k$. The Gram–Schmidt algorithm will determine which of these two cases holds. It cannot terminate in the first k steps, since we assume that $a_1, \ldots, a_k$ are linearly independent. It will terminate in the $(k+1)$st step with $\tilde{q}_{k+1} = 0$ if b is a linear combination of $a_1, \ldots, a_k$. It will not terminate in the $(k + 1)$st step (*i.e.*, $\tilde{q}_{k+1} \neq 0$), otherwise.

Checking if a collection of vectors is a basis.   To check if the n-vectors $a_1, \ldots, a_n$ are a basis, we run the Gram–Schmidt algorithm on them. If Gram–Schmidt terminates early, they are not a basis; if it runs to completion, we know they are a basis.

Complexity of the Gram–Schmidt algorithm.   We now derive an operation count for the Gram–Schmidt algorithm. In the first step of iteration i of the algorithm, $i-1$ inner products

$$q_1^T a_i, \ \ldots, \ q_{i-1}^T a_i$$

between vectors of length n are computed. This takes $(i-1)(2n-1)$ flops. We then use these inner products as the coefficients in $i-1$ scalar multiplications with the vectors $q_1, \ldots, q_{i-1}$. This requires $n(i-1)$ flops. We then subtract the $i-1$ resulting vectors from a_i, which requires another $n(i-1)$ flops. The total flop count for step 1 is

$$(i-1)(2n-1) + n(i-1) + n(i-1) = (4n-1)(i-1)$$

flops. In step 3 we compute the norm of $\tilde{q}_i$, which takes approximately $2n$ flops. We then divide $\tilde{q}_i$ by its norm, which requires n scalar divisions. So the total flop count for the ith iteration is $(4n-1)(i-1) + 3n$ flops.

   The total flop count for all k iterations of the algorithm is obtained by summing our counts for $i = 1, \ldots, k$:

$$\sum_{i=1}^{k} ((4n-1)(i-1) + 3n) = (4n-1)\frac{k(k-1)}{2} + 3nk \approx 2nk^2,$$

where we use the fact that

$$\sum_{i=1}^{k} (i-1) = 1 + 2 + \cdots + (k-2) + (k-1) = \frac{k(k-1)}{2}, \qquad (5.7)$$

which we justify below. The complexity of the Gram–Schmidt algorithm is $2nk^2$; its order is nk^2. We can guess that its running time grows linearly with the lengths of the vectors n, and quadratically with the number of vectors k.

   In the special case of $k = n$, the complexity of the Gram–Schmidt method is $2n^3$. For example, if the Gram–Schmidt algorithm is used to determine whether a collection of 1000 1000-vectors is linearly independent (and therefore a basis), the computational cost is around 2×10^9 flops. On a modern computer, can we can expect this to take on the order of one second.

   A famous anecdote alleges that the formula (5.7) was discovered by the mathematician Carl Friedrich Gauss when he was a child, although it was known before that time. Here is his argument, for the case when k is odd. Lump the first entry in the sum together with the last entry, the second entry together with the second-to-last entry, and so on. Each of these pairs of numbers adds up to k; since there are $(k-1)/2$ such pairs, the total is $k(k-1)/2$. A similar argument works when k is even.

Modified Gram–Schmidt algorithm.   When the Gram–Schmidt algorithm is implemented, a variation on it called the *modified Gram–Schmidt* algorithm is typically used. This algorithm produces the same results as the Gram–Schmidt algorithm (5.1), but is less sensitive to the small round-off errors that occur when arithmetic calculations are done using floating point numbers. (We do not consider round-off error in floating-point computations in this book.)

Exercises

5.1 *Linear independence of stacked vectors.* Consider the stacked vectors

$$c_1 = \left[\begin{array}{c} a_1 \\ b_1 \end{array} \right], \quad \ldots, \quad c_k = \left[\begin{array}{c} a_k \\ b_k \end{array} \right],$$

where $a_1, \ldots, a_k$ are n-vectors and $b_1, \ldots, b_k$ are m-vectors.

(a) Suppose $a_1, \ldots, a_k$ are linearly independent. (We make no assumptions about the vectors $b_1, \ldots, b_k$.) Can we conclude that the stacked vectors $c_1, \ldots, c_k$ are linearly independent?

(b) Now suppose that $a_1, \ldots, a_k$ are linearly dependent. (Again, with no assumptions about $b_1, \ldots, b_k$.) Can we conclude that the stacked vectors $c_1, \ldots, c_k$ are linearly dependent?

5.2 *A surprising discovery.* An intern at a quantitative hedge fund examines the daily returns of around 400 stocks over one year (which has 250 trading days). She tells her supervisor that she has discovered that the returns of one of the stocks, Google (GOOG), can be expressed as a linear combination of the others, which include many stocks that are unrelated to Google (say, in a different type of business or sector).

Her supervisor then says: "It is overwhelmingly unlikely that a linear combination of the returns of unrelated companies can reproduce the daily return of GOOG. So you've made a mistake in your calculations."

Is the supervisor right? Did the intern make a mistake? Give a very brief explanation.

5.3 *Replicating a cash flow with single-period loans.* We continue the example described on page 93. Let c be any n-vector representing a cash flow over n periods. Find the coefficients $\alpha_1, \ldots, \alpha_n$ of c in its expansion in the basis $e_1, l_1, \ldots, l_{n-1}$, *i.e.*,

$$c = \alpha_1 e_1 + \alpha_2 l_1 + \cdots + \alpha_n l_{n-1}.$$

Verify that α_1 is the net present value (NPV) of the cash flow c, defined on page 22, with interest rate r. *Hint.* Use the same type of argument that was used to show that $e_1, l_1, \ldots, l_{n-1}$ are linearly independent. Your method will find α_n first, then α_{n-1}, and so on.

5.4 *Norm of linear combination of orthonormal vectors.* Suppose $a_1, \ldots, a_k$ are orthonormal n-vectors, and $x = \beta_1 a_1 + \cdots + \beta_k a_k$, where $\beta_1, \ldots, \beta_k$ are scalars. Express $\|x\|$ in terms of $\beta = (\beta_1, \ldots, \beta_k)$.

5.5 *Orthogonalizing vectors.* Suppose that a and b are any n-vectors. Show that we can always find a scalar γ so that $(a - \gamma b) \perp b$, and that γ is unique if $b \neq 0$. (Give a formula for the scalar γ.) In other words, we can always subtract a multiple of a vector from another one, so that the result is orthogonal to the original vector. The orthogonalization step in the Gram–Schmidt algorithm is an application of this.

5.6 *Gram–Schmidt algorithm.* Consider the list of n n-vectors

$$a_1 = \left[\begin{array}{c} 1 \\ 0 \\ 0 \\ \vdots \\ 0 \end{array} \right], \quad a_2 = \left[\begin{array}{c} 1 \\ 1 \\ 0 \\ \vdots \\ 0 \end{array} \right], \quad \ldots, \quad a_n = \left[\begin{array}{c} 1 \\ 1 \\ 1 \\ \vdots \\ 1 \end{array} \right].$$

(The vector a_i has its first i entries equal to one, and the remaining entries zero.) Describe what happens when you run the Gram–Schmidt algorithm on this list of vectors, *i.e.*, say what $q_1, \ldots, q_n$ are. Is $a_1, \ldots, a_n$ a basis?

5.7 *Running Gram–Schmidt algorithm twice.* We run the Gram–Schmidt algorithm once on a given set of vectors $a_1, \ldots, a_k$ (we assume this is successful), which gives the vectors $q_1, \ldots, q_k$. Then we run the Gram–Schmidt algorithm on the vectors $q_1, \ldots, q_k$, which produces the vectors $z_1, \ldots, z_k$. What can you say about $z_1, \ldots, z_k$?

5.8 *Early termination of Gram–Schmidt algorithm.* When the Gram–Schmidt algorithm is run on a particular list of 10 15-vectors, it terminates in iteration 5 (since $\tilde{q}_5 = 0$). Which of the following must be true?

   (a) a_2, a_3, a_4 are linearly independent.

   (b) a_1, a_2, a_5 are linearly dependent.

   (c) a_1, a_2, a_3, a_4, a_5 are linearly dependent.

   (d) a_4 is nonzero.

5.9 *A* particular computer can carry out the Gram–Schmidt algorithm on a list of $k = 1000$ n-vectors, with $n = 10000$, in around 2 seconds. About how long would you expect it to take to carry out the Gram–Schmidt algorithm with $\tilde{k} = 500$ $\tilde{n}$-vectors, with $\tilde{n} = 1000$?

Part II

Matrices

Chapter 6

Matrices

In this chapter we introduce matrices and some basic operations on them. We give some applications in which they arise.

6.1 Matrices

A *matrix* is a rectangular array of numbers written between rectangular brackets, as in

$$\begin{bmatrix} 0 & 1 & -2.3 & 0.1 \\ 1.3 & 4 & -0.1 & 0 \\ 4.1 & -1 & 0 & 1.7 \end{bmatrix}.$$

It is also common to use large parentheses instead of rectangular brackets, as in

$$\begin{pmatrix} 0 & 1 & -2.3 & 0.1 \\ 1.3 & 4 & -0.1 & 0 \\ 4.1 & -1 & 0 & 1.7 \end{pmatrix}.$$

An important attribute of a matrix is its *size* or *dimensions*, *i.e.*, the numbers of rows and columns. The matrix above has 3 rows and 4 columns, so its size is 3×4. A matrix of size $m \times n$ is called an $m \times n$ matrix.

The *elements* (or *entries* or *coefficients*) of a matrix are the values in the array. The i, j element is the value in the ith row and jth column, denoted by double subscripts: the i, j element of a matrix A is denoted A_{ij} (or $A_{i,j}$, when i or j is more than one digit or character). The positive integers i and j are called the (row and column) *indices*. If A is an $m \times n$ matrix, then the row index i runs from 1 to m and the column index j runs from 1 to n. Row indices go from top to bottom, so row 1 is the top row and row m is the bottom row. Column indices go from left to right, so column 1 is the left column and column n is the right column.

If the matrix above is B, then we have $B_{13} = -2.3$, $B_{32} = -1$. The row index of the bottom left element (which has value 4.1) is 3; its column index is 1.

Two matrices are equal if they have the same size, and the corresponding entries are all equal. As with vectors, we normally deal with matrices with entries that are real numbers, which will be our assumption unless we state otherwise. The set of real $m \times n$ matrices is denoted $\mathbf{R}^{m \times n}$. But matrices with complex entries, for example, do arise in some applications.

Matrix indexing.   As with vectors, standard mathematical notation indexes the rows and columns of a matrix starting from 1. In computer languages, matrices are often (but not always) stored as 2-dimensional arrays, which can be indexed in a variety of ways, depending on the language. Lower level languages typically use indices starting from 0; higher level languages and packages that support matrix operations usually use standard mathematical indexing, starting from 1.

Square, tall, and wide matrices.   A *square* matrix has an equal number of rows and columns. A square matrix of size $n \times n$ is said to be of *order* n. A *tall* matrix has more rows than columns (size $m \times n$ with $m > n$). A *wide* matrix has more columns than rows (size $m \times n$ with $n > m$).

Column and row vectors.   An n-vector can be interpreted as an $n \times 1$ matrix; we do not distinguish between vectors and matrices with one column. A matrix with only one row, *i.e.*, with size $1 \times n$, is called a *row vector*; to give its size, we can refer to it as an *n-row-vector*. As an example,

$$\begin{bmatrix} -2.1 & -3 & 0 \end{bmatrix}$$

is a 3-row-vector (or 1×3 matrix). To distinguish them from row vectors, vectors are sometimes called *column vectors*. A 1×1 matrix is considered to be the same as a scalar.

Notational conventions.   Many authors (including us) tend to use capital letters to denote matrices, and lower case letters for (column or row) vectors. But this convention is not standardized, so you should be prepared to figure out whether a symbol represents a matrix, column vector, row vector, or a scalar, from context. (The more considerate authors will tell you what the symbols represent, for example, by referring to 'the matrix A' when introducing it.)

Columns and rows of a matrix.   An $m \times n$ matrix A has n columns, given by (the m-vectors)

$$a_j = \begin{bmatrix} A_{1j} \\ \vdots \\ A_{mj} \end{bmatrix},$$

for $j = 1, \ldots, n$. The same matrix has m rows, given by the (n-row-vectors)

$$b_i = \begin{bmatrix} A_{i1} & \cdots & A_{in} \end{bmatrix},$$

for $i = 1, \ldots, m$.

As a specific example, the 2×3 matrix

$$\begin{bmatrix} 1 & 2 & 3 \\ 4 & 5 & 6 \end{bmatrix}$$

has first row

$$\begin{bmatrix} 1 & 2 & 3 \end{bmatrix}$$

(which is a 3-row-vector or a 1×3 matrix), and second column

$$\begin{bmatrix} 2 \\ 5 \end{bmatrix}$$

(which is a 2-vector or 2×1 matrix), also written compactly as $(2, 5)$.

Block matrices and submatrices. It is useful to consider matrices whose entries are themselves matrices, as in

$$A = \begin{bmatrix} B & C \\ D & E \end{bmatrix},$$

where B, C, D, and E are matrices. Such matrices are called *block matrices*; the elements B, C, D, and E are called *blocks* or *submatrices* of A. The submatrices can be referred to by their block row and column indices; for example, C is the 1,2 block of A.

Block matrices must have the right dimensions to fit together. Matrices in the same (block) row must have the same number of rows (*i.e.*, the same 'height'); matrices in the same (block) column must have the same number of columns (*i.e.*, the same 'width'). In the example above, B and C must have the same number of rows, and C and E must have the same number of columns. Matrix blocks placed next to each other in the same row are said to be *concatenated*; matrix blocks placed above each other are called *stacked*.

As an example, consider

$$B = \begin{bmatrix} 0 & 2 & 3 \end{bmatrix}, \qquad C = \begin{bmatrix} -1 \end{bmatrix}, \qquad D = \begin{bmatrix} 2 & 2 & 1 \\ 1 & 3 & 5 \end{bmatrix}, \qquad E = \begin{bmatrix} 4 \\ 4 \end{bmatrix}.$$

Then the block matrix A above is given by

$$A = \begin{bmatrix} 0 & 2 & 3 & -1 \\ 2 & 2 & 1 & 4 \\ 1 & 3 & 5 & 4 \end{bmatrix}. \tag{6.1}$$

(Note that we have dropped the left and right brackets that delimit the blocks. This is similar to the way we drop the brackets in a 1×1 matrix to get a scalar.)

We can also divide a larger matrix (or vector) into 'blocks'. In this context the blocks are called *submatrices* of the big matrix. As with vectors, we can use colon notation to denote submatrices. If A is an $m \times n$ matrix, and p, q, r, s are integers with $1 \le p \le q \le m$ and $1 \le r \le s \le n$, then $A_{p:q,r:s}$ denotes the submatrix

$$A_{p:q,r:s} = \begin{bmatrix} A_{pr} & A_{p,r+1} & \cdots & A_{ps} \\ A_{p+1,r} & A_{p+1,r+1} & \cdots & A_{p+1,s} \\ \vdots & \vdots & & \vdots \\ A_{qr} & A_{q,r+1} & \cdots & A_{qs} \end{bmatrix}.$$

This submatrix has size $(q-p+1) \times (s-r+1)$ and is obtained by extracting from A the elements in rows p through q and columns r through s.

For the specific matrix A in (6.1), we have

$$A_{2:3,3:4} = \begin{bmatrix} 1 & 4 \\ 5 & 4 \end{bmatrix}.$$

Column and row representation of a matrix. Using block matrix notation we can write an $m \times n$ matrix A as a block matrix with one block row and n block columns,

$$A = \begin{bmatrix} a_1 & a_2 & \cdots & a_n \end{bmatrix},$$

where a_j, which is an m-vector, is the jth column of A. Thus, an $m \times n$ matrix can be viewed as its n columns, concatenated.

Similarly, an $m \times n$ matrix A can be written as a block matrix with one block column and m block rows:

$$A = \begin{bmatrix} b_1 \\ b_2 \\ \vdots \\ b_m \end{bmatrix},$$

where b_i, which is a row n-vector, is the ith row of A. In this notation, the matrix A is interpreted as its m rows, stacked.

Examples

Table interpretation. The most direct interpretation of a matrix is as a table of numbers that depend on two indices, i and j. (A vector is a list of numbers that depend on only one index.) In this case the rows and columns of the matrix usually have some simple interpretation. Some examples are given below.

- *Images.* A black and white image with $M \times N$ pixels is naturally represented as an $M \times N$ matrix. The row index i gives the vertical position of the pixel, the column index j gives the horizontal position of the pixel, and the i, j entry gives the pixel value.

- *Rainfall data.* An $m \times n$ matrix A gives the rainfall at m different locations on n consecutive days, so A_{42} (which is a number) is the rainfall at location 4 on day 2. The jth column of A, which is an m-vector, gives the rainfall at the m locations on day j. The ith row of A, which is an n-row-vector, is the time series of rainfall at location i.

- *Asset returns.* A $T \times n$ matrix R gives the returns of a collection of n assets (called the *universe* of assets) over T periods, with R_{ij} giving the return of asset j in period i. So $R_{12,7} = -0.03$ means that asset 7 had a 3% loss in period 12. The 4th column of R is a T-vector that is the return time series

Date	AAPL	GOOG	MMM	AMZN
March 1, 2016	0.00219	0.00006	−0.00113	0.00202
March 2, 2016	0.00744	−0.00894	−0.00019	−0.00468
March 3, 2016	0.01488	−0.00215	0.00433	−0.00407

Table 6.1 Daily returns of Apple (AAPL), Google (GOOG), 3M (MMM), and Amazon (AMZN), on March 1, 2, and 3, 2016 (based on closing prices).

for asset 4. The 3rd row of R is an n-row-vector that gives the returns of all assets in the universe in period 3.

An example of an asset return matrix, with a universe of $n = 4$ assets over $T = 3$ periods, is shown in table 6.1.

- *Prices from multiple suppliers.* An $m \times n$ matrix P gives the prices of n different goods from m different suppliers (or locations): P_{ij} is the price that supplier i charges for good j. The jth column of P is the m-vector of supplier prices for good j; the ith row gives the prices for all goods from supplier i.

- *Contingency table.* Suppose we have a collection of objects with two attributes, the first attribute with m possible values and the second with n possible values. An $m \times n$ matrix A can be used to hold the counts of the numbers of objects with the different pairs of attributes: A_{ij} is the number of objects with first attribute i and second attribute j. (This is the analog of a count n-vector, that records the counts of one attribute in a collection.) For example, a population of college students can be described by a 4×50 matrix, with the i, j entry the number of students in year i of their studies, from state j (with the states ordered in, say, alphabetical order). The ith row of A gives the geographic distribution of students in year i of their studies; the jth column of A is a 4-vector giving the numbers of student from state j in their first through fourth years of study.

- *Customer purchase history.* An $n \times N$ matrix P can be used to store a set of N customers' purchase histories of n products, items, or services, over some period. The entry P_{ij} represents the dollar value of product i that customer j purchased over the period (or as an alternative, the number or quantity of the product). The jth column of P is the purchase history vector for customer j; the ith row gives the sales report for product i across the N customers.

Matrix representation of a collection of vectors. Matrices are very often used as a compact way to give a set of indexed vectors of the same size. For example, if $x_1, \ldots, x_N$ are n-vectors that give the n feature values for each of N objects, we can collect them all into one $n \times N$ matrix

$$X = \begin{bmatrix} x_1 & x_2 & \cdots & x_N \end{bmatrix},$$

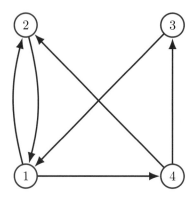

Figure 6.1 The relation (6.2) as a directed graph.

often called a *data matrix* or *feature matrix*. Its jth column is the feature n-vector for the jth object (in this context sometimes called the jth *example*). The ith row of the data matrix X is an N-row-vector whose entries are the values of the ith feature across the examples. We can also directly interpret the entries of the data matrix: X_{ij} (which is a number) is the value of the ith feature for the jth example.

As another example, a $3 \times M$ matrix can be used to represent a collection of M locations or positions in 3-D space, with its jth column giving the jth position.

Matrix representation of a relation or graph.   Suppose we have n objects labeled $1, \dots, n$. A *relation* $\mathcal{R}$ on the set of objects $\{1, \dots, n\}$ is a subset of ordered pairs of objects. As an example, $\mathcal{R}$ can represent a *preference relation* among n possible products or choices, with $(i, j) \in \mathcal{R}$ meaning that choice i is preferred to choice j.

A relation can also be viewed as a *directed graph*, with nodes (or vertices) labeled $1, \dots, n$, and a directed edge from j to i for each $(i, j) \in \mathcal{R}$. This is typically drawn as a graph, with arrows indicating the direction of the edge, as shown in figure 6.1, for the relation on 4 objects

$$\mathcal{R} = \{(1, 2),\ (1, 3),\ (2, 1),\ (2, 4),\ (3, 4),\ (4, 1)\}. \tag{6.2}$$

A relation $\mathcal{R}$ on $\{1, \dots, n\}$ is represented by the $n \times n$ matrix A with

$$A_{ij} = \begin{cases} 1 & (i, j) \in \mathcal{R} \\ 0 & (i, j) \notin \mathcal{R}. \end{cases}$$

This matrix is called the *adjacency matrix* associated with the graph. (Some authors define the adjacency matrix in the reverse sense, with $A_{ij} = 1$ meaning there is an edge from i to j.) The relation (6.2), for example, is represented by the matrix

$$A = \begin{bmatrix} 0 & 1 & 1 & 0 \\ 1 & 0 & 0 & 1 \\ 0 & 0 & 0 & 1 \\ 1 & 0 & 0 & 0 \end{bmatrix}.$$

This is the adjacency matrix of the associated graph, shown in figure 6.1. (We will encounter another matrix associated with a directed graph in §7.3.)

6.2   Zero and identity matrices

Zero matrix.   A zero matrix is a matrix with all elements equal to zero. The zero matrix of size $m \times n$ is sometimes written as $0_{m \times n}$, but usually a zero matrix is denoted just 0, the same symbol used to denote the number 0 or zero vectors. In this case the size of the zero matrix must be determined from the context.

Identity matrix.   An identity matrix is another common matrix. It is always square. Its *diagonal* elements, *i.e.*, those with equal row and column indices, are all equal to one, and its off-diagonal elements (those with unequal row and column indices) are zero. Identity matrices are denoted by the letter I. Formally, the identity matrix of size n is defined by

$$I_{ij} = \begin{cases} 1 & i = j \\ 0 & i \neq j, \end{cases}$$

for $i, j = 1, \ldots, n$. For example,

$$\begin{bmatrix} 1 & 0 \\ 0 & 1 \end{bmatrix}, \qquad \begin{bmatrix} 1 & 0 & 0 & 0 \\ 0 & 1 & 0 & 0 \\ 0 & 0 & 1 & 0 \\ 0 & 0 & 0 & 1 \end{bmatrix}$$

are the 2×2 and 4×4 identity matrices.

The column vectors of the $n \times n$ identity matrix are the unit vectors of size n. Using block matrix notation, we can write

$$I = \begin{bmatrix} e_1 & e_2 & \cdots & e_n \end{bmatrix},$$

where e_k is the kth unit vector of size n.

Sometimes a subscript is used to denote the size of an identity matrix, as in I_4 or $I_{2 \times 2}$. But more often the size is omitted and follows from the context. For example, if

$$A = \begin{bmatrix} 1 & 2 & 3 \\ 4 & 5 & 6 \end{bmatrix},$$

then

$$\begin{bmatrix} I & A \\ 0 & I \end{bmatrix} = \begin{bmatrix} 1 & 0 & 1 & 2 & 3 \\ 0 & 1 & 4 & 5 & 6 \\ 0 & 0 & 1 & 0 & 0 \\ 0 & 0 & 0 & 1 & 0 \\ 0 & 0 & 0 & 0 & 1 \end{bmatrix}.$$

The dimensions of the two identity matrices follow from the size of A. The identity matrix in the 1,1 position must be 2×2, and the identity matrix in the 2,2 position must be 3×3. This also determines the size of the zero matrix in the 2,1 position.

The importance of the identity matrix will become clear later, in §10.1.

Sparse matrices.   A matrix A is said to be *sparse* if many of its entries are zero, or (put another way) just a few of its entries are nonzero. Its *sparsity pattern* is the set of indices (i, j) for which $A_{ij} \neq 0$. The *number of nonzeros* of a sparse matrix A is the number of entries in its sparsity pattern, and denoted $\mathbf{nnz}(A)$. If A is $m \times n$ we have $\mathbf{nnz}(A) \leq mn$. Its *density* is $\mathbf{nnz}(A)/(mn)$, which is no more than one. Densities of sparse matrices that arise in applications are typically small or very small, as in 10^{-2} or 10^{-4}. There is no precise definition of how small the density must be for a matrix to qualify as sparse. A famous definition of sparse matrix due to the mathematician James H. Wilkinson is: A matrix is sparse if it has enough zero entries that it pays to take advantage of them. Sparse matrices can be stored and manipulated efficiently on a computer.

Many common matrices are sparse. An $n \times n$ identity matrix is sparse, since it has only n nonzeros, so its density is $1/n$. The zero matrix is the sparsest possible matrix, since it has no nonzero entries. Several special sparsity patterns have names; we describe some important ones below.

Like sparse vectors, sparse matrices arise in many applications. A typical customer purchase history matrix (see page 111) is sparse, since each customer has likely only purchased a small fraction of all the products available.

Diagonal matrices.   A square $n \times n$ matrix A is *diagonal* if $A_{ij} = 0$ for $i \neq j$. (The entries of a matrix with $i = j$ are called the *diagonal entries*; those with $i \neq j$ are its *off-diagonal* entries.) A diagonal matrix is one for which all off-diagonal entries are zero. Examples of diagonal matrices we have already seen are square zero matrices and identity matrices. Other examples are

$$
\begin{bmatrix} -3 & 0 \\ 0 & 0 \end{bmatrix}, \qquad \begin{bmatrix} 0.2 & 0 & 0 \\ 0 & -3 & 0 \\ 0 & 0 & 1.2 \end{bmatrix}.
$$

(Note that in the first example, one of the diagonal elements is also zero.)

The notation $\mathbf{diag}(a_1, \dots, a_n)$ is used to compactly describe the $n \times n$ diagonal matrix A with diagonal entries $A_{11} = a_1, \dots, A_{nn} = a_n$. This notation is not yet standard, but is coming into more prevalent use. As examples, the matrices above would be expressed as

$$
\mathbf{diag}(-3, 0), \qquad \mathbf{diag}(0.2, -3, 1.2),
$$

respectively. We also allow $\mathbf{diag}$ to take one n-vector argument, as in $I = \mathbf{diag}(\mathbf{1})$.

Triangular matrices.   A square $n \times n$ matrix A is *upper triangular* if $A_{ij} = 0$ for $i > j$, and it is *lower triangular* if $A_{ij} = 0$ for $i < j$. (So a diagonal matrix is one that is both lower and upper triangular.) If a matrix is either lower or upper triangular, it is called *triangular*. For example, the matrices

$$
\begin{bmatrix} 1 & -1 & 0.7 \\ 0 & 1.2 & -1.1 \\ 0 & 0 & 3.2 \end{bmatrix}, \qquad \begin{bmatrix} -0.6 & 0 \\ -0.3 & 3.5 \end{bmatrix},
$$

are upper and lower triangular, respectively.

A triangular $n \times n$ matrix A has up to $n(n+1)/2$ nonzero entries, *i.e.*, around half its entries are zero. Triangular matrices are generally not considered sparse matrices, since their density is around 50%, but their special sparsity pattern will be important in the sequel.

6.3 Transpose, addition, and norm

6.3.1 Matrix transpose

If A is an $m \times n$ matrix, its *transpose*, denoted A^T (or sometimes A' or A^*), is the $n \times m$ matrix given by $(A^T)_{ij} = A_{ji}$. In words, the rows and columns of A are transposed in A^T. For example,

$$\begin{bmatrix} 0 & 4 \\ 7 & 0 \\ 3 & 1 \end{bmatrix}^T = \begin{bmatrix} 0 & 7 & 3 \\ 4 & 0 & 1 \end{bmatrix}.$$

If we transpose a matrix twice, we get back the original matrix: $(A^T)^T = A$. (The superscript T in the transpose is the same one used to denote the inner product of two n-vectors; we will soon see how they are related.)

Row and column vectors. Transposition converts row vectors into column vectors and vice versa. It is sometimes convenient to express a row vector as a^T, where a is a column vector. For example, we might refer to the m rows of an $m \times n$ matrix A as $\tilde{a}_i^T, \ldots, \tilde{a}_m^T$, where $\tilde{a}_1, \ldots, \tilde{a}_m$ are (column) n-vectors. As an example, the second row of the matrix

$$\begin{bmatrix} 0 & 7 & 3 \\ 4 & 0 & 1 \end{bmatrix}$$

can be written as (the row vector) $(4, 0, 1)^T$.

It is common to extend concepts from (column) vectors to row vectors, by applying the concept to the transposed row vectors. We say that a collection of row vectors is linearly dependent (or independent) if their transposes (which are column vectors) are linearly dependent (or independent). For example, 'the rows of a matrix A are linearly independent' means that the columns of A^T are linearly independent. As another example, 'the rows of a matrix A are orthonormal' means that their transposes, the columns of A^T, are orthonormal. 'Clustering the rows of a matrix X' means clustering the columns of X^T.

Transpose of block matrix. The transpose of a block matrix has the simple form (shown here for a 2×2 block matrix)

$$\begin{bmatrix} A & B \\ C & D \end{bmatrix}^T = \begin{bmatrix} A^T & C^T \\ B^T & D^T \end{bmatrix},$$

where A, B, C, and D are matrices with compatible sizes. The transpose of a block matrix is the transposed block matrix, with each element transposed.

Document-term matrix. Consider a corpus (collection) of N documents, with word count vectors for a dictionary with n words. The *document-term* matrix associated with the corpus is the $N \times n$ matrix A, with A_{ij} the number of times word j appears in document i. The rows of the document-term matrix are $a_1^T, \ldots, a_N^T$, where the n-vectors $a_1, \ldots, a_N$ are the word count vectors for documents $1, \ldots, N$, respectively. The columns of the document-term matrix are also interesting. The jth column of A, which is an N-vector, gives the number of times word j appears in the corpus of N documents.

Data matrix. A collection of N n-vectors, for example feature n-vectors associated with N objects, can be given as an $n \times N$ matrix whose N columns are the vectors, as described on page 111. It is also common to describe this collection of vectors using the transpose of that matrix. In this case, we give the vectors as an $N \times n$ matrix X. Its ith row is x_i^T, the transpose of the ith vector. Its jth column gives the value of the jth entry (or feature) across the collection of N vectors. When an author refers to a *data matrix* or *feature matrix*, it can usually be determined from context (for example, its dimensions) whether they mean the data matrix organized by rows or columns.

Symmetric matrix. A square matrix A is *symmetric* if $A = A^T$, *i.e.*, $A_{ij} = A_{ji}$ for all i, j. Symmetric matrices arise in several applications. For example, suppose that A is the adjacency matrix of a graph or relation (see page 112). The matrix A is symmetric when the relation is symmetric, *i.e.*, whenever $(i, j) \in \mathcal{R}$, we also have $(j, i) \in \mathcal{R}$. An example is the *friend relation* on a set of n people, where $(i, j) \in \mathcal{R}$ means that person i and person j are friends. (In this case the associated graph is called the 'social network graph'.)

6.3.2   Matrix addition

Two matrices of the same size can be added together. The result is another matrix of the same size, obtained by adding the corresponding elements of the two matrices. For example,

$$
\begin{bmatrix} 0 & 4 \\ 7 & 0 \\ 3 & 1 \end{bmatrix} + \begin{bmatrix} 1 & 2 \\ 2 & 3 \\ 0 & 4 \end{bmatrix} = \begin{bmatrix} 1 & 6 \\ 9 & 3 \\ 3 & 5 \end{bmatrix}.
$$

Matrix subtraction is similar. As an example,

$$
\begin{bmatrix} 1 & 6 \\ 9 & 3 \end{bmatrix} - I = \begin{bmatrix} 0 & 6 \\ 9 & 2 \end{bmatrix}.
$$

(This gives another example where we have to figure out the size of the identity matrix. Since we can only add or subtract matrices of the same size, I refers to a 2×2 identity matrix.)

Properties of matrix addition. The following important properties of matrix addition can be verified directly from the definition. We assume here that A, B, and C are matrices of the same size.

- *Commutativity.* $A + B = B + A$.

- *Associativity.* $(A + B) + C = A + (B + C)$. We therefore write both as $A + B + C$.

- *Addition with zero matrix.* $A + 0 = 0 + A = A$. Adding the zero matrix to a matrix has no effect.

- *Transpose of sum.* $(A + B)^T = A^T + B^T$. The transpose of a sum of two matrices is the sum of their transposes.

6.3.3 Scalar-matrix multiplication

Scalar multiplication of matrices is defined in a similar way as for vectors, and is done by multiplying every element of the matrix by the scalar. For example

$$(-2) \begin{bmatrix} 1 & 6 \\ 9 & 3 \\ 6 & 0 \end{bmatrix} = \begin{bmatrix} -2 & -12 \\ -18 & -6 \\ -12 & 0 \end{bmatrix}.$$

As with scalar-vector multiplication, the scalar can also appear on the right. Note that $0\,A = 0$ (where the left-hand zero is the scalar zero, and the right-hand 0 is the zero matrix).

Several useful properties of scalar multiplication follow directly from the definition. For example, $(\beta A)^T = \beta(A^T)$ for a scalar β and a matrix A. If A is a matrix and β, γ are scalars, then

$$(\beta + \gamma)A = \beta A + \gamma A, \qquad (\beta\gamma)A = \beta(\gamma A).$$

It is useful to identify the symbols appearing in these two equations. The $+$ symbol on the left of the left-hand equation is addition of scalars, while the $+$ symbol on the right of the left-hand equation denotes matrix addition. On the left side of the right-hand equation we see scalar-scalar multiplication ($\alpha\beta$) and scalar-matrix multiplication; on the right we see two cases of scalar-matrix multiplication.

Finally, we mention that scalar-matrix multiplication has higher precedence than matrix addition, which means that we should carry out multiplication before addition (when there are no parentheses to fix the order). So the right-hand side of the left equation above is to be interpreted as $(\beta A) + (\gamma A)$.

6.3.4 Matrix norm

The norm of an $m \times n$ matrix A, denoted $\|A\|$, is the squareroot of the sum of the squares of its entries,

$$\|A\| = \sqrt{\sum_{i=1}^{m} \sum_{j=1}^{n} A_{ij}^2}. \tag{6.3}$$

This agrees with our definition for vectors when A is a vector, $i.e.$, $n = 1$. The norm of an $m \times n$ matrix is the norm of an mn-vector formed from the entries of the matrix (in any order). Like the vector norm, the matrix norm is a quantitative measure of the magnitude of a matrix. In some applications it is more natural to use the RMS values of the matrix entries, $\|A\|/\sqrt{mn}$, as a measure of matrix size. The RMS value of the matrix entries tells us the typical size of the entries, independent of the matrix dimensions.

The matrix norm (6.3) satisfies the properties of any norm, given on page 46. For any $m \times n$ matrix A, we have $\|A\| \geq 0$ ($i.e.$, the norm is nonnegative), and $\|A\| = 0$ only if $A = 0$ (definiteness). The matrix norm is nonnegative homogeneous: For any scalar γ and $m \times n$ matrix A, we have $\|\gamma A\| = |\gamma| \|A\|$. Finally, for any two $m \times n$ matrices A and B, we have the triangle inequality,

$$\|A + B\| \leq \|A\| + \|B\|.$$

(The plus symbol on the left-hand side is matrix addition, and the plus symbol on the right-hand side is addition of numbers.)

The matrix norm allows us to define the distance between two matrices as $\|A - B\|$. As with vectors, we can say that one matrix is close to, or near, another one if their distance is small. (What qualifies as small depends on the application.)

In this book we will only use the matrix norm (6.3). Several other norms of a matrix are commonly used, but are beyond the scope of this book. In contexts where other norms of a matrix are used, the norm (6.3) is called the *Frobenius norm*, after the mathematician Ferdinand Georg Frobenius, and is usually denoted with a subscript, as $\|A\|_F$.

One simple property of the matrix norm is $\|A\| = \|A^T\|$, $i.e.$, the norm of a matrix is the same as the norm of its transpose. Another one is

$$\|A\|^2 = \|a_1\|^2 + \cdots + \|a_n\|^2,$$

where $a_1, \ldots, a_n$ are the columns of A. In other words: The squared norm of a matrix is the sum of the squared norms of its columns.

6.4   Matrix-vector multiplication

If A is an $m \times n$ matrix and x is an n-vector, then the *matrix-vector product* $y = Ax$ is the m-vector y with elements

$$y_i = \sum_{k=1}^{n} A_{ik} x_k = A_{i1} x_1 + \cdots + A_{in} x_n, \quad i = 1, \ldots, m. \tag{6.4}$$

As a simple example, we have

$$\begin{bmatrix} 0 & 2 & -1 \\ -2 & 1 & 1 \end{bmatrix} \begin{bmatrix} 2 \\ 1 \\ -1 \end{bmatrix} = \begin{bmatrix} (0)(2) + (2)(1) + (-1)(-1) \\ (-2)(2) + (1)(1) + (1)(-1) \end{bmatrix} = \begin{bmatrix} 3 \\ -4 \end{bmatrix}.$$

Row and column interpretations.   We can express the matrix-vector product in terms of the rows or columns of the matrix. From (6.4) we see that y_i is the inner product of x with the ith row of A:

$$y_i = b_i^T x, \quad i = 1, \dots, m,$$

where b_i^T is the row i of A. The matrix-vector product can also be interpreted in terms of the columns of A. If a_k is the kth column of A, then $y = Ax$ can be written

$$y = x_1 a_1 + x_2 a_2 + \cdots + x_n a_n.$$

This shows that $y = Ax$ is a linear combination of the columns of A; the coefficients in the linear combination are the elements of x.

General examples.   In the examples below, A is an $m \times n$ matrix and x is an n-vector.

- *Zero matrix.* When $A = 0$, we have $Ax = 0$. In other words, $0x = 0$. (The left-hand 0 is an $m \times n$ matrix, and the right-hand zero is an m-vector.)

- *Identity.* We have $Ix = x$ for any vector x. (The identity matrix here has dimension $n \times n$.) In other words, multiplying a vector by the identity matrix gives the same vector.

- *Picking out columns and rows.* An important identity is $Ae_j = a_j$, the jth column of A. Multiplying a unit vector by a matrix 'picks out' one of the columns of the matrix. $A^T e_i$, which is an n-vector, is the ith row of A, transposed. (In other words, $(A^T e_i)^T$ is the ith row of A.)

- *Summing or averaging columns or rows.* The m-vector $A\mathbf{1}$ is the sum of the columns of A; its ith entry is the sum of the entries in the ith row of A. The m-vector $A(\mathbf{1}/n)$ is the average of the columns of A; its ith entry is the average of the entries in the ith row of A. In a similar way, $A^T \mathbf{1}$ is an n-vector, whose jth entry is the sum of the entries in the jth column of A.

- *Difference matrix.* The $(n-1) \times n$ matrix

$$D = \begin{bmatrix} -1 & 1 & 0 & \cdots & 0 & 0 & 0 \\ 0 & -1 & 1 & \cdots & 0 & 0 & 0 \\ & & \ddots & \ddots & & & \\ & & & \ddots & \ddots & & \\ 0 & 0 & 0 & \cdots & -1 & 1 & 0 \\ 0 & 0 & 0 & \cdots & 0 & -1 & 1 \end{bmatrix} \tag{6.5}$$

(where entries not shown are zero, and entries with diagonal dots are 1 or -1, continuing the pattern) is called the *difference matrix*. The vector Dx is the $(n-1)$-vector of differences of consecutive entries of x:

$$Dx = \begin{bmatrix} x_2 - x_1 \\ x_3 - x_2 \\ \vdots \\ x_n - x_{n-1} \end{bmatrix}.$$

- *Running sum matrix.* The $n \times n$ matrix

$$
S = \begin{bmatrix}
1 & 0 & 0 & \cdots & & 0 & 0 \\
1 & 1 & 0 & \cdots & & 0 & 0 \\
& & \ddots & & \ddots & & \\
& & & \ddots & & \ddots & \\
1 & 1 & 1 & \cdots & & 1 & 0 \\
1 & 1 & 1 & \cdots & & 1 & 1
\end{bmatrix}
\tag{6.6}
$$

  is called the *running sum matrix*. The ith entry of the n-vector Sx is the sum of the first i entries of x:

$$
Sx = \begin{bmatrix}
x_1 \\
x_1 + x_2 \\
x_1 + x_2 + x_3 \\
\vdots \\
x_1 + \cdots + x_n
\end{bmatrix}.
$$

Application examples.

- *Feature matrix and weight vector.* Suppose X is a feature matrix, where its N columns $x_1, \ldots, x_N$ are feature n-vectors for N objects or examples. Let the n-vector w be a *weight vector*, and let $s_i = x_i^T w$ be the score associated with object i using the weight vector w. Then we can write $s = X^T w$, where s is the N-vector of scores of the objects.

- *Portfolio return time series.* Suppose that R is a $T \times n$ asset return matrix, that gives the returns of n assets over T periods. A common trading strategy maintains constant investment weights given by the n-vector w over the T periods. For example, $w_4 = 0.15$ means that 15% of the total portfolio value is held in asset 4. (Short positions are denoted by negative entries in w.) Then Rw, which is a T-vector, is the time series of the portfolio returns over the periods $1, \ldots, T$.

  As an example, consider a portfolio of the 4 assets in table 6.1, with weights $w = (0.4, 0.3, -0.2, 0.5)$. The product $Rw = (0.00213, -0.00201, 0.00241)$ gives the portfolio returns over the three periods in the example.

- *Polynomial evaluation at multiple points.* Suppose the entries of the n-vector c are the coefficients of a polynomial p of degree $n - 1$ or less:

$$
p(t) = c_1 + c_2 t + \cdots + c_{n-1} t^{n-2} + c_n t^{n-1}.
$$

  Let $t_1, \ldots, t_m$ be m numbers, and define the m-vector y as $y_i = p(t_i)$. Then we have $y = Ac$, where A is the $m \times n$ matrix

$$
A = \begin{bmatrix}
1 & t_1 & \cdots & t_1^{n-2} & t_1^{n-1} \\
1 & t_2 & \cdots & t_2^{n-2} & t_2^{n-1} \\
\vdots & \vdots & & \vdots & \vdots \\
1 & t_m & \cdots & t_m^{n-2} & t_m^{n-1}
\end{bmatrix}.
\tag{6.7}
$$

So multiplying a vector c by the matrix A is the same as evaluating a polynomial with coefficients c at m points. The matrix A in (6.7) comes up often, and is called a *Vandermonde matrix* (of degree $n-1$, at the points $t_1, \ldots, t_m$), named for the mathematician Alexandre-Théophile Vandermonde.

- *Total price from multiple suppliers.* Suppose the $m \times n$ matrix P gives the prices of n goods from m suppliers (or in m different locations). If q is an n-vector of quantities of the n goods (sometimes called a *basket* of goods), then $c = Pq$ is an N-vector that gives the total cost of the goods, from each of the N suppliers.

- *Document scoring.* Suppose A in an $N \times n$ document-term matrix, which gives the word counts of a corpus of N documents using a dictionary of n words, so the rows of A are the word count vectors for the documents. Suppose that w in an n-vector that gives a set of weights for the words in the dictionary. Then $s = Aw$ is an N-vector that gives the scores of the documents, using the weights and the word counts. A search engine, for example, might choose w (based on the search query) so that the scores are predictions of relevance of the documents (to the search).

- *Audio mixing.* Suppose the k columns of A are vectors representing audio signals or tracks of length T, and w is a k-vector. Then $b = Aw$ is a T-vector representing the mix of the audio signals, with track weights given by the vector w.

Inner product.   When a and b are n-vectors, $a^T b$ is exactly the inner product of a and b, obtained from the rules for transposing matrices and forming a matrix-vector product. We start with the (column) n-vector a, consider it as an $n \times 1$ matrix, and transpose it to obtain the n-row-vector a^T. Now we multiply this $1 \times n$ matrix by the n-vector b, to obtain the 1-vector $a^T b$, which we also consider a scalar. So the notation $a^T b$ for the inner product is just a special case of matrix-vector multiplication.

Linear dependence of columns.   We can express the concepts of linear dependence and independence in a compact form using matrix-vector multiplication. The columns of a matrix A are linearly dependent if $Ax = 0$ for some $x \neq 0$. The columns of a matrix A are linearly independent if $Ax = 0$ implies $x = 0$.

Expansion in a basis.   If the columns of A are a basis, which means A is square with linearly independent columns $a_1, \ldots, a_n$, then for any n-vector b there is a unique n-vector x that satisfies $Ax = b$. In this case the vector x gives the coefficients in the expansion of b in the basis $a_1, \ldots, a_n$.

Properties of matrix-vector multiplication.   Matrix-vector multiplication satisfies several properties that are readily verified. First, it distributes across the vector argument: For any $m \times n$ matrix A and any n-vectors u and v, we have

$$A(u + v) = Au + Av.$$

Matrix-vector multiplication, like ordinary multiplication of numbers, has higher precedence than addition, which means that when there are no parentheses to force the order of evaluation, multiplications are to be carried out before additions. This means that the right-hand side above is to be interpreted as $(Au) + (Av)$. The equation above looks innocent and natural, but must be read carefully. On the left-hand side, we first add the vectors u and v, which is the addition of n-vectors. We then multiply the resulting n-vector by the matrix A. On the right-hand side, we first multiply each of n-vectors by the matrix A (this is two matrix-vector multiplies); and then add the two resulting m-vectors together. The left- and right-hand sides of the equation above involve very different steps and operations, but the final result of each is the same m-vector.

Matrix-vector multiplication also distributes across the matrix argument: For any $m \times n$ matrices A and B, and any n-vector u, we have

$$(A + B)u = Au + Bu.$$

On the left-hand side the plus symbol is matrix addition; on the right-hand side it is vector addition.

Another basic property is, for any $m \times n$ matrix A, any n-vector u, and any scalar α, we have

$$(\alpha A)u = \alpha(Au)$$

(and so we can write this as αAu). On the left-hand side, we have scalar-matrix multiplication, followed by matrix-vector multiplication; on the right-hand side, we start with matrix-vector multiplication, and then perform scalar-vector multiplication. (Note that we also have $\alpha Au = A(\alpha u)$.)

Input-output interpretation. We can interpret the relation $y = Ax$, with A an $m \times n$ matrix, as a mapping from the n-vector x to the m-vector y. In this context we might think of x as an input, and y as the corresponding output. From equation (6.4), we can interpret A_{ij} as the factor by which y_i depends on x_j. Some examples of conclusions we can draw are given below.

- If A_{23} is positive and large, then y_2 depends strongly on x_3, and increases as x_3 increases.

- If A_{32} is much larger than the other entries in the third row of A, then y_3 depends much more on x_2 than the other inputs.

- If A is square and lower triangular, then y_i only depends on $x_1, \ldots, x_i$.

6.5   Complexity

Computer representation of matrices.   An $m \times n$ matrix is usually represented on a computer as an $m \times n$ array of floating point numbers, which requires $8mn$ bytes. In some software systems symmetric matrices are represented in a more efficient way, by only storing the upper triangular elements in the matrix, in some specific order. This reduces the memory requirement by around a factor of two. Sparse matrices are represented by various methods that encode for each nonzero element its row index i (an integer), its column index j (an integer) and its value A_{ij} (a floating point number). When the row and column indices are represented using 4 bytes (which allows m and n to range up to around 4.3 billion) this requires a total of around $16 \mathbf{nnz}(A)$ bytes.

Complexity of matrix addition, scalar multiplication, and transposition. The addition of two $m \times n$ matrices or a scalar multiplication of an $m \times n$ matrix each take mn flops. When A is sparse, scalar multiplication requires $\mathbf{nnz}(A)$ flops. When at least one of A and B is sparse, computing $A + B$ requires at most $\min\{\mathbf{nnz}(A), \mathbf{nnz}(B)\}$ flops. (For any entry i, j for which one of A_{ij} or B_{ij} is zero, no arithmetic operations are needed to find $(A+B)_{ij}$.) Matrix transposition, i.e., computing A^T, requires zero flops, since we simply copy entries of A to those of A^T. (Copying the entries does take time to carry out, but this is not reflected in the flop count.)

Complexity of matrix-vector multiplication. A matrix-vector multiplication of an $m \times n$ matrix A with an n-vector x requires $m(2n - 1)$ flops, which we simplify to $2mn$ flops. This can be seen as follows. The result $y = Ax$ of the product is an m-vector, so there are m numbers to compute. The ith element of y is the inner product of the ith row of A and the vector x, which takes $2n - 1$ flops.

If A is sparse, computing Ax requires $\mathbf{nnz}(A)$ multiplies (of A_{ij} and x_j, for each nonzero entry of A) and a number of additions that is no more than $\mathbf{nnz}(A)$. Thus, the complexity is between $\mathbf{nnz}(A)$ and $2 \mathbf{nnz}(A)$ flops. As a special example, suppose A is $n \times n$ and diagonal. Then Ax can be computed with n multiplies (A_{ii} times x_i) and no additions, a total of $n = \mathbf{nnz}(A)$ flops.

Exercises

6.1 *Matrix and vector notation.* Suppose $a_1, \ldots, a_n$ are m-vectors. Determine whether each expression below makes sense (*i.e.*, uses valid notation). If the expression does make sense, give its dimensions.

(a) $\begin{bmatrix} a_1 \\ \vdots \\ a_n \end{bmatrix}$

(b) $\begin{bmatrix} a_1^T \\ \vdots \\ a_n^T \end{bmatrix}$

(c) $\begin{bmatrix} a_1 & \cdots & a_n \end{bmatrix}$

(d) $\begin{bmatrix} a_1^T & \cdots & a_n^T \end{bmatrix}$

6.2 *Matrix notation.* Suppose the block matrix

$$\begin{bmatrix} A & I \\ I & C \end{bmatrix}$$

makes sense, where A is a $p \times q$ matrix. What are the dimensions of C?

6.3 *Block matrix.* Assuming the matrix

$$K = \begin{bmatrix} I & A^T \\ A & 0 \end{bmatrix}$$

makes sense, which of the following statements must be true? ('Must be true' means that it follows with no additional assumptions.)

(a) K is square.

(b) A is square or wide.

(c) K is symmetric, *i.e.*, $K^T = K$.

(d) The identity and zero submatrices in K have the same dimensions.

(e) The zero submatrix is square.

6.4 *Adjacency matrix row and column sums.* Suppose A is the adjacency matrix of a directed graph (see page 112). What are the entries of the vector $A\mathbf{1}$? What are the entries of the vector $A^T\mathbf{1}$?

6.5 *Adjacency matrix of reversed graph.* Suppose A is the adjacency matrix of a directed graph (see page 112). The *reversed graph* is obtained by reversing the directions of all the edges of the original graph. What is the adjacency matrix of the reversed graph? (Express your answer in terms of A.)

6.6 *Matrix-vector multiplication.* For each of the following matrices, describe in words how x and $y = Ax$ are related. In each case x and y are n-vectors, with $n = 3k$.

(a) $A = \begin{bmatrix} 0 & 0 & I_k \\ 0 & I_k & 0 \\ I_k & 0 & 0 \end{bmatrix}$.

(b) $A = \begin{bmatrix} E & 0 & 0 \\ 0 & E & 0 \\ 0 & 0 & E \end{bmatrix}$, where E is the $k \times k$ matrix with all entries $1/k$.

6.7 *Currency exchange matrix.* We consider a set of n currencies, labeled $1, \ldots, n$. (These might correspond to USD, RMB, EUR, and so on.) At a particular time the exchange or conversion rates among the n currencies are given by an $n \times n$ (exchange rate) matrix R, where R_{ij} is the amount of currency i that you can buy for one unit of currency j. (All entries of R are positive.) The exchange rates include commission charges, so we have $R_{ji} R_{ij} < 1$ for all $i \neq j$. You can assume that $R_{ii} = 1$.

Suppose $y = Rx$, where x is a vector (with nonnegative entries) that represents the amounts of the currencies that we hold. What is y_i? Your answer should be in English.

6.8 *Cash flow to bank account balance.* The T-vector c represents the cash flow for an interest bearing bank account over T time periods. Positive values of c indicate a deposit, and negative values indicate a withdrawal. The T-vector b denotes the bank account balance in the T periods. We have $b_1 = c_1$ (the initial deposit or withdrawal) and

$$b_t = (1 + r)b_{t-1} + c_t, \quad t = 2, \ldots, T,$$

where $r > 0$ is the (per-period) interest rate. (The first term is the previous balance plus the interest, and the second term is the deposit or withdrawal.)

Find the $T \times T$ matrix A for which $b = Ac$. That is, the matrix A maps a cash flow sequence into a bank account balance sequence. Your description must make clear what all entries of A are.

6.9 *Multiple channel marketing campaign.* Potential customers are divided into m market segments, which are groups of customers with similar demographics, *e.g.*, college educated women aged 25–29. A company markets its products by purchasing advertising in a set of n channels, *i.e.*, specific TV or radio shows, magazines, web sites, blogs, direct mail, and so on. The ability of each channel to deliver impressions or views by potential customers is characterized by the *reach matrix*, the $m \times n$ matrix R, where R_{ij} is the number of views of customers in segment i for each dollar spent on channel j. (We assume that the total number of views in each market segment is the sum of the views from each channel, and that the views from each channel scale linearly with spending.) The n-vector c will denote the company's purchases of advertising, in dollars, in the n channels. The m-vector v gives the total number of impressions in the m market segments due to the advertising in all channels. Finally, we introduce the m-vector a, where a_i gives the profit in dollars per impression in market segment i. The entries of R, c, v, and a are all nonnegative.

(a) Express the total amount of money the company spends on advertising using vector/matrix notation.

(b) Express v using vector/matrix notation, in terms of the other vectors and matrices.

(c) Express the total profit from all market segments using vector/matrix notation.

(d) How would you find the single channel most effective at reaching market segment 3, in terms of impressions per dollar spent?

(e) What does it mean if R_{35} is very small (compared to other entries of R)?

6.10 *Resource requirements.* We consider an application with n different job (types), each of which consumes m different resources. We define the $m \times n$ resource matrix R, with entry R_{ij} giving the amount of resource i that is needed to run one unit of job j, for $i = 1, \ldots, m$ and $j = 1, \ldots, n$. (These numbers are typically positive.) The number (or amount) of each of the different jobs to be processed or run is given by the entries of the n-vector x. (These entries are typically nonnegative integers, but they can be fractional if the jobs are divisible.) The entries of the m-vector p give the price per unit of each of the resources.

(a) Let y be the m-vector whose entries give the total of each of the m resources needed to process the jobs given by x. Express y in terms of R and x using matrix and vector notation.

(b) Let c be an n-vector whose entries gives the cost per unit for each job type. (This is the total cost of the resources required to run one unit of the job type.) Express c in terms of R and p using matrix and vector notation.

Remark. One example is a data center, which runs many instances of each of n types of application programs. The resources include number of cores, amount of memory, disk, and network bandwidth.

6.11 Let A and B be two $m \times n$ matrices. Under each of the assumptions below, determine whether $A = B$ must always hold, or whether $A = B$ holds only sometimes.

(a) Suppose $Ax = Bx$ holds for all n-vectors x.

(b) Suppose $Ax = Bx$ for some nonzero n-vector x.

6.12 *Skew-symmetric matrices.* An $n \times n$ matrix A is called *skew-symmetric* if $A^T = -A$, i.e., its transpose is its negative. (A symmetric matrix satisfies $A^T = A$.)

(a) Find all 2×2 skew-symmetric matrices.

(b) Explain why the diagonal entries of a skew-symmetric matrix must be zero.

(c) Show that for a skew-symmetric matrix A, and any n-vector x, $(Ax) \perp x$. This means that Ax and x are orthogonal. *Hint.* First show that for any $n \times n$ matrix A and n-vector x, $x^T(Ax) = \sum_{i,j=1}^n A_{ij}x_ix_j$.

(d) Now suppose A is any matrix for which $(Ax) \perp x$ for any n-vector x. Show that A must be skew-symmetric. *Hint.* You might find the formula

$$(e_i + e_j)^T(A(e_i + e_j)) = A_{ii} + A_{jj} + A_{ij} + A_{ji},$$

valid for any $n \times n$ matrix A, useful. For $i = j$, this reduces to $e_i^T(Ae_i) = A_{ii}$.

6.13 *Polynomial differentiation.* Suppose p is a polynomial of degree $n - 1$ or less, given by $p(t) = c_1 + c_2t + \cdots + c_nt^{n-1}$. Its derivative (with respect to t) $p'(t)$ is a polynomial of degree $n - 2$ or less, given by $p'(t) = d_1 + d_2t + \cdots + d_{n-1}t^{n-2}$. Find a matrix D for which $d = Dc$. (Give the entries of D, and be sure to specify its dimensions.)

6.14 *Norm of matrix-vector product.* Suppose A is an $m \times n$ matrix and x is an n-vector. A famous inequality relates $\|x\|$, $\|A\|$, and $\|Ax\|$:

$$\|Ax\| \leq \|A\|\|x\|.$$

The left-hand side is the (vector) norm of the matrix-vector product; the right-hand side is the (scalar) product of the matrix and vector norms. Show this inequality. *Hints.* Let a_i^T be the ith row of A. Use the Cauchy–Schwarz inequality to get $(a_i^T x)^2 \leq \|a_i\|^2\|x\|^2$. Then add the resulting m inequalities.

6.15 *Distance between adjacency matrices.* Let A and B be the $n \times n$ adjacency matrices of two directed graphs with n vertices (see page 112). The squared distance $\|A - B\|^2$ can be used to express how different the two graphs are. Show that $\|A - B\|^2$ is the total number of directed edges that are in one of the two graphs but not in the other.

6.16 *Columns of difference matrix.* Are the columns of the difference matrix D, defined in (6.5), linearly independent?

6.17 *Stacked matrix.* Let A be an $m \times n$ matrix, and consider the stacked matrix S defined by

$$S = \begin{bmatrix} A \\ I \end{bmatrix}.$$

When does S have linearly independent columns? When does S have linearly independent rows? Your answer can depend on m, n, or whether or not A has linearly independent columns or rows.

6.18 *Vandermonde matrices.* A Vandermonde matrix is an $m \times n$ matrix of the form

$$V = \begin{bmatrix} 1 & t_1 & t_1^2 & \cdots & t_1^{n-1} \\ 1 & t_2 & t_2^2 & \cdots & t_2^{n-1} \\ \vdots & \vdots & \vdots & \ddots & \vdots \\ 1 & t_m & t_m^2 & \cdots & t_m^{n-1} \end{bmatrix}$$

where $t_1, \ldots, t_m$ are numbers. Multiplying an n-vector c by the Vandermonde matrix V is the same as evaluating the polynomial of degree less than n, with coefficients $c_1, \ldots, c_n$, at the points $t_1, \ldots, t_m$; see page 120. Show that the columns of a Vandermonde matrix are linearly independent if the numbers $t_1, \ldots, t_m$ are distinct, *i.e.*, different from each other. *Hint.* Use the following fact from algebra: If a polynomial p with degree less than n has n or more roots (points t for which $p(t) = 0$) then all its coefficients are zero.

6.19 *Back-test timing.* The $T \times n$ asset returns matrix R gives the returns of n assets over T periods. (See page 120.) When the n-vector w gives a set of portfolio weights, the T-vector Rw gives the time series of portfolio return over the T time periods. Evaluating portfolio return with past returns data is called *back-testing*.

Consider a specific case with $n = 5000$ assets, and $T = 2500$ returns. (This is 10 years of daily returns, since there are around 250 trading days in each year.) About how long would it take to carry out this back-test, on a 1 Gflop/s computer?

6.20 *Complexity of matrix-vector multiplication.* On page 123 we worked out the complexity of computing the m-vector Ax, where A is an $m \times n$ matrix and x is an n-vector, when each entry of Ax is computed as an inner product of a row of A and the vector x. Suppose instead that we compute Ax as a linear combination of the columns of A, with coefficients $x_1, \ldots, x_n$. How many flops does this method require? How does it compare to the method described on page 123?

6.21 *Complexity of matrix-sparse-vector multiplication.* On page 123 we consider the complexity of computing Ax, where A is a sparse $m \times n$ matrix and x is an n-vector x (not assumed to be sparse). Now consider the complexity of computing Ax when the $m \times n$ matrix A is not sparse, but the n-vector x is sparse, with $\mathbf{nnz}(x)$ nonzero entries. Give the total number of flops in terms of m, n, and $\mathbf{nnz}(x)$, and simplify it by dropping terms that are dominated by others when the dimensions are large. *Hint.* The vector Ax is a linear combination of $\mathbf{nnz}(x)$ columns of A.

6.22 *Distribute or not?* Suppose you need to compute $z = (A + B)(x + y)$, where A and B are $m \times n$ matrices and x and y are n-vectors.

(a) What is the approximate flop count if you evaluate z as expressed, *i.e.*, by adding A and B, adding x and y, and then carrying out the matrix-vector multiplication?

(b) What is the approximate flop count if you evaluate z as $z = Ax + Ay + Bx + By$, *i.e.*, with four matrix-vector multiplies and three vector additions?

(c) Which method requires fewer flops? Your answer can depend on m and n. *Remark.* When comparing two computation methods, we usually do not consider a factor of 2 or 3 in flop counts to be significant, but in this exercise you can.

Chapter 7

Matrix examples

In this chapter we describe some special matrices that occur often in applications.

7.1 Geometric transformations

Suppose the 2-vector (or 3-vector) x represents a position in 2-D (or 3-D) space. Several important geometric transformations or mappings from points to points can be expressed as matrix-vector products $y = Ax$, with A a 2×2 (or 3×3) matrix. In the examples below, we consider the mapping from x to y, and focus on the 2-D case (for which some of the matrices are simpler to describe).

Scaling. Scaling is the mapping $y = ax$, where a is a scalar. This can be expressed as $y = Ax$ with $A = aI$. This mapping stretches a vector by the factor $|a|$ (or shrinks it when $|a| < 1$), and it flips the vector (reverses its direction) if $a < 0$.

Dilation. Dilation is the mapping $y = Dx$, where D is a diagonal matrix, $D = \mathbf{diag}(d_1, d_2)$. This mapping stretches the vector x by different factors along the two different axes. (Or shrinks, if $|d_i| < 1$, and flips, if $d_i < 0$.)

Rotation. Suppose that y is the vector obtained by rotating x by θ radians counterclockwise. Then we have

$$y = \begin{bmatrix} \cos\theta & -\sin\theta \\ \sin\theta & \cos\theta \end{bmatrix} x. \tag{7.1}$$

This matrix is called (for obvious reasons) a *rotation matrix*.

Reflection. Suppose that y is the vector obtained by reflecting x through the line that passes through the origin, inclined θ radians with respect to horizontal. Then we have

$$y = \begin{bmatrix} \cos(2\theta) & \sin(2\theta) \\ \sin(2\theta) & -\cos(2\theta) \end{bmatrix} x.$$

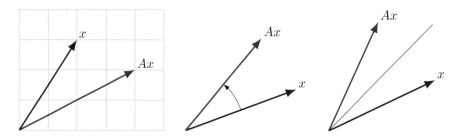

Figure 7.1 From left to right: A dilation with $A = \mathbf{diag}(2, 2/3)$, a counterclockwise rotation by $\pi/6$ radians, and a reflection through a line that makes an angle of $\pi/4$ radians with the horizontal line.

Projection onto a line.   The projection of the point x onto a set is the point in the set that is closest to x. Suppose y is the projection of x onto the line that passes through the origin, inclined θ radians with respect to horizontal. Then we have

$$y = \left[\begin{array}{cc} (1/2)(1 + \cos(2\theta)) & (1/2)\sin(2\theta) \\ (1/2)\sin(2\theta) & (1/2)(1 - \cos(2\theta)) \end{array} \right] x.$$

Some of these geometric transformations are illustrated in figure 7.1.

Finding the matrix.   When a geometric transformation is represented by matrix-vector multiplication (as in the examples above), a simple method to find the matrix is to find its columns. The ith column is the vector obtained by applying the transformation to e_i. As a simple example consider clockwise rotation by $90°$ in 2-D. Rotating the vector $e_1 = (1, 0)$ by $90°$ gives $(0, -1)$; rotating $e_2 = (0, 1)$ by $90°$ gives $(1, 0)$. So rotation by $90°$ is given by

$$y = \left[\begin{array}{cc} 0 & 1 \\ -1 & 0 \end{array} \right] x.$$

Change of coordinates.   In many applications multiple coordinate systems are used to describe locations or positions in 2-D or 3-D. For example in aerospace engineering we can describe a position using *earth-fixed* coordinates or *body-fixed* coordinates, where the body refers to an aircraft. Earth-fixed coordinates are with respect to a specific origin, with the three axes pointing East, North, and straight up, respectively. The origin of the body-fixed coordinates is a specific location on the aircraft (typically the center of gravity), and the three axes point forward (along the aircraft body), left (with respect to the aircraft body), and up (with respect to the aircraft body). Suppose the 3-vector x^{body} describes a location using the body coordinates, and x^{earth} describes the same location in earth-fixed coordinates. These are related by

$$x^{\mathrm{earth}} = p + Q x^{\mathrm{body}},$$

where p is the location of the airplane center (in earth-fixed coordinates) and Q is a 3×3 matrix. The ith column of Q gives the earth-fixed coordinates for the ith axis of the airplane. For an airplane in level flight, heading due South, we have

$$Q = \begin{bmatrix} 0 & 1 & 0 \\ -1 & 0 & 0 \\ 0 & 0 & 1 \end{bmatrix}.$$

7.2 Selectors

An $m \times n$ *selector matrix* A is one in which each row is a unit vector (transposed):

$$A = \begin{bmatrix} e_{k_1}^T \\ \vdots \\ e_{k_m}^T \end{bmatrix},$$

where $k_1, \ldots, k_m$ are integers in the range $1, \ldots, n$. When it multiplies a vector, it simply copies the k_ith entry of x into the ith entry of $y = Ax$:

$$y = (x_{k_1}, x_{k_2}, \ldots, x_{k_m}).$$

In words, each entry of Ax is a selection of an entry of x.

The identity matrix, and the *reverser matrix*

$$A = \begin{bmatrix} e_n^T \\ \vdots \\ e_1^T \end{bmatrix} = \begin{bmatrix} 0 & 0 & \cdots & 0 & 1 \\ 0 & 0 & \cdots & 1 & 0 \\ \vdots & \vdots & \ddots & \vdots & \vdots \\ 0 & 1 & \cdots & 0 & 0 \\ 1 & 0 & \cdots & 0 & 0 \end{bmatrix}$$

are special cases of selector matrices. (The reverser matrix reverses the order of the entries of a vector: $Ax = (x_n, x_{n-1}, \ldots, x_2, x_1)$.) Another one is the $r:s$ *slicing matrix*, which can be described as the block matrix

$$A = \begin{bmatrix} 0_{m \times (r-1)} & I_{m \times m} & 0_{m \times (n-s)} \end{bmatrix},$$

where $m = s - r + 1$. (We show the dimensions of the blocks for clarity.) We have $Ax = x_{r:s}$, i.e., multiplying by A gives the $r:s$ slice of a vector.

Down-sampling. Another example is the $(n/2) \times n$ matrix (with n even)

$$A = \begin{bmatrix} 1 & 0 & 0 & 0 & 0 & 0 & \cdots & 0 & 0 & 0 & 0 \\ 0 & 0 & 1 & 0 & 0 & 0 & \cdots & 0 & 0 & 0 & 0 \\ 0 & 0 & 0 & 0 & 1 & 0 & \cdots & 0 & 0 & 0 & 0 \\ \vdots & \vdots & \vdots & \vdots & \vdots & \vdots & & \vdots & \vdots & \vdots & \vdots \\ 0 & 0 & 0 & 0 & 0 & 0 & \cdots & 1 & 0 & 0 & 0 \\ 0 & 0 & 0 & 0 & 0 & 0 & \cdots & 0 & 0 & 1 & 0 \end{bmatrix}.$$

If $y = Ax$, we have $y = (x_1, x_3, x_5, \ldots, x_{n-3}, x_{n-1})$. When x is a time series, y is called the $2\times$ *down-sampled* version of x. If x is a quantity sampled every hour, then y is the same quantity, sampled every 2 hours.

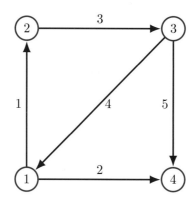

Figure 7.2 Directed graph with four vertices and five edges.

Image cropping.   As a more interesting example, suppose that x is an image with $M \times N$ pixels, with M and N even. (That is, x is an MN-vector, with its entries giving the pixel values in some specific order.) Let y be the $(M/2) \times (N/2)$ image that is the upper left corner of the image x, *i.e.*, a cropped version. Then we have $y = Ax$, where A is an $(MN/4) \times (MN)$ selector matrix. The ith row of A is $e_{k_i}^T$, where k_i is the index of the pixel in x that corresponds to the ith pixel in y.

Permutation matrices.   An $n \times n$ *permutation matrix* is one in which each column is a unit vector, and each row is the transpose of a unit vector. (In other words, A and A^T are both selector matrices.) Thus, exactly one entry of each row is one, and exactly one entry of each column is one. This means that $y = Ax$ can be expressed as $y_i = x_{\pi_i}$, where π is a permutation of $1, 2, \ldots, n$, *i.e.*, each integer from 1 to n appears exactly once in $\pi_1, \ldots, \pi_n$.

As a simple example consider the permutation $\pi = (3, 1, 2)$. The associated permutation matrix is

$$A = \begin{bmatrix} 0 & 0 & 1 \\ 1 & 0 & 0 \\ 0 & 1 & 0 \end{bmatrix}.$$

Multiplying a 3-vector by A re-orders its entries: $Ax = (x_3, x_1, x_2)$.

7.3   Incidence matrix

Directed graph.   A *directed graph* consists of a set of *vertices* (or nodes), labeled $1, \ldots, n$, and a set of *directed edges* (or branches), labeled $1, \ldots, m$. Each edge is connected from one of the nodes and into another one, in which case we say the two nodes are connected or adjacent. Directed graphs are often drawn with the vertices as circles or dots, and the edges as arrows, as in figure 7.2. A directed graph can be described by its $n \times m$ *incidence matrix*, defined as

$$A_{ij} = \begin{cases} 1 & \text{edge } j \text{ points to node } i \\ -1 & \text{edge } j \text{ points from node } i \\ 0 & \text{otherwise.} \end{cases}$$

The incidence matrix is evidently sparse, since it has only two nonzero entries in each column (one with value 1 and other with value -1). The jth column is associated with the jth edge; the indices of its two nonzero entries give the nodes that the edge connects. The ith row of A corresponds to node i; its nonzero entries tell us which edges connect to the node, and whether they point into or away from the node. The incidence matrix for the graph shown in figure 7.2 is

$$A = \begin{bmatrix} -1 & -1 & 0 & 1 & 0 \\ 1 & 0 & -1 & 0 & 0 \\ 0 & 0 & 1 & -1 & -1 \\ 0 & 1 & 0 & 0 & 1 \end{bmatrix}.$$

A directed graph can also be described by its adjacency matrix, described on page 112. The adjacency and incidence matrices for a directed graph are closely related, but not the same. The adjacency matrix does not explicitly label the edges $j = 1, \ldots, m$. There are also some small differences in the graphs that can be represented using incidence and adjacency matrices. For example, self edges (that connect from and to the same vertex) cannot be represented in an incidence matrix.

7.3.1   Networks

In many applications a graph is used to represent a *network*, through which some commodity or quantity such as electricity, water, heat, or vehicular traffic flows. The edges of the graph represent the *paths* or *links* over which the quantity can move or flow, in either direction. If x is an m-vector representing a flow in the network, we interpret x_j as the flow (rate) along the edge j, with a positive value meaning the flow is in the direction of edge j, and negative meaning the flow is in the opposite direction of edge j. In a network, the direction of the edge or link does not specify the direction of flow; it only specifies which direction of flow we consider to be positive.

Flow conservation.   When x represents a flow in a network, the matrix-vector product $y = Ax$ can be given a very simple interpretation. The n-vector $y = Ax$ can be interpreted as the vector of net flows, from the edges, into the nodes: y_i is equal to the total of the flows that come in to node i, minus the total of the flows that go out from node i. The quantity y_i is sometimes called the *flow surplus* at node i.

If $Ax = 0$, we say that *flow conservation* occurs, since at each node, the total in-flow matches the total out-flow. In this case the flow vector x is called a *circulation*. This could be used as a model of traffic flow (in a closed system), with the nodes

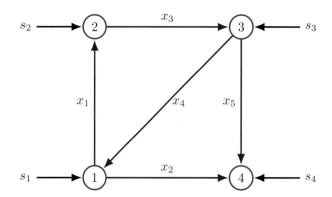

Figure 7.3 Network with four nodes and five edges, with source flows shown.

representing intersections and the edges representing road segments (one for each direction).

For a network described by the directed graph example above, the vector

$$x = (1, -1, 1, 0, 1)$$

is a circulation, since $Ax = 0$. This flow corresponds to a unit clockwise flow on the outer edges (1, 3, 5, and 2) and no flow on the diagonal edge (4). (Visualizing this explains why such vectors are called circulations.)

Sources. In many applications it is useful to include additional flows called *source flows* or *exogenous flows*, that enter or leave the network at the nodes, but not along the edges, as shown in figure 7.3. We denote these flows with an n-vector s. We can think of s_i as a flow that enters the network at node i from outside, *i.e.*, not from any edge. When $s_i > 0$ the exogenous flow is called a *source*, since it is injecting the quantity into the network at the node. When $s_i < 0$ the exogenous flow is called a *sink*, since it is removing the quantity from the network at the node.

Flow conservation with sources. The equation $Ax + s = 0$ means that the flow is conserved at each node, counting the source flow: The total of all incoming flow, from the incoming edges and exogenous source, minus the total outgoing flow from outgoing edges and exogenous sinks, is zero.

As an example, flow conservation with sources can be used as an approximate model of a power grid (ignoring losses), with x being the vector of power flows along the transmission lines, $s_i > 0$ representing a generator injecting power into the grid at node i, $s_i < 0$ representing a load that consumes power at node i, and $s_i = 0$ representing a substation where power is exchanged among transmission lines, with no generation or load attached.

For the example above, consider the source vector $s = (1, 0, -1, 0)$, which corresponds to an injection of one unit of flow into node 1, and the removal of one unit of flow at node 3. In other words, node 1 is a source, node 3 is a sink, and flow is conserved at nodes 2 and 4. For this source, the flow vector

$$x = (0.6,\ 0.3,\ 0.6,\ -0.1,\ -0.3)$$

satisfies flow conservation, *i.e.*, $Ax + s = 0$. This flow can be explained in words: The unit external flow entering node 1 splits three ways, with 0.6 flowing up, 0.3 flowing right, and 0.1 flowing diagonally up (on edge 4). The upward flow on edge 1 passes through node 2, where flow is conserved, and proceeds right on edge 3 towards node 3. The rightward flow on edge 2 passes through node 4, where flow is conserved, and proceeds up on edge 5 to node 3. The one unit of excess flow arriving at node 3 is removed as external flow.

Node potentials.   A graph is also useful when we focus on the values of some quantity at each graph vertex or node. Let v be an n-vector, often interpreted as a *potential*, with v_i the potential value at node i. We can give a simple interpretation to the matrix-vector product $u = A^T v$. The m-vector $u = A^T v$ gives the potential differences across the edges: $u_j = v_l - v_k$, where edge j goes from node k to node l.

Dirichlet energy.   When the m-vector $A^T v$ is small, it means that the potential differences across the edges are small. Another way to say this is that the potentials of connected vertices are near each other. A quantitative measure of this is the function of v given by

$$\mathcal{D}(v) = \|A^T v\|^2.$$

This function arises in many applications, and is called the *Dirichlet energy* (or *Laplacian quadratic form*) associated with the graph. It can be expressed as

$$\mathcal{D}(v) = \sum_{\text{edges } (k,l)} (v_l - v_k)^2,$$

which is the sum of the squares of the potential differences of v across all edges in the graph. The Dirichlet energy is small when the potential differences across the edges of the graph are small, *i.e.*, nodes that are connected by edges have similar potential values.

The Dirichlet energy is used as a measure the non-smoothness (roughness) of a set of node potentials on a graph. A set of node potentials with small Dirichlet energy can be thought of as smoothly varying across the graph. Conversely, a set of potentials with large Dirichlet energy can be thought of as non-smooth or rough. The Dirichlet energy will arise as a measure of roughness in several applications we will encounter later.

As a simple example, consider the potential vector $v = (1, -1, 2, -1)$ for the graph shown in figure 7.2. For this set of potentials, the potential differences across the edges are relatively large, with $A^T v = (-2, -2, 3, -1, -3)$, and the associated Dirichlet energy is $\|A^T v\|^2 = 27$. Now consider the potential vector $v = (1, 2, 2, 1)$. The associated edge potential differences are $A^T v = (1, 0, 0, -1, -1)$, and the Dirichlet energy has the much smaller value $\|A^T v\|^2 = 3$.

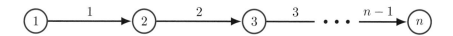

Figure 7.4 Chain graph.

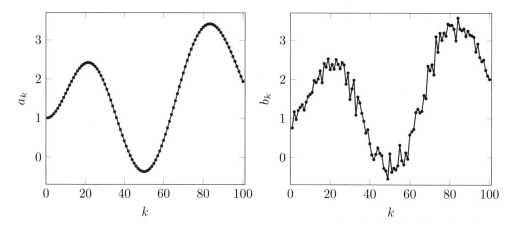

Figure 7.5 Two vectors of length 100, with Dirichlet energy $\mathcal{D}(a) = 1.14$ and $\mathcal{D}(b) = 8.99$.

Chain graph. The incidence matrix and the Dirichlet energy function have a particularly simple form for the *chain graph* shown in figure 7.4, with n vertices and $n-1$ edges. The $n \times (n-1)$ incidence matrix is the transpose of the difference matrix D described on page 119, in (6.5). The Dirichlet energy is then

$$\mathcal{D}(v) = \|Dv\|^2 = (v_2 - v_1)^2 + \cdots + (v_n - v_{n-1})^2,$$

the sum of squares of the differences between consecutive entries of the n-vector v. This is used as a measure of the non-smoothness of the vector v, considered as a time series. Figure 7.5 shows an example.

7.4   Convolution

The *convolution* of an n-vector a and an m-vector b is the $(n + m - 1)$-vector denoted $c = a * b$, with entries

$$c_k = \sum_{i+j=k+1} a_i b_j, \quad k = 1, \ldots, n + m - 1, \qquad (7.2)$$

where the subscript in the sum means that we should sum over all values of i and j in their index ranges $1, \ldots, n$ and $1, \ldots, m$, for which the sum $i + j$ is $k + 1$. For example, with $n = 4$, $m = 3$, we have

$$
\begin{aligned}
c_1 &= a_1 b_1 \\
c_2 &= a_1 b_2 + a_2 b_1 \\
c_3 &= a_1 b_3 + a_2 b_2 + a_3 b_1 \\
c_4 &= a_2 b_3 + a_3 b_2 + a_4 b_1 \\
c_5 &= a_3 b_3 + a_4 b_2 \\
c_6 &= a_4 b_3.
\end{aligned}
$$

Convolution reduces to ordinary multiplication of numbers when $n = m = 1$, and to scalar-vector multiplication when either $n = 1$ or $m = 1$. Convolution arises in many applications and contexts.

As a specific numerical example, we have $(1, 0, -1) * (2, 1, -1) = (2, 1, -3, -1, 1)$, where the entries of the convolution result are found from

$$
\begin{aligned}
2 &= (1)(2) \\
1 &= (1)(1) + (0)(2) \\
-3 &= (1)(-1) + (0)(1) + (-1)(2) \\
-1 &= (0)(-1) + (-1)(1) \\
1 &= (-1)(-1).
\end{aligned}
$$

Polynomial multiplication.   If a and b represent the coefficients of two polynomials

$$
p(x) = a_1 + a_2 x + \cdots + a_n x^{n-1}, \qquad q(x) = b_1 + b_2 x + \cdots + b_m x^{m-1},
$$

then the coefficients of the product polynomial $p(x)q(x)$ are represented by $c = a * b$:

$$
p(x)q(x) = c_1 + c_2 x + \cdots + c_{n+m-1} x^{n+m-2}.
$$

To see this we will show that c_k is the coefficient of x^{k-1} in $p(x)q(x)$. We expand the product polynomial into mn terms, and collect those terms associated with x^{k-1}. These terms have the form $a_i b_j x^{i+j-2}$, for i and j that satisfy $i + j - 2 = k - 1$, i.e., $i + j = k - 1$. It follows that $c_k = \sum_{i+j=k+1} a_i b_j$, which agrees with the convolution formula (7.2).

Properties of convolution.   Convolution is symmetric: We have $a * b = b * a$. It is also associative: We have $(a * b) * c = a * (b * c)$, so we can write both as $a * b * c$. Another property is that $a * b = 0$ implies that either $a = 0$ or $b = 0$. These properties follow from the polynomial coefficient property above, and can also be directly shown. As an example, let us show that $a * b = b * a$. Suppose p is the polynomial with coefficients a, and q is the polynomial with coefficients b. The two polynomials $p(x)q(x)$ and $q(x)p(x)$ are the same (since multiplication of numbers is commutative), so they have the same coefficients. The coefficients of $p(x)q(x)$ are $a * b$ and the coefficients of $q(x)p(x)$ are $b * a$. These must be the same.

A basic property is that for fixed a, the convolution $a * b$ is a linear function of b; and for fixed b, it is a linear function of a. This means we can express $a * b$ as a matrix-vector product:

$$
a * b = T(b)a = T(a)b,
$$

where $T(b)$ is the $(n + m - 1) \times n$ matrix with entries

$$T(b)_{ij} = \begin{cases} b_{i-j+1} & 1 \le i - j + 1 \le m \\ 0 & \text{otherwise} \end{cases} \tag{7.3}$$

and similarly for $T(a)$. For example, with $n = 4$ and $m = 3$, we have

$$T(b) = \begin{bmatrix} b_1 & 0 & 0 & 0 \\ b_2 & b_1 & 0 & 0 \\ b_3 & b_2 & b_1 & 0 \\ 0 & b_3 & b_2 & b_1 \\ 0 & 0 & b_3 & b_2 \\ 0 & 0 & 0 & b_3 \end{bmatrix}, \qquad T(a) = \begin{bmatrix} a_1 & 0 & 0 \\ a_2 & a_1 & 0 \\ a_3 & a_2 & a_1 \\ a_4 & a_3 & a_2 \\ 0 & a_4 & a_3 \\ 0 & 0 & a_4 \end{bmatrix}.$$

The matrices $T(b)$ and $T(a)$ are called *Toeplitz* matrices (named after the mathematician Otto Toeplitz), which means the entries on any diagonal (*i.e.*, indices with $i - j$ constant) are the same. The columns of the Toeplitz matrix $T(a)$ are simply shifted versions of the vector a, padded with zero entries.

Variations.   Several slightly different definitions of convolution are used in different applications. In one variation, a and b are infinite two-sided sequences (and not vectors) with indices ranging from $-\infty$ to ∞. In another variation, the rows of $T(a)$ at the top and bottom that do not contain all the coefficients of a are dropped. (In this version, the rows of $T(a)$ are shifted versions of the vector a, reversed.) For consistency, we will use the one definition (7.2).

Examples.

- *Time series smoothing.* Suppose the n-vector x is a time series, and $a = (1/3, 1/3, 1/3)$. Then the $(n + 2)$-vector $y = a * x$ can be interpreted as a *smoothed* version of the original time series: for $i = 3, \dots, n$, y_i is the average of x_i, x_{i-1}, x_{i-2}. The time series y is called the (3-period) *moving average* of the time series x. Figure 7.6 shows an example.

- *First order differences.* If the n-vector x is a time series and $a = (1, -1)$, the time series $y = a * x$ gives the first order differences in the series x:

$$y = (x_1, \, x_2 - x_1, \, x_3 - x_2, \, \dots, \, x_n - x_{n-1}, \, -x_n).$$

  (The first and last entries here would be the first order difference if we take $x_0 = x_{n+1} = 0$.)

- *Audio filtering.* If the n-vector x is an audio signal, and a is a vector (typically with length less than around 0.1 second of real time) the vector $y = a * x$ is called the *filtered* audio signal, with *filter coefficients* a. Depending on the coefficients a, y will be perceived as enhancing or suppressing different frequencies, like the familiar audio tone controls.

- *Communication channel.* In a modern data communication system, a time series u is transmitted or sent over some channel (*e.g.*, electrical, optical, or radio) to a receiver, which receives the time series y. A very common model is that y and u are related via convolution: $y = c * u$, where the vector c is the *channel impulse response*.

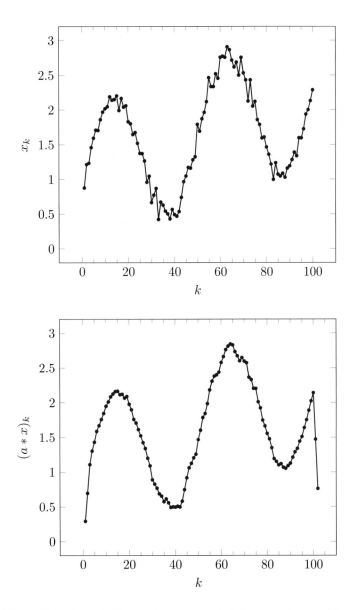

Figure 7.6 *Top.* A time series represented by a vector x of length 100. *Bottom.* The 3-period moving average of the time series as a vector of length 102. This vector is the convolution of x with $a = (1/3, 1/3, 1/3)$.

Input-output convolution system. Many physical systems with an *input* (time series) m-vector u and *output* (time series) y are well modeled as $y = h*u$, where the n-vector h is called the *system impulse response*. For example, u_t might represent the power level of a heater at time period t, and y_t might represent the resulting temperature rise (above the surrounding temperature). The lengths of u and y, n and $m+n-1$, are typically large, and not relevant in these applications. We can express the ith entry of the output y as

$$y_i = \sum_{j=1}^{n} u_{i-j+1} h_j,$$

where we interpret u_k as zero for $k < 0$ or $k > n$. This formula states that y at time i is a linear combination of $u_i, u_{i-1}, \ldots, u_{i-n+1}$, *i.e.*, a linear combination of the current input value u_i, and the past $n-1$ input values $u_{i-1}, \ldots, u_{i-n+1}$. The coefficients are precisely $h_1, \ldots, h_n$. Thus, h_3 can be interpreted as the factor by which the current output depends on what the input was, 2 time steps before. Alternatively, we can say that h_3 is the factor by which the input at any time will affect the output 2 steps in the future.

Complexity of convolution. The naïve method to compute the convolution $c = a * b$ of an n-vector a and an m-vector b, using the basic formula (7.2) to calculate each c_k, requires around $2mn$ flops. The same number of flops is required to compute the matrix-vector products $T(a)b$ or $T(b)a$, taking into account the zeros at the top right and bottom left in the Toeplitz matrices $T(b)$ and $T(a)$. Forming these matrices requires us to store mn numbers, even though the original data contains only $m + n$ numbers.

It turns out that the convolution of two vectors can be computed far faster, using a so-called *fast convolution algorithm*. By exploiting the special structure of the convolution equations, this algorithm can compute the convolution of an n-vector and an m-vector in around $5(m + n) \log_2(m + n)$ flops, and with no additional memory requirement beyond the original $m + n$ numbers. The fast convolution algorithm is based on the *fast Fourier transform* (FFT), which is beyond the scope of this book. (The Fourier transform is named for the mathematician Jean-Baptiste Fourier.)

7.4.1 2-D convolution

Convolution has a natural extension to multiple dimensions. Suppose that A is an $m \times n$ matrix and B is a $p \times q$ matrix. Their convolution is the $(m+p-1) \times (n+q-1)$ matrix

$$C_{rs} = \sum_{i+k=r+1, \; j+l=s+1} A_{ij} B_{kl}, \quad r = 1, \ldots, m+p-1, \quad s = 1, \ldots, n+q-1,$$

where the indices are restricted to their ranges (or alternatively, we assume that A_{ij} and B_{kl} are zero, when the indices are out of range). This is *not* denoted

$C = A * B$, however, in standard mathematical notation. So we will use the notation $C = A \star B$.

The same properties that we observed for 1-D convolution hold for 2-D convolution: We have $A \star B = B \star A$, $(A \star B) \star C = A \star (B \star C)$, and for fixed B, $A \star B$ is a linear function of A.

Image blurring. If the $m \times n$ matrix X represents an image, $Y = X \star B$ represents the effect of *blurring* the image by the *point spread function* (PSF) given by the entries of the matrix B. If we represent X and Y as vectors, we have $y = T(B)x$, for some $(m + p - 1)(n + q - 1) \times mn$-matrix $T(B)$.

As an example, with

$$B = \begin{bmatrix} 1/4 & 1/4 \\ 1/4 & 1/4 \end{bmatrix}, \tag{7.4}$$

$Y = X \star B$ is an image where each pixel value is the average of a 2×2 block of 4 adjacent pixels in X. The image Y would be perceived as the image X, with some blurring of the fine details. This is illustrated in figure 7.7 for the 8×9 matrix

$$X = \begin{bmatrix} 1 & 1 & 1 & 1 & 1 & 1 & 1 & 1 & 1 \\ 1 & 1 & 1 & 1 & 1 & 1 & 1 & 1 & 1 \\ 1 & 1 & 0 & 0 & 0 & 0 & 0 & 1 & 1 \\ 1 & 1 & 1 & 0 & 1 & 1 & 0 & 1 & 1 \\ 1 & 1 & 1 & 0 & 1 & 1 & 0 & 1 & 1 \\ 1 & 1 & 1 & 0 & 1 & 1 & 0 & 1 & 1 \\ 1 & 1 & 1 & 1 & 1 & 1 & 1 & 1 & 1 \\ 1 & 1 & 1 & 1 & 1 & 1 & 1 & 1 & 1 \end{bmatrix} \tag{7.5}$$

and its convolution with B,

$$X \star B = \begin{bmatrix} 1/4 & 1/2 & 1/2 & 1/2 & 1/2 & 1/2 & 1/2 & 1/2 & 1/2 & 1/4 \\ 1/2 & 1 & 1 & 1 & 1 & 1 & 1 & 1 & 1 & 1/2 \\ 1/2 & 1 & 3/4 & 1/2 & 1/2 & 1/2 & 1/2 & 3/4 & 1 & 1/2 \\ 1/2 & 1 & 3/4 & 1/4 & 1/4 & 1/2 & 1/4 & 1/2 & 1 & 1/2 \\ 1/2 & 1 & 1 & 1/2 & 1/2 & 1 & 1/2 & 1/2 & 1 & 1/2 \\ 1/2 & 1 & 1 & 1/2 & 1/2 & 1 & 1/2 & 1/2 & 1 & 1/2 \\ 1/2 & 1 & 1 & 3/4 & 3/4 & 1 & 3/4 & 3/4 & 1 & 1/2 \\ 1/2 & 1 & 1 & 1 & 1 & 1 & 1 & 1 & 1 & 1/2 \\ 1/4 & 1/2 & 1/2 & 1/2 & 1/2 & 1/2 & 1/2 & 1/2 & 1/2 & 1/4 \end{bmatrix}.$$

With the point spread function

$$D^{\mathrm{hor}} = \begin{bmatrix} 1 & -1 \end{bmatrix},$$

the pixel values in the image $Y = X \star D^{\mathrm{hor}}$ are the horizontal first order differences of those in X:

$$Y_{ij} = X_{ij} - X_{i,j-1}, \quad i = 1, \dots, m, \quad j = 2, \dots, n$$

(and $Y_{i1} = X_{i1}$, $X_{i,n+1} = -X_{in}$ for $i = 1, \dots, m$). With the point spread function

$$D^{\mathrm{ver}} = \begin{bmatrix} 1 \\ -1 \end{bmatrix},$$

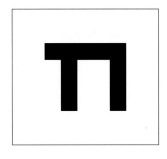
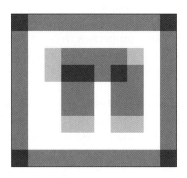

Figure 7.7 An 8×9 image and its convolution with the point spread function (7.4).

the pixel values in the image $Y = X \star D^{\mathrm{ver}}$ are the vertical first order differences of those in X:

$$Y_{ij} = X_{ij} - X_{i-1,j}, \quad i = 2, \ldots, m, \quad j = 1, \ldots, n$$

(and $Y_{1j} = X_{1j}$, $X_{m+1,j} = -X_{mj}$ for $j = 1, \ldots, n$). As an example, the convolutions of the matrix (7.5) with D^{hor} and D^{ver} are

$$X \star D^{\mathrm{hor}} = \begin{bmatrix}
1 & 0 & 0 & 0 & 0 & 0 & 0 & 0 & 0 & -1 \\
1 & 0 & 0 & 0 & 0 & 0 & 0 & 0 & 0 & -1 \\
1 & 0 & -1 & 0 & 0 & 0 & 0 & 1 & 0 & -1 \\
1 & 0 & 0 & -1 & 1 & 0 & -1 & 1 & 0 & -1 \\
1 & 0 & 0 & -1 & 1 & 0 & -1 & 1 & 0 & -1 \\
1 & 0 & 0 & -1 & 1 & 0 & -1 & 1 & 0 & -1 \\
1 & 0 & 0 & 0 & 0 & 0 & 0 & 0 & 0 & -1 \\
1 & 0 & 0 & 0 & 0 & 0 & 0 & 0 & 0 & -1
\end{bmatrix}$$

and

$$X \star D^{\mathrm{ver}} = \begin{bmatrix}
1 & 1 & 1 & 1 & 1 & 1 & 1 & 1 & 1 \\
0 & 0 & 0 & 0 & 0 & 0 & 0 & 0 & 0 \\
0 & 0 & -1 & -1 & -1 & -1 & -1 & 0 & 0 \\
0 & 0 & 1 & 0 & 1 & 1 & 0 & 0 & 0 \\
0 & 0 & 0 & 0 & 0 & 0 & 0 & 0 & 0 \\
0 & 0 & 0 & 0 & 0 & 0 & 0 & 0 & 0 \\
0 & 0 & 0 & 1 & 0 & 0 & 1 & 0 & 0 \\
0 & 0 & 0 & 0 & 0 & 0 & 0 & 0 & 0 \\
-1 & -1 & -1 & -1 & -1 & -1 & -1 & -1 & -1
\end{bmatrix}.$$

Figure 7.8 shows the effect of convolution on a larger image. The figure shows an image of size 512×512 and its convolution with the 8×8 matrix B with constant entries $B_{ij} = 1/64$.

Figure 7.8 512×512 image and the 519×519 image that results from the convolution of the first image with an 8×8 matrix with constant entries $1/64$. Image credit: NASA.

Exercises

7.1 *Projection on a line.* Let $P(x)$ denote the projection of the 2-D point (2-vector) x onto the line that passes through $(0, 0)$ and $(1, 3)$. (This means that $P(x)$ is the point on the line that is closest to x; see exercise 3.12.) Show that P is a linear function, and give the matrix A for which $P(x) = Ax$ for any x.

7.2 *3-D rotation.* Let x and y be 3-vectors representing positions in 3-D. Suppose that the vector y is obtained by rotating the vector x about the vertical axis (*i.e.*, e_3) by $45°$ (counterclockwise, *i.e.*, from e_1 toward e_2). Find the 3×3 matrix A for which $y = Ax$. *Hint.* Determine the three columns of A by finding the result of the transformation on the unit vectors e_1, e_2, e_3.

7.3 *Trimming a vector.* Find a matrix A for which $Ax = (x_2, \ldots, x_{n-1})$, where x is an n-vector. (Be sure to specify the size of A, and describe all its entries.)

7.4 *Down-sampling and up-conversion.* We consider n-vectors x that represent signals, with x_k the value of the signal at time k for $k = 1, \ldots, n$. Below we describe two functions of x that produce new signals $f(x)$. For each function, give a matrix A such that $f(x) = Ax$ for all x.

(a) $2\times$ *downsampling.* We assume n is even and define $f(x)$ as the $n/2$-vector y with elements $y_k = x_{2k}$. To simplify your notation you can assume that $n = 8$, *i.e.*,

$$f(x) = (x_2, \, x_4, \, x_6, \, x_8).$$

(On page 131 we describe a different type of down-sampling, that uses the average of pairs of original values.)

(b) $2\times$ *up-conversion with linear interpolation.* We define $f(x)$ as the $(2n - 1)$-vector y with elements $y_k = x_{(k+1)/2}$ if k is odd and $y_k = (x_{k/2} + x_{k/2+1})/2$ if k is even. To simplify your notation you can assume that $n = 5$, *i.e.*,

$$f(x) = \left(x_1, \, \frac{x_1 + x_2}{2}, \, x_2, \, \frac{x_2 + x_3}{2}, \, x_3, \, \frac{x_3 + x_4}{2}, \, x_4, \, \frac{x_4 + x_5}{2}, \, x_5\right).$$

7.5 *Transpose of selector matrix.* Suppose the $m \times n$ matrix A is a selector matrix. Describe the relation between the m-vector u and the n-vector $v = A^T u$.

7.6 *Rows of incidence matrix.* Show that the rows of the incidence matrix of a graph are always linearly dependent. *Hint.* Consider the sum of the rows.

7.7 *Incidence matrix of reversed graph.* (See exercise 6.5.) Suppose A is the incidence matrix of a graph. The reversed graph is obtained by reversing the directions of all the edges of the original graph. What is the incidence matrix of the reversed graph? (Express your answer in terms of A.)

7.8 *Flow conservation with sources.* Suppose that A is the incidence matrix of a graph, x is the vector of edge flows, and s is the external source vector, as described in §7.3. Assuming that flow is conserved, *i.e.*, $Ax + s = 0$, show that $\mathbf{1}^T s = 0$. This means that the total amount injected into the network by the sources ($s_i > 0$) must exactly balance the total amount removed from the network at the sink nodes ($s_i < 0$). For example if the network is a (lossless) electrical power grid, the total amount of electrical power generated (and injected into the grid) must exactly balance the total electrical power consumed (from the grid).

7.9 *Social network graph.* Consider a group of n people or users, and some symmetric social relation among them. This means that some pairs of users are *connected*, or *friends* (say). We can create a directed graph by associating a node with each user, and an edge between each pair of friends, arbitrarily choosing the direction of the edge. Now consider an n-vector v, where v_i is some quantity for user i, for example, age or education level (say, given in years). Let $\mathcal{D}(v)$ denote the Dirichlet energy associated with the graph and v, thought of as a potential on the nodes.

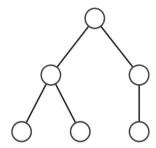

Figure 7.9 Tree with six vertices.

(a) Explain why the number $\mathcal{D}(v)$ does not depend on the choice of directions for the edges of the graph.

(b) Would you guess that $\mathcal{D}(v)$ is small or large? This is an open-ended, vague question; there is no right answer. Just make a guess as to what you might expect, and give a short English justification of your guess.

7.10 *Circle graph.* A *circle graph* (also called a *cycle graph*) has n vertices, with edges pointing from vertex 1 to vertex 2, from vertex 2 to vertex 3, ..., from vertex $n - 1$ to vertex n, and finally, from vertex n to vertex 1. (This last edge completes the circle.)

(a) Draw a diagram of a circle graph, and give its incidence matrix A.

(b) Suppose that x is a circulation for a circle graph. What can you say about x?

(c) Suppose the n-vector v is a potential on a circle graph. What is the Dirichlet energy $\mathcal{D}(v) = \|A^T v\|^2$?

Remark. The circle graph arises when an n-vector v represents a periodic time series. For example, v_1 could be the value of some quantity on Monday, v_2 its value on Tuesday, and v_7 its value on Sunday. The Dirichlet energy is a measure of the roughness of such an n-vector v.

7.11 *Tree.* An undirected graph is called a tree if it is connected (there is a path from every vertex to every other vertex) and contains no cycles, *i.e.*, there is no path that begins and ends at the same vertex. Figure 7.9 shows a tree with six vertices. For the tree in the figure, find a numbering of the vertices and edges, and an orientation of the edges, so that the incidence matrix A of the resulting directed graph satisfies $A_{ii} = 1$ for $i = 1, \ldots, 5$ and $A_{ij} = 0$ for $i < j$. In other words, the first 5 rows of A form a lower triangular matrix with ones on the diagonal.

7.12 *Some properties of convolution.* Suppose that a is an n-vector.

(a) *Convolution with* 1. What is $1 * a$? (Here we interpret 1 as a 1-vector.)

(b) *Convolution with a unit vector.* What is $e_k * a$, where e_k is the kth unit vector of dimension q? Describe this vector mathematically (*i.e.*, give its entries), and via a brief English description. You might find vector slice notation useful.

7.13 *Sum property of convolution.* Show that for any vectors a and b, we have $\mathbf{1}^T (a * b) = (\mathbf{1}^T a)(\mathbf{1}^T b)$. In words: The sum of the coefficients of the convolution of two vectors is the product of the sums of the coefficients of the vectors. *Hint.* If the vector a represents the coefficients of a polynomial p, $\mathbf{1}^T a = p(1)$.

7.14 *Rainfall and river height.* The T-vector r gives the daily rainfall in some region over a period of T days. The vector h gives the daily height of a river in the region (above its normal height). By careful modeling of water flow, or by fitting a model to past data, it is found that these vectors are (approximately) related by convolution: $h = g * r$, where

$$g = (0.1, 0.4, 0.5, 0.2).$$

Give a short story in English (with no mathematical terms) to approximately describe this relation. For example, you might mention how many days after a one day heavy rainfall the river height is most affected. Or how many days it takes for the river height to return to the normal height once the rain stops.

7.15 *Channel equalization.* We suppose that $u_1, \ldots, u_m$ is a signal (time series) that is transmitted (for example by radio). A receiver receives the signal $y = c * u$, where the n-vector c is called the channel impulse response. (See page 138.) In most applications n is small, *e.g.*, under 10, and m is much larger. An *equalizer* is a k-vector h that satisfies $h * c \approx e_1$, the first unit vector of length $n + k - 1$. The receiver equalizes the received signal y by convolving it with the equalizer to obtain $z = h * y$.

(a) How are z (the equalized received signal) and u (the original transmitted signal) related? *Hint.* Recall that $h * (c * u) = (h * c) * u$.

(b) *Numerical example.* Generate a signal u of length $m = 50$, with each entry a random value that is either -1 or $+1$. Plot u and $y = c * u$, with $c = (1, 0.7, -0.3)$. Also plot the equalized signal $z = h * y$, with

$$h = (0.9, -0.5, 0.5, -0.4, 0.3, -0.3, 0.2, -0.1).$$

Chapter 8

Linear equations

In this chapter we consider vector-valued linear and affine functions, and systems of linear equations.

8.1 Linear and affine functions

Vector-valued functions of vectors. The notation $f : \mathbf{R}^n \to \mathbf{R}^m$ means that f is a function that maps real n-vectors to real m-vectors. The value of the function f, evaluated at an n-vector x, is an m-vector $f(x) = (f_1(x), f_2(x), \ldots, f_m(x))$. Each of the components f_i of f is itself a scalar-valued function of x. As with scalar-valued functions, we sometimes write $f(x) = f(x_1, x_2, \ldots, x_n)$ to emphasize that f is a function of n scalar arguments. We use the same notation for each of the components of f, writing $f_i(x) = f_i(x_1, x_2, \ldots, x_n)$ to emphasize that f_i is a function mapping the scalar arguments $x_1, \ldots, x_n$ into a scalar.

The matrix-vector product function. Suppose A is an $m \times n$ matrix. We can define a function $f : \mathbf{R}^n \to \mathbf{R}^m$ by $f(x) = Ax$. The inner product function $f : \mathbf{R}^n \to \mathbf{R}$, defined as $f(x) = a^T x$, discussed in §2.1, is the special case with $m = 1$.

Superposition and linearity. The function $f : \mathbf{R}^n \to \mathbf{R}^m$, defined by $f(x) = Ax$, is *linear*, *i.e.*, it satisfies the superposition property:

$$f(\alpha x + \beta y) = \alpha f(x) + \beta f(y) \tag{8.1}$$

holds for all n-vectors x and y and all scalars α and β. It is a good exercise to parse this simple looking equation, since it involves overloading of notation. On the left-hand side, the scalar-vector multiplications αx and βy involve n-vectors, and the sum $\alpha x + \beta y$ is the sum of two n-vectors. The function f maps n-vectors to m-vectors, so $f(\alpha x + \beta y)$ is an m-vector. On the right-hand side, the scalar-vector multiplications and the sum are those for m-vectors. Finally, the equality sign is equality between two m-vectors.

We can verify that superposition holds for f using properties of matrix-vector and scalar-vector multiplication:

$$
\begin{aligned}
f(\alpha x + \beta y) &= A(\alpha x + \beta y) \\
&= A(\alpha x) + A(\beta y) \\
&= \alpha(Ax) + \beta(Ay) \\
&= \alpha f(x) + \beta f(y)
\end{aligned}
$$

Thus we can associate with every matrix A a linear function $f(x) = Ax$.

The converse is also true. Suppose f is a function that maps n-vectors to m-vectors, and is linear, $i.e.$, (8.1) holds for all n-vectors x and y and all scalars α and β. Then there exists an $m \times n$ matrix A such that $f(x) = Ax$ for all x. This can be shown in the same way as for scalar-valued functions in §2.1, by showing that if f is linear, then

$$
f(x) = x_1 f(e_1) + x_2 f(e_2) + \cdots + x_n f(e_n), \tag{8.2}
$$

where e_k is the kth unit vector of size n. The right-hand side can also be written as a matrix-vector product Ax, with

$$
A = \begin{bmatrix} f(e_1) & f(e_2) & \cdots & f(e_n) \end{bmatrix}.
$$

The expression (8.2) is the same as (2.3), but here $f(x)$ and $f(e_k)$ are vectors. The implications are exactly the same: A linear vector-valued function f is completely characterized by evaluating f at the n unit vectors $e_1, \ldots, e_n$.

As in §2.1 it is easily shown that the matrix-vector representation of a linear function is unique. If $f : \mathbf{R}^n \to \mathbf{R}^m$ is a linear function, then there exists exactly one matrix A such that $f(x) = Ax$ for all x.

Examples of linear functions.   In the examples below we define functions f that map n-vectors x to n-vectors $f(x)$. Each function is described in words, in terms of its effect on an arbitrary x. In each case we give the associated matrix multiplication representation.

- *Negation.* f changes the sign of x: $f(x) = -x$.

  Negation can be expressed as $f(x) = Ax$ with $A = -I$.

- *Reversal.* f reverses the order of the elements of x: $f(x) = (x_n, x_{n-1}, \ldots, x_1)$.

  The reversal function can be expressed as $f(x) = Ax$ with

$$
A = \begin{bmatrix}
0 & \cdots & 0 & 1 \\
0 & \cdots & 1 & 0 \\
\vdots & \cdot^{\cdot^{\cdot}} & \vdots & \vdots \\
1 & \cdots & 0 & 0
\end{bmatrix}.
$$

  (This is the $n \times n$ identity matrix with the order of its columns reversed. It is the *reverser matrix* introduced in §7.2.)

- *Running sum.* f forms the running sum of the elements in x:

$$f(x) = (x_1, \; x_1 + x_2, \; x_1 + x_2 + x_3, \; \ldots, \; x_1 + x_2 + \cdots + x_n).$$

The running sum function can be expressed as $f(x) = Ax$ with

$$A = \begin{bmatrix} 1 & 0 & \cdots & 0 & 0 \\ 1 & 1 & \cdots & 0 & 0 \\ \vdots & \vdots & \ddots & \vdots & \vdots \\ 1 & 1 & \cdots & 1 & 0 \\ 1 & 1 & \cdots & 1 & 1 \end{bmatrix},$$

i.e., $A_{ij} = 1$ if $i \geq j$ and $A_{ij} = 0$ otherwise. This is the running sum matrix defined in (6.6).

- *De-meaning.* f subtracts the mean from each entry of a vector x: $f(x) = x - \mathbf{avg}(x)\mathbf{1}$.

The de-meaning function can be expressed as $f(x) = Ax$ with

$$A = \begin{bmatrix} 1 - 1/n & -1/n & \cdots & -1/n \\ -1/n & 1 - 1/n & \cdots & -1/n \\ \vdots & \vdots & \ddots & \vdots \\ -1/n & -1/n & \cdots & 1 - 1/n \end{bmatrix}.$$

Examples of functions that are not linear.   Here we list some examples of functions f that map n-vectors x to n-vectors $f(x)$ that are *not* linear. In each case we show a superposition counterexample.

- *Absolute value.* f replaces each element of x with its absolute value: $f(x) = (|x_1|, |x_2|, \ldots, |x_n|)$.

The absolute value function is not linear. For example, with $n = 1$, $x = 1$, $y = 0$, $\alpha = -1$, $\beta = 0$, we have

$$f(\alpha x + \beta y) = 1 \neq \alpha f(x) + \beta f(y) = -1,$$

so superposition does not hold.

- *Sort.* f sorts the elements of x in decreasing order.

The sort function is not linear (except when $n = 1$, in which case $f(x) = x$). For example, if $n = 2$, $x = (1, 0)$, $y = (0, 1)$, $\alpha = \beta = 1$, then

$$f(\alpha x + \beta y) = (1, 1) \neq \alpha f(x) + \beta f(y) = (2, 0).$$

Affine functions.   A vector-valued function $f : \mathbf{R}^n \to \mathbf{R}^m$ is called affine if it can be expressed as $f(x) = Ax + b$, where A is an $m \times n$ matrix and b is an m-vector. It can be shown that a function $f : \mathbf{R}^n \to \mathbf{R}^m$ is affine if and only if

$$f(\alpha x + \beta y) = \alpha f(x) + \beta f(y)$$

holds for all n-vectors x, y, and all scalars α, β that satisfy $\alpha + \beta = 1$. In other words, superposition holds for affine combinations of vectors. (For linear functions, superposition holds for any linear combinations of vectors.)

The matrix A and the vector b in the representation of an affine function as $f(x) = Ax + b$ are unique. These parameters can be obtained by evaluating f at the vectors $0, e_1, \ldots, e_n$, where e_k is the kth unit vector in $\mathbf{R}^n$. We have

$$A = \begin{bmatrix} f(e_1) - f(0) & f(e_2) - f(0) & \cdots & f(e_n) - f(0) \end{bmatrix}, \qquad b = f(0).$$

Just like affine scalar-valued functions, affine vector-valued functions are often called linear, even though they are linear only when the vector b is zero.

8.2   Linear function models

Many functions or relations between variables that arise in natural science, engineering, and social sciences can be *approximated* as linear or affine functions. In these cases we refer to the linear function relating the two sets of variables as a *model* or an *approximation*, to remind us that the relation is only an approximation, and not exact. We give a few examples here.

- *Price elasticity of demand.* Consider n goods or services with prices given by the n-vector p, and demands for the goods given by the n-vector d. A change in prices will induce a change in demands. We let δ^{price} be the n-vector that gives the fractional change in the prices, *i.e.*, $\delta_i^{\text{price}} = (p_i^{\text{new}} - p_i)/p_i$, where p^{new} is the n-vector of new (changed) prices. We let δ^{dem} be the n-vector that gives the fractional change in the product demands, *i.e.*, $\delta_i^{\text{dem}} = (d_i^{\text{new}} - d_i)/d_i$, where d^{new} is the n-vector of new demands. A linear demand elasticity model relates these vectors as $\delta^{\text{dem}} = E^{\text{d}} \delta^{\text{price}}$, where E^{d} is the $n \times n$ *demand elasticity matrix*. For example, suppose $E_{11}^{\text{d}} = -0.4$ and $E_{21}^{\text{d}} = 0.2$. This means that a 1% increase in the price of the first good, with other prices kept the same, will cause demand for the first good to drop by 0.4%, and demand for the second good to increase by 0.2%. (In this example, the second good is acting as a *partial substitute* for the first good.)

- *Elastic deformation.* Consider a steel structure like a bridge or the structural frame of a building. Let f be an n-vector that gives the forces applied to the structure at n specific places (and in n specific directions), sometimes called a *loading*. The structure will deform slightly due to the loading. Let d be an m-vector that gives the displacements (in specific directions) of m points in the structure, due to the load, *e.g.*, the amount of sag at a specific point on a bridge. For small displacements, the relation between displacement and loading is well approximated as linear: $d = Cf$, where C is the $m \times n$ *compliance matrix*. The units of the entries of C are m/N.

8.2.1   Taylor approximation

Suppose $f : \mathbf{R}^n \to \mathbf{R}^m$ is differentiable, *i.e.*, has partial derivatives, and z is an n-vector. The first-order Taylor approximation of f near z is given by

$$
\begin{aligned}
\hat{f}(x)_i &= f_i(z) + \frac{\partial f_i}{\partial x_1}(z)(x_1 - z_1) + \cdots + \frac{\partial f_i}{\partial x_n}(z)(x_n - z_n) \\
&= f_i(z) + \nabla f_i(z)^T (x - z),
\end{aligned}
$$

for $i = 1, \ldots, m$. (This is just the first-order Taylor approximation of each of the scalar-valued functions f_i, described in §2.2.) For x near z, $\hat{f}(x)$ is a very good approximation of $f(x)$. We can express this approximation in compact notation, using matrix-vector multiplication, as

$$
\hat{f}(x) = f(z) + Df(z)(x - z), \tag{8.3}
$$

where the $m \times n$ matrix $Df(z)$ is the *derivative* or *Jacobian* matrix of f at z (see §C.1). Its components are the partial derivatives of f,

$$
Df(z)_{ij} = \frac{\partial f_i}{\partial x_j}(z), \quad i = 1, \ldots, m, \quad j = 1, \ldots, n,
$$

evaluated at the point z. The rows of the Jacobian are $\nabla f_i(z)^T$, for $i = 1, \ldots, m$. The Jacobian matrix is named for the mathematician Carl Gustav Jacob Jacobi.

As in the scalar-valued case, Taylor approximation is sometimes written with a second argument as $\hat{f}(x; z)$ to show the point z around which the approximation is made. Evidently the Taylor series approximation $\hat{f}$ is an affine function of x. (It is often called a linear approximation of f, even though it is not, in general, a linear function.)

8.2.2   Regression model

Recall the regression model (2.7)

$$
\hat{y} = x^T \beta + v, \tag{8.4}
$$

where the n-vector x is a feature vector for some object, β is an n-vector of weights, v is a constant (the offset), and $\hat{y}$ is the (scalar) value of the regression model prediction.

Now suppose we have a set of N objects (also called *samples* or *examples*), with feature vectors $x^{(1)}, \ldots, x^{(N)}$. The regression model predictions associated with the examples are given by

$$
\hat{y}^{(i)} = (x^{(i)})^T \beta + v, \quad i = 1, \ldots, N.
$$

These numbers usually correspond to predictions of the value of the outputs or responses. If in addition to the example feature vectors $x^{(i)}$ we are also given the actual value of the associated response variables, $y^{(1)}, \ldots, y^{(N)}$, then our *prediction errors* or *residuals* are

$$r^{(i)} = y^{(i)} - \hat{y}^{(i)}, \quad i = 1, \ldots, N.$$

(Some authors define the prediction errors as $\hat{y}^{(i)} - y^{(i)}$.)

We can express this using compact matrix-vector notation. We form the $n \times N$ feature matrix X with columns $x^{(1)}, \ldots, x^{(N)}$. We let y^{d} denote the N-vector whose entries are the actual values of the response for the N examples. (The superscript 'd' stands for 'data'.) We let $\hat{y}^{\mathrm{d}}$ denote the N-vector of regression model predictions for the N examples, and we let r^{d} denote the N-vector of residuals or prediction errors. We can then express the regression model predictions for this data set in matrix-vector form as

$$\hat{y}^{\mathrm{d}} = X^T \beta + v\mathbf{1}.$$

The vector of N prediction errors for the examples is given by

$$r^{\mathrm{d}} = y^{\mathrm{d}} - \hat{y}^{\mathrm{d}} = y^{\mathrm{d}} - X^T \beta - v\mathbf{1}.$$

We can include the offset v in the regression model by including an additional feature equal to one as the first entry of each feature vector:

$$\hat{y}^{\mathrm{d}} = \left[\begin{array}{c} \mathbf{1}^T \\ X \end{array} \right]^T \left[\begin{array}{c} v \\ \beta \end{array} \right] = \tilde{X}^T \tilde{\beta},$$

where $\tilde{X}$ is the new feature matrix, with a new first row of ones, and $\tilde{\beta} = (v, \beta)$ is the vector of regression model parameters. This is often written without the tildes, as $\hat{y}^{\mathrm{d}} = X^T \beta$, by simply including the feature one as the first feature.

The equation above shows that the N-vector of predictions for the N examples is a linear function of the model parameters (v, β). The N-vector of prediction errors is an affine function of the model parameters.

8.3   Systems of linear equations

Consider a set (also called a system) of m linear equations in n variables or unknowns $x_1, \ldots, x_n$:

$$\begin{aligned}
A_{11}x_1 + A_{12}x_2 + \cdots + A_{1n}x_n &= b_1 \\
A_{21}x_1 + A_{22}x_2 + \cdots + A_{2n}x_n &= b_2 \\
&\vdots \\
A_{m1}x_1 + A_{m2}x_2 + \cdots + A_{mn}x_n &= b_m.
\end{aligned}$$

The numbers A_{ij} are called the *coefficients* in the linear equations, and the numbers b_i are called the *right-hand sides* (since by tradition, they appear on the right-hand side of the equation). These equations can be written succinctly in matrix notation as

$$Ax = b. \tag{8.5}$$

In this context, the $m \times n$ matrix A is called the *coefficient matrix*, and the m-vector b is called the *right-hand side*. An n-vector x is called a *solution* of the linear equations if $Ax = b$ holds. A set of linear equations can have no solutions, one solution, or multiple solutions.

Examples.

- The set of linear equations

$$x_1 + x_2 = 1, \quad x_1 = -1, \quad x_1 - x_2 = 0$$

  is written as $Ax = b$ with

$$A = \begin{bmatrix} 1 & 1 \\ 1 & 0 \\ 1 & -1 \end{bmatrix}, \qquad b = \begin{bmatrix} 1 \\ -1 \\ 0 \end{bmatrix}.$$

  It has no solutions.

- The set of linear equations

$$x_1 + x_2 = 1, \quad x_2 + x_3 = 2$$

  is written as $Ax = b$ with

$$A = \begin{bmatrix} 1 & 1 & 0 \\ 0 & 1 & 1 \end{bmatrix}, \qquad b = \begin{bmatrix} 1 \\ 2 \end{bmatrix}.$$

  It has multiple solutions, including $x = (1, 0, 2)$ and $x = (0, 1, 1)$.

Over-determined and under-determined systems of linear equations. The set of linear equations is called *over-determined* if $m > n$, *under-determined* if $m < n$, and *square* if $m = n$; these correspond to the coefficient matrix being tall, wide, and square, respectively. When the system of linear equations is over-determined, there are more equations than variables or unknowns. When the system of linear equations is under-determined, there are more unknowns than equations. When the system of linear equations is square, the numbers of unknowns and equations is the same. A set of equations with zero right-hand side, $Ax = 0$, is called a *homogeneous* set of equations. Any homogeneous set of equations has $x = 0$ as a solution.

In chapter 11 we will address the question of how to determine if a system of linear equations has a solution, and how to find one when it does. For now, we give a few interesting examples.

8.3.1   Examples

Coefficients of linear combinations.   Let $a_1, \dots, a_n$ denote the columns of A. The system of linear equations $Ax = b$ can be expressed as

$$x_1 a_1 + \cdots + x_n a_n = b,$$

i.e., b is a linear combination of $a_1, \dots, a_n$ with coefficients $x_1, \dots, x_n$. So solving $Ax = b$ is the same as finding coefficients that express b as a linear combination of the vectors $a_1, \dots, a_n$.

Polynomial interpolation.   We seek a polynomial p of degree at most $n - 1$ that interpolates a set of m given points (t_i, y_i), $i = 1, \dots, m$. (This means that $p(t_i) = y_i$.) We can express this as a set of m linear equations in the n unknowns c, where c is the n-vector of coefficients: $Ac = y$. Here the matrix A is the Vandermonde matrix (6.7), and the vector c is the vector of polynomial coefficients, as described in the example on page 120.

Balancing chemical reactions.   A chemical reaction involves p reactants (molecules) and q products, and can be written as

$$a_1 R_1 + \cdots + a_p R_p \longrightarrow b_1 P_1 + \cdots + b_q P_q.$$

Here $R_1, \dots, R_p$ are the reactants, $P_1, \dots, P_q$ are the products, and the numbers $a_1, \dots, a_p$ and $b_1, \dots, b_q$ are positive numbers that tell us how many of each of these molecules is involved in the reaction. They are typically integers, but can be scaled arbitrarily; we could double all of these numbers, for example, and we still have the same reaction. As a simple example, we have the electrolysis of water,

$$2\mathrm{H_2O} \longrightarrow 2\mathrm{H_2} + \mathrm{O_2},$$

which has one reactant, water ($\mathrm{H_2O}$), and two products, molecular hydrogen ($\mathrm{H_2}$) and molecular oxygen ($\mathrm{O_2}$). The coefficients tell us that 2 water molecules create 2 hydrogen molecules and 1 oxygen molecule. The coefficients in a reaction can be multiplied by any nonzero numbers; for example, we could write the reaction above as $3\mathrm{H_2O} \longrightarrow 3\mathrm{H_2} + (3/2)\mathrm{O_2}$. By convention reactions are written with all coefficients integers, with least common divisor one.

   In a chemical reaction the numbers of constituent atoms must balance. This means that for each atom appearing in any of the reactants or products, the total amount on the left-hand side must equal the total amount on the right-hand side. (If any of the reactants or products is charged, *i.e.*, an ion, then the total charge must also balance.) In the simple water electrolysis reaction above, for example, we have 4 hydrogen atoms on the left (2 water molecules, each with 2 hydrogen atoms), and 4 on the right (2 hydrogen molecules, each with 2 hydrogen atoms). The oxygen atoms also balance, so this reaction is balanced.

   Balancing a chemical reaction with specified reactants and products, *i.e.*, finding the numbers $a_1, \dots, a_p$ and $b_1, \dots, b_q$, can be expressed as a system of linear equations. We can express the requirement that the reaction balances as a set of

m equations, where m is the number of different atoms appearing in the chemical reaction. We define the $m \times p$ matrix R by

$$R_{ij} = \text{number of atoms of type } i \text{ in } R_j, \quad i = 1, \ldots, m, \quad j = 1, \ldots, p.$$

(The entries of R are nonnegative integers.) The matrix R is interesting; for example, its jth column gives the chemical formula for reactant R_j. We let a denote the p-vector with entries $a_1, \ldots, a_p$. Then, the m-vector Ra gives the total number of atoms of each type appearing in the reactants. We define an $m \times q$ matrix P in a similar way, so the m-vector Pb gives the total number of atoms of each type that appears in the products.

We write the balance condition using vectors and matrices as $Ra = Pb$. We can express this as

$$\begin{bmatrix} R & -P \end{bmatrix} \begin{bmatrix} a \\ b \end{bmatrix} = 0,$$

which is a set of m homogeneous linear equations.

A simple solution of these equations is $a = 0$, $b = 0$. But we seek a nonzero solution. We can set one of the coefficients, say a_1, to be one. (This might cause the other quantities to be fractional-valued.) We can add the condition that $a_1 = 1$ to our system of linear equations as

$$\begin{bmatrix} R & -P \\ e_1^T & 0 \end{bmatrix} \begin{bmatrix} a \\ b \end{bmatrix} = e_{m+1}.$$

Finally, we have a set of $m + 1$ equations in $p + q$ variables that expresses the requirement that the chemical reaction balances. Finding a solution of this set of equations is called *balancing* the chemical reaction.

For the example of electrolysis of water described above, we have $p = 1$ reactant (water) and $q = 2$ products (molecular hydrogen and oxygen). The reaction involves $m = 2$ atoms, hydrogen and oxygen. The reactant and product matrices are

$$R = \begin{bmatrix} 2 \\ 1 \end{bmatrix}, \qquad P = \begin{bmatrix} 2 & 0 \\ 0 & 2 \end{bmatrix}.$$

The balancing equations are then

$$\begin{bmatrix} 2 & -2 & 0 \\ 1 & 0 & -2 \\ 1 & 0 & 0 \end{bmatrix} \begin{bmatrix} a_1 \\ b_1 \\ b_2 \end{bmatrix} = \begin{bmatrix} 0 \\ 0 \\ 1 \end{bmatrix}.$$

These equations are easily solved, and have the solution $(1, 1, 1/2)$. (Multiplying these coefficients by 2 gives the reaction given above.)

Diffusion systems.   A *diffusion system* is a common model that arises in many areas of physics to describe *flows* and *potentials*. We start with a directed graph with n nodes and m edges. (See §6.1.) Some quantity (like electricity, heat, energy, or mass) can flow across the edges, from one node to another.

With edge j we associate a flow (rate) f_j, which is a scalar; the vector of all m flows is the flow m-vector f. The flows f_j can be positive or negative: Positive

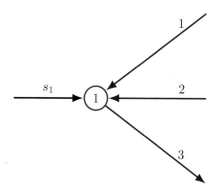

Figure 8.1 A node in a diffusion system with label 1, exogenous flow s_1 and three incident edges.

f_j means the quantity flows in the direction of edge j, and negative f_j means the quantity flows in the opposite direction of edge j. The flows can represent, for example, heat flow (in units of Watts) in a thermal model, electrical current (Amps) in an electrical circuit, or movement (diffusion) of mass (such as, for example, a pollutant). We also have a source (or exogenous) flow s_i at each node, with $s_i > 0$ meaning that an exogenous flow is injected into node i, and $s_i < 0$ means that an exogenous flow is removed from node i. (In some contexts, a node where flow is removed is called a *sink*.) In a thermal system, the sources represent thermal (heat) sources; in an electrical circuit, they represent electrical current sources; in a system with diffusion, they represent external injection or removal of the mass.

In a diffusion system, the flows must satisfy (flow) *conservation*, which means that at each node, the total flow entering each node from adjacent edges and the exogenous source, must be zero. This is illustrated in figure 8.1, which shows three edges adjacent to node 1, two entering node 1 (flows 1 and 2), and one (flow 3) leaving node 1, and an exogenous flow. Flow conservation at this node is expressed as

$$f_1 + f_2 - f_3 + s_1 = 0.$$

Flow conservation at every node can be expressed by the simple matrix-vector equation

$$Af + s = 0, \tag{8.6}$$

where A is the incidence matrix described in §7.3. (This is called *Kirchhoff's current law* in an electrical circuit, after the physicist Gustav Kirchhoff; when the flows represent movement of mass, it is called *conservation of mass*.)

With node i we associate a potential e_i; the n-vector e gives the potential at all nodes. (Note that here, e represents the n-vector of potentials; e_i is the scalar potential at node i, and not the standard ith unit vector.) The potential might represent the node temperature in a thermal model, the electrical potential (voltage) in an electrical circuit, and the concentration in a system that involves mass diffusion.

Figure 8.2 The flow through edge 8 is equal to $f_8 = (e_2 - e_3)/r_8$.

In a diffusion system *the flow on an edge is proportional to the potential differ-ence across its adjacent nodes.* This is typically written as $r_j f_j = e_k - e_l$, where edge j goes from node k to node l, and r_j (which is typically positive) is called the *resistance* of edge j. In a thermal model, r_j is called the thermal resistance of the edge; in an electrical circuit, it is called the electrical resistance. This is illustrated in figure 8.2, which shows edge 8, connecting node 2 and node 3, corresponding to an edge flow equation

$$r_8 f_8 = e_2 - e_3.$$

We can write the edge flow equations in a compact way as

$$Rf = -A^T e, \tag{8.7}$$

where $R = \mathbf{diag}(r)$ is called the *resistance matrix*.

The diffusion model can be expressed as one set of block linear equations in the variables f, s, and e:

$$\begin{bmatrix} A & I & 0 \\ R & 0 & A^T \end{bmatrix} \begin{bmatrix} f \\ s \\ e \end{bmatrix} = 0.$$

This is a set of $n + m$ homogeneous equations in $m + 2n$ variables. To these under-determined equations we can add others, for example, by specifying some of the entries of f, s, and e.

Leontief input-ouput model. We consider an economy with n different industrial sectors. We let x_i be the economic activity level, or total production output, of sector i, for $i = 1, \ldots, n$, measured in a common unit, such as (billions of) dollars. The output of each sector flows to other sectors, to support their production, and also to consumers. We denote the total consumer demand for sector i as d_i, for $i = 1, \ldots, n$.

Supporting the output level x_j for sector j requires $A_{ij} x_j$ output for sector i. We refer to $A_{ij} x_j$ as the sector i *input* that flows to sector j. (We can have $A_{ii} \neq 0$; for example, it requires some energy to support the production of energy.) Thus, $A_{i1} x_1 + \cdots + A_{in} x_n$ is the total sector i output required by, or flowing into, the n industrial sectors. The matrix A is called the *input-output matrix* of the economy, since it describes the flows of sector outputs to the inputs of itself and other sectors. The vector Ax gives the sector outputs required to support the production levels given by x. (This sounds circular, but isn't.)

Finally, we require that for each sector, the total production level matches the demand plus the total amount required to support production. This leads to the balance equations,

$$x = Ax + d.$$

Suppose the demand vector d is given, and we wish to find the sector output levels that will support it. We can write this as a set of n equations in n unknowns,

$$(I - A)x = d.$$

This model of the sector inputs and outputs of an economy was developed by Wassily Leontief in the late 1940s, and is now known as Leontief input-output analysis. He was awarded the Nobel Prize in economics for this work in 1973.

Exercises

8.1 *Sum of linear functions.* Suppose $f : \mathbf{R}^n \to \mathbf{R}^m$ and $g : \mathbf{R}^n \to \mathbf{R}^m$ are linear functions. Their *sum* is the function $h : \mathbf{R}^n \to \mathbf{R}^m$, defined as $h(x) = f(x) + g(x)$ for any n-vector x. The sum function is often denoted as $h = f + g$. (This is another case of overloading the $+$ symbol, in this case to the sum of functions.) If f has matrix representation $f(x) = Fx$, and g has matrix representation $f(x) = Gx$, where F and G are $m \times n$ matrices, what is the matrix representation of the sum function $h = f + g$? Be sure to identify any $+$ symbols appearing in your justification.

8.2 *Averages and affine functions.* Suppose that $G : \mathbf{R}^n \to \mathbf{R}^m$ is an affine function. Let $x_1, \dots, x_k$ be n-vectors, and define the m-vectors $y_1 = G(x_1), \dots, y_k = G(x_k)$. Let

$$\overline{x} = (x_1 + \cdots + x_k)/k, \qquad \overline{y} = (y_1 + \cdots + y_k)/k$$

be the averages of these two lists of vectors. (Here $\overline{x}$ is an n-vector and $\overline{y}$ is an m-vector.) Show that we always have $\overline{y} = G(\overline{x})$. In words: The average of an affine function applied to a list of vectors is the same as the affine function applied to the average of the list of vectors.

8.3 *Cross-product.* The cross product of two 3-vectors $a = (a_1, a_2, a_3)$ and $x = (x_1, x_2, x_3)$ is defined as the vector

$$a \times x = \begin{bmatrix} a_2 x_3 - a_3 x_2 \\ a_3 x_1 - a_1 x_3 \\ a_1 x_2 - a_2 x_1 \end{bmatrix}.$$

The cross product comes up in physics, for example in electricity and magnetism, and in dynamics of mechanical systems like robots or satellites. (You do not need to know this for this exercise.)

Assume a is fixed. Show that the function $f(x) = a \times x$ is a linear function of x, by giving a matrix A that satisfies $f(x) = Ax$ for all x.

8.4 *Linear functions of images.* In this problem we consider several linear functions of a monochrome image with $N \times N$ pixels. To keep the matrices small enough to work out by hand, we will consider the case with $N = 3$ (which would hardly qualify as an image). We represent a 3×3 image as a 9-vector using the ordering of pixels shown below.

1	4	7
2	5	8
3	6	9

(This ordering is called *column-major*.) Each of the operations or transformations below defines a function $y = f(x)$, where the 9-vector x represents the original image, and the 9-vector y represents the resulting or transformed image. For each of these operations, give the 9×9 matrix A for which $y = Ax$.

(a) Turn the original image x upside-down.

(b) Rotate the original image x clockwise $90°$.

(c) Translate the image up by 1 pixel and to the right by 1 pixel. In the translated image, assign the value $y_i = 0$ to the pixels in the first column and the last row.

(d) Set each pixel value y_i to be the average of the neighbors of pixel i in the original image. By neighbors, we mean the pixels immediately above and below, and immediately to the left and right. The center pixel has 4 neighbors; corner pixels have 2 neighbors, and the remaining pixels have 3 neighbors.

8.5 *Symmetric and anti-symmetric part.* An n-vector x is *symmetric* if $x_k = x_{n-k+1}$ for $k = 1, \ldots, n$. It is *anti-symmetric* if $x_k = -x_{n-k+1}$ for $k = 1, \ldots, n$.

(a) Show that every vector x can be decomposed in a unique way as a sum $x = x_\mathrm{s} + x_\mathrm{a}$ of a symmetric vector x_s and an anti-symmetric vector x_a.

(b) Show that the symmetric and anti-symmetric parts x_s and x_a are linear functions of x. Give matrices A_s and A_a such that $x_\mathrm{s} = A_\mathrm{s} x$ and $x_\mathrm{a} = A_\mathrm{a} x$ for all x.

8.6 *Linear functions.* For each description of y below, express it as $y = Ax$ for some A. (You should specify A.)

(a) y_i is the difference between x_i and the average of $x_1, \ldots, x_{i-1}$. (We take $y_1 = x_1$.)

(b) y_i is the difference between x_i and the average value of all other x_js, *i.e.*, the average of $x_1, \ldots, x_{i-1}, x_{i+1}, \ldots, x_n$.

8.7 *Interpolation of polynomial values and derivatives.* The 5-vector c represents the coefficients of a quartic polynomial $p(x) = c_1 + c_2 x + c_3 x^2 + c_4 x^3 + c_5 x^4$. Express the conditions

$$p(0) = 0, \quad p'(0) = 0, \quad p(1) = 1, \quad p'(1) = 0,$$

as a set of linear equations of the form $Ac = b$. Is the system of equations under-determined, over-determined, or square?

8.8 *Interpolation of rational functions.* A *rational function* of degree two has the form

$$f(t) = \frac{c_1 + c_2 t + c_3 t^2}{1 + d_1 t + d_2 t^2},$$

where c_1, c_2, c_3, d_1, d_2 are coefficients. ('Rational' refers to the fact that f is a ratio of polynomials. Another name for f is *bi-quadratic.*) Consider the interpolation conditions

$$f(t_i) = y_i, \quad i = 1, \ldots, K,$$

where t_i and y_i are given numbers. Express the interpolation conditions as a set of linear equations in the vector of coefficients $\theta = (c_1, c_2, c_3, d_1, d_2)$, as $A\theta = b$. Give A and b, and their dimensions.

8.9 *Required nutrients.* We consider a set of n basic foods (such as rice, beans, apples) and a set of m nutrients or components (such as protein, fat, sugar, vitamin C). Food j has a cost given by c_j (say, in dollars per gram), and contains an amount N_{ij} of nutrient i (per gram). (The nutrients are given in some appropriate units, which can depend on the particular nutrient.) A daily diet is represented by an n-vector d, with d_i the daily intake (in grams) of food i. Express the condition that a diet d contains the total nutrient amounts given by the m-vector n^des, and has a total cost B (the budget) as a set of linear equations in the variables $d_1, \ldots, d_n$. (The entries of d must be nonnegative, but we ignore this issue here.)

8.10 *Blending crude oil.* A set of K different types of crude oil are blended (mixed) together in proportions $\theta_1, \ldots, \theta_K$. These numbers sum to one; they must also be nonnegative, but we will ignore that requirement here. Associated with crude oil type k is an n-vector c_k that gives its concentration of n different constituents, such as specific hydrocarbons. Find a set of linear equations on the blending coefficients, $A\theta = b$, that expresses the requirement that the blended crude oil achieves a target set of constituent concentrations, given by the n-vector c^tar. (Include the condition that $\theta_1 + \cdots + \theta_K = 1$ in your equations.)

8.11 *Location from range measurements.* The 3-vector x represents a location in 3-D. We measure the distances (also called the range) of x to four points at known locations a_1, a_2, a_3, a_4:

$$\rho_1 = \|x - a_1\|, \qquad \rho_2 = \|x - a_2\|, \qquad \rho_3 = \|x - a_3\|, \qquad \rho_4 = \|x - a_4\|.$$

Express these distance conditions as a set of three linear equations in the vector x. *Hint.* Square the distance equations, and subtract one from the others.

8.12 *Quadrature.* Consider a function $f : \mathbf{R} \to \mathbf{R}$. We are interested in *estimating* the definite integral $\alpha = \int_{-1}^{1} f(x)\,dx$ based on the value of f at some points $t_1, \ldots, t_n$. (We typically have $-1 \le t_1 < t_2 < \cdots < t_n \le 1$, but this is not needed here.) The standard method for estimating α is to form a weighted sum of the values $f(t_i)$:

$$\hat{\alpha} = w_1 f(t_1) + \cdots + w_n f(t_n),$$

where $\hat{\alpha}$ is our estimate of α, and $w_1, \ldots, w_n$ are the weights. This method of estimating the value of an integral of a function from its values at some points is a classical method in applied mathematics called *quadrature*. There are many quadrature methods (*i.e.*, choices of the points t_i and weights w_i). The most famous one is due to the mathematician Carl Friedrich Gauss, and bears his name.

(a) A typical requirement in quadrature is that the approximation should be exact (*i.e.*, $\hat{\alpha} = \alpha$) when f is any polynomial up to degree d, where d is given. In this case we say that the quadrature method has *order* d. Express this condition as a set of linear equations on the weights, $Aw = b$, assuming the points $t_1, \ldots, t_n$ are given.
Hint. If $\hat{\alpha} = \alpha$ holds for the specific cases $f(x) = 1$, $f(x) = x$, ..., $f(x) = x^d$, then it holds for any polynomial of degree up to d.

(b) Show that the following quadrature methods have order 1, 2, and 3 respectively.

- *Trapezoid rule:* $n = 2$, $t_1 = -1$, $t_2 = 1$, and

$$w_1 = 1/2, \qquad w_2 = 1/2.$$

- *Simpson's rule:* $n = 3$, $t_1 = -1$, $t_2 = 0$, $t_3 = 1$, and

$$w_1 = 1/3, \qquad w_2 = 4/3, \qquad w_3 = 1/3.$$

(Named after the mathematician Thomas Simpson.)
- *Simpson's 3/8 rule:* $n = 4$, $t_1 = -1$, $t_2 = -1/3$, $t_3 = 1/3$, $t_4 = 1$,

$$w_1 = 1/4, \qquad w_2 = 3/4, \qquad w_3 = 3/4, \qquad w_4 = 1/4.$$

8.13 *Portfolio sector exposures.* (See exercise 1.14.) The n-vector h denotes a portfolio of investments in n assets, with h_i the dollar value invested in asset i. We consider a set of m industry sectors, such as pharmaceuticals or consumer electronics. Each asset is assigned to one of these sectors. (More complex models allow for an asset to be assigned to more than one sector.) The *exposure* of the portfolio to sector i is defined as the sum of investments in the assets in that sector. We denote the sector exposures using the m-vector s, where s_i is the portfolio exposure to sector i. (When $s_i = 0$, the portfolio is said to be *neutral* to sector i.) An investment advisor specifies a set of desired sector exposures, given as the m-vector s^{des}. Express the requirement $s = s^{\mathrm{des}}$ as a set of linear equations of the form $Ah = b$. (You must describe the matrix A and the vector b.)
Remark. A typical practical case involves $n = 1000$ assets and $m = 50$ sectors. An advisor might specify $s_i^{\mathrm{des}} = 0$ if she does not have an opinion as how companies in that sector will do in the future; she might specify a positive value for s_i^{des} if she thinks the companies in that sector will do well (*i.e.*, generate positive returns) in the future, and a negative value if she thinks they will do poorly.

8.14 *Affine combinations of solutions of linear equations.* Consider the set of m linear equations in n variables $Ax = b$, where A is an $m \times n$ matrix, b is an m-vector, and x is the n-vector of variables. Suppose that the n-vectors $z_1, \ldots, z_k$ are solutions of this set of equations, *i.e.*, satisfy $Az_i = b$. Show that if the coefficients $\alpha_1, \ldots, \alpha_k$ satisfy $\alpha_1 + \cdots + \alpha_k = 1$, then the affine combination

$$w = \alpha_1 z_1 + \cdots + \alpha_k z_k$$

is a solution of the linear equations, *i.e.*, satisfies $Aw = b$. In words: Any affine combination of solutions of a set of linear equations is also a solution of the equations.

8.15 *Stoichiometry and equilibrium reaction rates.* We consider a system (such as a single cell) containing m metabolites (chemical species), with n reactions among the metabolites occurring at rates given by the n-vector r. (A negative reaction rate means the reaction runs in reverse.) Each reaction consumes some metabolites and produces others, in known rates proportional to the reaction rate. This is specified in the $m \times n$ *stoichiometry matrix* S, where S_{ij} is the rate of metabolite i production by reaction j, running at rate one. (When S_{ij} is negative, it means that when reaction j runs at rate one, metabolite i is consumed.) The system is said to be in equilibrium if the total production rate of each metabolite, due to all the reactions, is zero. This means that for each metabolite, the total production rate balances the total consumption rate, so the total quantities of the metabolites in the system do not change. Express the condition that the system is in equilibrium as a set of linear equations in the reaction rates.

8.16 *Bi-linear interpolation.* We are given a scalar value at each of the four corners of a square in 2-D, (x_1, y_1), (x_1, y_2), (x_2, y_1), and (x_2, y_2), where $x_1 < x_2$ and $y_1 < y_2$. We refer to these four values as F_{11}, F_{12}, F_{21}, and F_{22}, respectively. A *bi-linear interpolation* is a function of the form

$$f(u, v) = \theta_1 + \theta_2 u + \theta_3 v + \theta_4 uv,$$

where $\theta_1, \ldots, \theta_4$ are coefficients, that satisfies

$$f(x_1, y_1) = F_{11}, \quad f(x_1, y_2) = F_{12}, \quad f(x_2, y_1) = F_{21}, \quad f(x_2, y_2) = F_{22},$$

i.e., it agrees with (or interpolates) the given values on the four corners of the square. (The function f is usually evaluated only for points (u, v) inside the square. It is called bi-linear since it is affine in u when v is fixed, and affine in v when u is fixed.)

Express the interpolation conditions as a set of linear equations of the form $A\theta = b$, where A is a 4×4 matrix and b is a 4-vector. Give the entries of A and b in terms of x_1, x_2, y_1, y_2, F_{11}, F_{12}, F_{21}, and F_{22}.

Remark. Bi-linear interpolation is used in many applications to guess or approximate the values of a function at an arbitrary point in 2-D, given the function values on a grid of points. To approximate the value at a point (x, y), we first find the square of grid points that the point lies in. Then we use bi-linear interpolation to get the approximate value at (x, y).

Chapter 9

Linear dynamical systems

In this chapter we consider a useful application of matrix-vector multiplication, which is used to describe many systems or phenomena that change or evolve over time.

9.1 Linear dynamical systems

Suppose $x_1, x_2, \ldots$ is a sequence of n-vectors. The index (subscript) denotes time or period, and is written as t; x_t, the value of the sequence at time (or period) t, is called the *state* at time t. We can think of x_t as a vector that changes over time, *i.e.*, one that changes dynamically. In this context, the sequence $x_1, x_2, \ldots$ is sometimes called a *trajectory* or *state trajectory*. We sometimes refer to x_t as the *current state* of the system (implicitly assuming the current time is t), and x_{t+1} as the *next state*, x_{t-1} as the *previous state*, and so on.

The state x_t can represent a portfolio that changes daily, or the positions and velocities of the parts of a mechanical system, or the quarterly activity of an economy. If x_t represents a portfolio that changes daily, $(x_5)_3$ is the amount of asset 3 held in the portfolio on (trading) day 5.

A *linear dynamical system* is a simple model for the sequence, in which each x_{t+1} is a linear function of x_t:

$$x_{t+1} = A_t x_t, \quad t = 1, 2, \ldots. \tag{9.1}$$

Here the $n \times n$ matrices A_t are called the *dynamics matrices*. The equation above is called the *dynamics* or *update* equation, since it gives us the next value of x, *i.e.*, x_{t+1}, as a function of the current value x_t. Often the dynamics matrix does not depend on t, in which case the linear dynamical system is called *time-invariant*.

If we know x_t (and $A_t, A_{t+1}, \ldots$) we can determine $x_{t+1}, x_{t+2}, \ldots$ simply by iterating the dynamics equation (9.1). In other words: If we know the *current* value of x, we can find all *future* values. In particular, we do not need to know the *past* states. This is why x_t is called the *state* of the system. It contains all the information needed at time t to determine the future evolution of the system.

Linear dynamical system with input.   There are many variations on and extensions of the basic linear dynamical system model (9.1), some of which we will encounter later. As an example, we can add additional terms to the update equation:

$$x_{t+1} = A_t x_t + B_t u_t + c_t, \quad t = 1, 2, \ldots. \tag{9.2}$$

Here u_t is an m-vector called the *input*, B_t is the $n \times m$ *input matrix*, and the n-vector c_t is called the *offset*, all at time t. The input and offset are used to model other factors that affect the time evolution of the state. Another name for the input u_t is *exogenous variable*, since, roughly speaking, it comes from outside the system.

Markov model.   The linear dynamical system (9.1) is sometimes called a *Markov model* (after the mathematician Andrey Markov). Markov studied systems in which the next state value depends on the current one, and not on the previous state values $x_{t-1}, x_{t-2}, \ldots$. The linear dynamical system (9.1) is the special case of a Markov system where the next state is a linear function of the current state.

In a variation on the Markov model, called a (linear) K-Markov model, the next state x_{t+1} depends on the current state and $K - 1$ previous states. Such a system has the form

$$x_{t+1} = A_1 x_t + \cdots + A_K x_{t-K+1}, \quad t = K, K + 1, \ldots. \tag{9.3}$$

Models of this form are used in time series analysis and econometrics, where they are called (vector) *auto-regressive models*. When $K = 1$, the Markov model (9.3) is the same as a linear dynamical system (9.1). When $K > 1$, the Markov model (9.3) can be reduced to a standard linear dynamical system (9.1), with an appropriately chosen state; see exercise 9.4.

Simulation.   If we know the dynamics (and input) matrices, and the state at time t, we can find the future state trajectory $x_{t+1}, x_{t+2}, \ldots$ by iterating the equation (9.1) (or (9.2), provided we also know the input sequence $u_t, u_{t+1}, \ldots$). This is called *simulating* the linear dynamical system. Simulation makes predictions about the future state of a system. (To the extent that (9.1) is only an approximation or model of some real system, we must be careful when interpreting the results.) We can carry out what-if simulations, to see what would happen if the system changes in some way, or if a particular set of inputs occurs.

9.2   Population dynamics

Linear dynamical systems can be used to describe the evolution of the age distribution in some population over time. Suppose x_t is a 100-vector, with $(x_t)_i$ denoting the number of people in some population (say, a country) with age $i - 1$ (say, on January 1) in year t, where t is measured starting from some base year, for $i = 1, \ldots, 100$. While $(x_t)_i$ is an integer, it is large enough that we simply consider

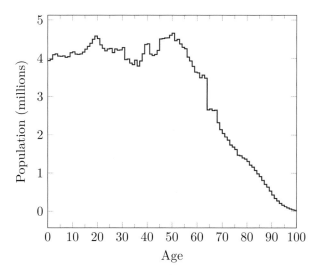

Figure 9.1 Age distribution in the US in 2010. (United States Census Bureau, census.gov).

it a real number. In any case, our model certainly is not accurate at the level of individual people. Also, note that the model does not track people 100 and older. The distribution of ages in the US in 2010 is shown in figure 9.1.

The birth rate is given by a 100-vector b, where b_i is the average number of births per person with age $i - 1$, $i = 1, \ldots, 100$. (This is half the average number of births per woman with age $i - 1$, assuming equal numbers of men and women in the population.) Of course b_i is approximately zero for $i < 13$ and $i > 50$. The approximate birth rates for the US in 2010 are shown in figure 9.2. The death rate is given by a 100-vector d, where d_i is the portion of those aged $i - 1$ who will die this year. The death rates for the US in 2010 are shown in figure 9.3.

To derive the dynamics equation (9.1), we find x_{t+1} in terms of x_t, taking into account only births and deaths, and not immigration. The number of 0-year olds next year is the total number of births this year:

$$(x_{t+1})_1 = b^T x_t.$$

The number of i-year olds next year is the number of $(i - 1)$-year-olds this year, minus those who die:

$$(x_{t+1})_{i+1} = (1 - d_i)(x_t)_i, \quad i = 1, \ldots, 99.$$

We can assemble these equations into the time-invariant linear dynamical system

$$x_{t+1} = Ax_t, \quad t = 1, 2, \ldots, \tag{9.4}$$

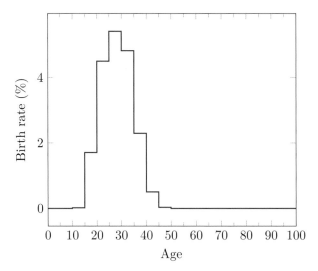

Figure 9.2 Approximate birth rate versus age in the US in 2010. The figure is based on statistics for age groups of five years (hence, the piecewise-constant shape) and assumes an equal number of men and women in each age group. (Martin J.A., Hamilton B.E., Ventura S.J. *et al.*, Births: Final data for 2010. National Vital Statistics Reports; vol. 61, no. 1. National Center for Health Statistics, 2012.)

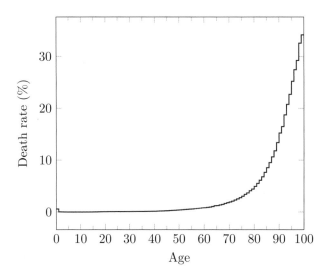

Figure 9.3 Death rate versus age, for ages 0–99, in the US in 2010. (Centers for Disease Control and Prevention, National Center for Health Statistics, wonder.cdc.gov.)

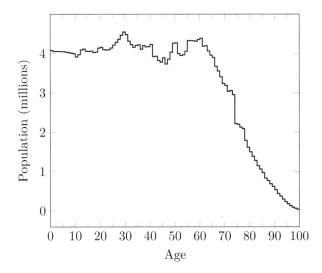

Figure 9.4 Predicted age distribution in the US in 2020.

where A is given by

$$
A = \begin{bmatrix}
b_1 & b_2 & b_3 & \cdots & b_{98} & b_{99} & b_{100} \\
1-d_1 & 0 & 0 & \cdots & 0 & 0 & 0 \\
0 & 1-d_2 & 0 & \cdots & 0 & 0 & 0 \\
\vdots & \vdots & \vdots & & \vdots & \vdots & \vdots \\
0 & 0 & 0 & \cdots & 1-d_{98} & 0 & 0 \\
0 & 0 & 0 & \cdots & 0 & 1-d_{99} & 0
\end{bmatrix}.
$$

We can use this model to predict the total population in 10 years (not including immigration), or to predict the number of school age children, or retirement age adults. Figure 9.4 shows the predicted age distribution in 2020, computed by iterating the model $x_{t+1} = Ax_t$ for $t = 1, \ldots, 10$, with initial value x_1 given by the 2010 age distribution of figure 9.1. Note that the distribution is based on an approximate model, since we neglect the effect of immigration, and assume that the death and birth rates remain constant and equal to the values shown in figures 9.2 and 9.3.

Population dynamics models are used to carry out projections of the future age distribution, which in turn is used to predict how many retirees there will be in some future year. They are also used to carry out various 'what if' analyses, to predict the effect of changes in birth or death rates on the future age distribution.

It is easy to include the effects of immigration and emigration in the population dynamics model (9.4), by simply adding a 100-vector u_t:

$$
x_{t+1} = Ax_t + u_t,
$$

which is a time-invariant linear dynamical system of the form (9.2), with input u_t and $B = I$. The vector u_t gives the net immigration in year t over all ages; $(u_t)_i$ is the number of immigrants in year t of age $i - 1$. (Negative entries mean net emigration.)

9.3   Epidemic dynamics

The dynamics of infection and the spread of an epidemic can be modeled using a linear dynamical system. (More sophisticated nonlinear epidemic dynamic models are also used.) In this section we describe a simple example.

A disease is introduced into a population. In each period (say, days) we count the fraction of the population that is in four different infection states:

- *Susceptible.* These individuals can acquire the disease the next day.

- *Infected.* These individuals have the disease.

- *Recovered* (and immune). These individuals had the disease and survived, and now have immunity.

- *Deceased.* These individuals had the disease, and unfortunately died from it.

We denote the fractions of each of these as a 4-vector x_t, so, for example, $x_t = (0.75, 0.10, 0.10, 0.05)$ means that in day t, 75% of the population is susceptible, 10% is infected, 10% is recovered and immune, and 5% has died from the disease.

There are many mathematical models that predict how the disease state fractions x_t evolve over time. One simple model can be expressed as a linear dynamical system. The model assumes the following happens over each day.

- 5% of the susceptible population will acquire the disease. (The other 95% will remain susceptible.)

- 1% of the infected population will die from the disease, 10% will recover and acquire immunity, and 4% will recover and not acquire immunity (and therefore, become susceptible). The remaining 85% will remain infected.

(Those who have have recovered with immunity and those who have died remain in those states.)

We first determine $(x_{t+1})_1$, the fraction of susceptible individuals in the next day. These include the susceptible individuals from today, who did not become infected, which is $0.95(x_t)_1$, plus the infected individuals today who recovered without immunity, which is $0.04(x_t)_2$. All together we have $(x_{t+1})_1 = 0.95(x_t)_1 + 0.04(x_t)_2$. We have $(x_{t+1})_2 = 0.85(x_t)_2 + 0.05(x_t)_1$; the first term counts those who are infected and remain infected, and the second term counts those who are susceptible and acquire the disease. Similar arguments give $(x_{t+1})_3 = (x_t)_3 + 0.10(x_t)_2$, and $(x_{t+1})_4 = (x_t)_4 + 0.01(x_t)_2$. We put these together to get

$$
x_{t+1} = \begin{bmatrix} 0.95 & 0.04 & 0 & 0 \\ 0.05 & 0.85 & 0 & 0 \\ 0 & 0.10 & 1 & 0 \\ 0 & 0.01 & 0 & 1 \end{bmatrix} x_t,
$$

which is a time-invariant linear dynamical system of the form (9.1).

Figure 9.5 shows the evolution of the four groups from the initial condition $x_0 = (1, 0, 0, 0)$. The simulation shows that after around 100 days, the state converges to one with a little under 10% of the population deceased, and the remaining population immune.

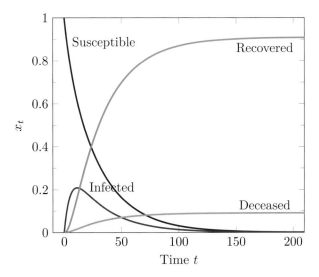

Figure 9.5 Simulation of epidemic dynamics.

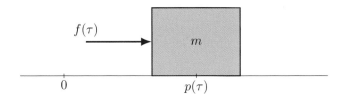

Figure 9.6 Mass moving along a line.

9.4   Motion of a mass

Linear dynamical systems can be used to (approximately) describe the motion of many mechanical systems, for example, an airplane (that is not undergoing extreme maneuvers), or the (hopefully not too large) movement of a building during an earthquake. Here we describe the simplest example: A single mass moving in 1-D (*i.e.*, a straight line), with an external force and a drag force acting on it. This is illustrated in figure 9.6. The (scalar) position of the mass at time τ is given by $p(\tau)$. (Here τ is continuous, *i.e.*, a real number.) The position satisfies Newton's law of motion, the differential equation

$$m\frac{d^2p}{d\tau^2}(\tau) = -\eta\frac{dp}{d\tau}(\tau) + f(\tau),$$

where $m > 0$ is the mass, $f(\tau)$ is the external force acting on the mass at time τ, and $\eta > 0$ is the drag coefficient. The right-hand side is the total force acting on the mass; the first term is the drag force, which is proportional to the velocity and in the opposite direction.

Introducing the velocity of the mass, $v(\tau) = dp(\tau)/d\tau$, we can write the equation above as two coupled differential equations,

$$\frac{dp}{d\tau}(\tau) = v(\tau), \qquad m\frac{dv}{d\tau}(\tau) = -\eta v(\tau) + f(\tau).$$

The first equation relates the position and velocity; the second is from the law of motion.

Discretization.   To develop an (approximate) linear dynamical system model from the differential equations above, we first discretize time. We let $h > 0$ be a time interval (called the 'sampling interval') that is small enough that the velocity and forces do not change very much over h seconds. We define

$$p_k = p(kh), \qquad v_k = v(kh), \qquad f_k = f(kh),$$

which are the continuous quantities 'sampled' at multiples of h seconds. We now use the approximations

$$\frac{dp}{d\tau}(kh) \approx \frac{p_{k+1} - p_k}{h}, \qquad \frac{dv}{d\tau}(kh) \approx \frac{v_{k+1} - v_k}{h}, \tag{9.5}$$

which are justified since h is small. This leads to the (approximate) equations (replacing $\approx$ with $=$)

$$\frac{p_{k+1} - p_k}{h} = v_k, \qquad m\frac{v_{k+1} - v_k}{h} = f_k - \eta v_k.$$

Finally, using state $x_k = (p_k, v_k)$, we write this as

$$x_{k+1} = \begin{bmatrix} 1 & h \\ 0 & 1 - h\eta/m \end{bmatrix} x_k + \begin{bmatrix} 0 \\ h/m \end{bmatrix} f_k, \quad k = 1, 2, \ldots,$$

which is a linear dynamical system of the form (9.2), with input f_k and dynamics and input matrices

$$A = \begin{bmatrix} 1 & h \\ 0 & 1 - h\eta/m \end{bmatrix}, \qquad B = \begin{bmatrix} 0 \\ h/m \end{bmatrix}.$$

This linear dynamical system gives an approximation of the true motion, due to our approximation (9.5) of the derivatives. But for h small enough, it is accurate. This linear dynamical system can be used to simulate the motion of the mass, if we know the external force applied to it, $i.e.$, u_t for $t = 1, 2, \ldots$.

The approximation (9.5), which turns a set of differential equations into a recursion that approximates it, is called the *Euler method*, named after the mathematician Leonhard Euler. (There are other, more sophisticated, methods for approximating differential equations as recursions.)

Example.   As a simple example, we consider the case with $m = 1$ (kilogram), $\eta = 1$ (Newtons per meter per second), and sampling period $h = 0.01$ (seconds). The external force is

$$f(\tau) = \begin{cases} 0.0 & 0.0 \le \tau < 0.5 \\ 1.0 & 0.5 \le \tau < 1.0 \\ -1.3 & 1.0 \le \tau < 1.4 \\ 0.0 & 1.4 \le \tau. \end{cases}$$

We simulate this system for a period of 2.5 seconds, starting from initial state $x_1 = (0, 0)$, which corresponds to the mass starting at rest (zero velocity) at position 0. The simulation involves iterating the dynamics equation from $k = 1$ to $k = 250$. Figure 9.7 shows the force, position, and velocity of the mass, with the axes labeled using continuous time τ.

9.5   Supply chain dynamics

The dynamics of a supply chain can often be modeled using a linear dynamical system. (This simple model does not include some important aspects of a real supply chain, for example limits on storage at the warehouses, or the fact that demand fluctuates.) We give a simple example here.

   We consider a supply chain for a single divisible commodity (say, oil or gravel, or discrete quantities so small that their quantities can be considered real numbers). The commodity is stored at n warehouses or storage locations. Each of these locations has a target (desired) level or amount of the commodity, and we let the n-vector x_t denote the *deviations* of the levels of the commodities from their target levels. For example, $(x_5)_3$ is the actual commodity level at location 3, in period 5, minus the target level for location 3. If this is positive it means we have more than the target level at the location; if it is negative, we have less than the target level at the location.

   The commodity is moved or transported in each period over a set of m transportation links between the storage locations, and also enters and exits the nodes through purchases (from suppliers) and sales (to end-users). The purchases and sales are given by the n-vectors p_t and s_t, respectively. We expect these to be positive; but they can be negative if we include returns. The net effect of the purchases and sales is that we add $(p_t - s_t)_i$ of the commodity at location i. (This number is negative if we sell more than we purchase at the location.)

   We describe the links by the $n \times m$ incidence matrix A^{sc} (see §7.3). The direction of each link does not indicate the direction of commodity flow; it only sets the *reference direction* for the flow: Commodity flow in the direction of the link is considered positive and commodity flow in the opposite direction is considered negative. We describe the commodity flow in period t by the m-vector f_t. For example, $(f_6)_2 = -1.4$ means that in time period 6, 1.4 units of the commodity are moved along link 2 in the direction opposite the link direction (since the flow is negative). The n-vector $A^{\mathrm{sc}} f_t$ gives the net flow of the commodity into the n locations, due to the transport across the links.

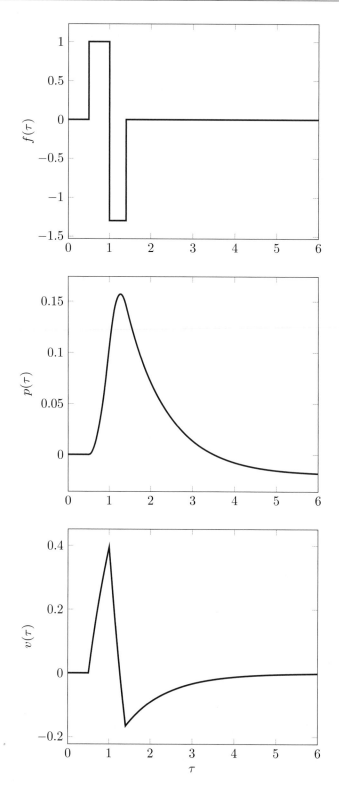

Figure 9.7 Simulation of mass moving along a line. Applied force (top), position (middle), and velocity (bottom).

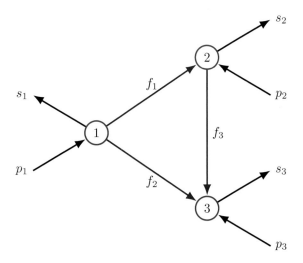

Figure 9.8 A simple supply chain with $n = 3$ storage locations and $m = 3$ transportation links.

Taking into account the movement of the commodity across the network, and the purchase and sale of the commodity, we get the dynamics

$$x_{t+1} = x_t + A^{\mathrm{sc}} f_t + p_t - s_t, \quad t = 1, 2, \ldots.$$

In applications where we control or run a supply chain, s_t is beyond our control, but we can manipulate f_t (the flow of goods between storage locations) and p_t (purchases at locations). This suggests treating s_t as the offset, and $u_t = (f_t, p_t)$ as the input in a linear dynamical system with input (9.2). We can write the dynamics equations above in this form, with dynamics and input matrices

$$A = I, \qquad B = \begin{bmatrix} A^{\mathrm{sc}} & I \end{bmatrix}.$$

(Note that A^{sc} refers to the supply chain graph incidence matrix, while A is the dynamics matrix in (9.2).) This gives

$$x_{t+1} = Ax_t + B(f_t, p_t) - s_t, \quad t = 1, 2, \ldots,.$$

A simple example is shown in figure 9.8. The supply chain dynamics equation is

$$x_{t+1} = x_t + \begin{bmatrix} -1 & -1 & 0 & 1 & 0 & 0 \\ 1 & 0 & -1 & 0 & 1 & 0 \\ 0 & 1 & 1 & 0 & 0 & 1 \end{bmatrix} \begin{bmatrix} f_t \\ p_t \end{bmatrix} - s_t, \quad t = 1, 2, \ldots.$$

It is a good exercise to check that the matrix-vector product (the middle term of the right-hand side) gives the amount of commodity added at each location, as a result of shipment and purchasing.

Exercises

9.1 *Compartmental system.* A *compartmental system* is a model used to describe the movement of some material over time among a set of n compartments of a system, and the outside world. It is widely used in *pharmaco-kinetics*, the study of how the concentration of a drug varies over time in the body. In this application, the material is a drug, and the compartments are the bloodstream, lungs, heart, liver, kidneys, and so on. Compartmental systems are special cases of linear dynamical systems.

In this problem we will consider a very simple compartmental system with 3 compartments. We let $(x_t)_i$ denote the amount of the material (say, a drug) in compartment i at time period t. Between period t and period $t + 1$, the material moves as follows.

- 10% of the material in compartment 1 moves to compartment 2. (This decreases the amount in compartment 1 and increases the amount in compartment 2.)

- 5% of the material in compartment 2 moves to compartment 3.

- 5% of the material in compartment 3 moves to compartment 1.

- 5% of the material in compartment 3 is eliminated.

Express this compartmental system as a linear dynamical system, $x_{t+1} = Ax_t$. (Give the matrix A.) Be sure to account for all the material entering and leaving each compartment.

9.2 *Dynamics of an economy.* An economy (of a country or region) is described by an n-vector a_t, where $(a_t)_i$ is the economic output in sector i in year t (measured in billions of dollars, say). The total output of the economy in year t is $\mathbf{1}^T a_t$. A very simple model of how the economic output changes over time is $a_{t+1} = Ba_t$, where B is an $n \times n$ matrix. (This is closely related to the Leontief input-output model described on page 157 of the book. But the Leontief model is static, *i.e.*, doesn't consider how an economy changes over time.) The entries of a_t and B are positive in general.

In this problem we will consider the specific model with $n = 4$ sectors and

$$B = \begin{bmatrix} 0.10 & 0.06 & 0.05 & 0.70 \\ 0.48 & 0.44 & 0.10 & 0.04 \\ 0.00 & 0.55 & 0.52 & 0.04 \\ 0.04 & 0.01 & 0.42 & 0.51 \end{bmatrix}.$$

(a) Briefly interpret B_{23}, in English.

(b) *Simulation.* Suppose $a_1 = (0.6, 0.9, 1.3, 0.5)$. Plot the four sector outputs (*i.e.*, $(a_t)_i$ for $i = 1, \ldots, 4$) and the total economic output (*i.e.*, $\mathbf{1}^T a_t$) versus t, for $t = 1, \ldots, 20$.

9.3 *Equilibrium point for linear dynamical system.* Consider a time-invariant linear dynamical system with offset, $x_{t+1} = Ax_t + c$, where x_t is the state n-vector. We say that a vector z is an *equilibrium point* of the linear dynamical system if $x_1 = z$ implies $x_2 = z$, $x_3 = z$, $\ldots$. (In words: If the system starts in state z, it stays in state z.)

Find a matrix F and vector g for which the set of linear equations $Fz = g$ characterizes equilibrium points. (This means: If z is an equilibrium point, then $Fz = g$; conversely if $Fz = g$, then z is an equilibrium point.) Express F and g in terms of A, c, any standard matrices or vectors (*e.g.*, I, $\mathbf{1}$, or 0), and matrix and vector operations.

Remark. Equilibrium points often have interesting interpretations. For example, if the linear dynamical system describes the population dynamics of a country, with the vector c denoting immigration (emigration when entries of c are negative), an equilibrium point is a population distribution that does not change, year to year. In other words, immigration exactly cancels the changes in population distribution caused by aging, births, and deaths.

9.4 *Reducing a Markov model to a linear dynamical system.* Consider the 2-Markov model

$$x_{t+1} = A_1 x_t + A_2 x_{t-1}, \quad t = 2, 3, \ldots,$$

where x_t is an n-vector. Define $z_t = (x_t, x_{t-1})$. Show that z_t satisfies the linear dynamical system equation $z_{t+1} = B z_t$, for $t = 2, 3, \ldots$, where B is a $(2n) \times (2n)$ matrix. This idea can be used to express any K-Markov model as a linear dynamical system, with state $(x_t, \ldots, x_{t-K+1})$.

9.5 *Fibonacci sequence.* The Fibonacci sequence $y_0, y_1, y_2, \ldots$ starts with $y_0 = 0$, $y_1 = 1$, and for $t = 2, 3, \ldots$, y_t is the sum of the previous two entries, *i.e.*, $y_{t-1} + y_{t-2}$. (Fibonacci is the name used by the 13th century mathematician Leonardo of Pisa.) Express this as a time-invariant linear dynamical system with state $x_t = (y_t, y_{t-1})$ and output y_t, for $t = 1, 2, \ldots$. Use your linear dynamical system to simulate (compute) the Fibonacci sequence up to $t = 20$. Also simulate a modified Fibonacci sequence $z_0, z_1, z_2, \ldots$, which starts with the same values $z_0 = 0$ and $z_1 = 1$, but for $t = 2, 3, \ldots$, z_t is the difference of the two previous values, *i.e.*, $z_{t-1} - z_{t-2}$.

9.6 *Recursive averaging.* Suppose that $u_1, u_2, \ldots$ is a sequence of n-vectors. Let $x_1 = 0$, and for $t = 2, 3, \ldots$, let x_t be the average of $u_1, \ldots, u_{t-1}$, *i.e.*, $x_t = (u_1 + \cdots + u_{t-1})/(t-1)$. Express this as a linear dynamical system with input, *i.e.*, $x_{t+1} = A_t x_t + B_t u_t$, $t = 1, 2, \ldots$ (with initial state $x_1 = 0$). *Remark.* This can be used to compute the average of an extremely large collection of vectors, by accessing them one-by-one.

9.7 *Complexity of linear dynamical system simulation.* Consider the time-invariant linear dynamical system with n-vector state x_t and m-vector input u_t, and dynamics $x_{t+1} = A x_t + B u_t$, $t = 1, 2, \ldots$. You are given the matrices A and B, the initial state x_1, and the inputs $u_1, \ldots, u_{T-1}$. What is the complexity of carrying out a simulation, *i.e.*, computing $x_2, x_3, \ldots, x_T$? About how long would it take to carry out a simulation with $n = 15$, $m = 5$, and $T = 10^5$, using a 1 Gflop/s computer?

Chapter 10

Matrix multiplication

In this chapter we introduce matrix multiplication, a generalization of matrix-vector multiplication, and describe several interpretations and applications.

10.1 Matrix-matrix multiplication

It is possible to multiply two matrices using *matrix multiplication*. You can multiply two matrices A and B provided their dimensions are *compatible*, which means the number of columns of A equals the number of rows of B. Suppose A and B are compatible, *e.g.*, A has size $m \times p$ and B has size $p \times n$. Then the product matrix $C = AB$ is the $m \times n$ matrix with elements

$$C_{ij} = \sum_{k=1}^{p} A_{ik}B_{kj} = A_{i1}B_{1j}+\cdots+A_{ip}B_{pj}, \qquad i = 1,\ldots,m, \quad j = 1,\ldots,n. \quad (10.1)$$

There are several ways to remember this rule. To find the i,j element of the product $C = AB$, you need to know the ith row of A and the jth column of B. The summation above can be interpreted as 'moving left to right along the ith row of A' while moving 'top to bottom' down the jth column of B. As you go, you keep a running sum of the product of elements, one from A and one from B.

As a specific example, we have

$$\begin{bmatrix} -1.5 & 3 & 2 \\ 1 & -1 & 0 \end{bmatrix} \begin{bmatrix} -1 & -1 \\ 0 & -2 \\ 1 & 0 \end{bmatrix} = \begin{bmatrix} 3.5 & -4.5 \\ -1 & 1 \end{bmatrix}.$$

To find the $1,2$ entry of the right-hand matrix, we move along the first row of the left-hand matrix, and down the second column of the middle matrix, to get $(-1.5)(-1) + (3)(-2) + (2)(0) = -4.5$.

Matrix-matrix multiplication includes as special cases several other types of multiplication (or product) we have encountered so far.

Scalar-vector product.   If x is an n-vector and a is a number, we can interpret the scalar-vector product xa, with the scalar appearing on the right, as matrix-matrix multiplication. We consider the n-vector x to be an $n \times 1$ matrix, and the scalar a to be a 1×1 matrix. The matrix product xa then makes sense, and is an $n \times 1$ matrix, which we consider the same as an n-vector. It coincides with the scalar-vector product xa, which we usually write (by convention) as ax. But note that ax cannot be interpreted as matrix-matrix multiplication (except when $n = 1$), since the number of columns of a (which is one) is not equal to the number of rows of x (which is n).

Inner product.   An important special case of matrix-matrix multiplication is the multiplication of a row vector with a column vector. If a and b are n-vectors, then the inner product

$$a^T b = a_1 b_1 + a_2 b_2 + \cdots + a_n b_n$$

can be interpreted as the matrix-matrix product of the $1 \times n$ matrix a^T and the $n \times 1$ matrix b. The result is a 1×1 matrix, which we consider to be a scalar. (This explains the notation $a^T b$ for the inner product of vectors a and b, defined in §1.4.)

Matrix-vector multiplication.   The matrix-vector product $y = Ax$ defined in (6.4) can be interpreted as a matrix-matrix product of A with the $n \times 1$ matrix x.

Vector outer product.   The *outer product* of an m-vector a and an n-vector b is given by ab^T, which is an $m \times n$ matrix

$$ab^T = \begin{bmatrix} a_1 b_1 & a_1 b_2 & \cdots & a_1 b_n \\ a_2 b_1 & a_2 b_2 & \cdots & a_2 b_n \\ \vdots & \vdots & & \vdots \\ a_m b_1 & a_m b_2 & \cdots & a_m b_n \end{bmatrix},$$

whose entries are all products of the entries of a and the entries of b. Note that the outer product does not satisfy $ab^T = ba^T$, *i.e.*, it is not symmetric (like the inner product). Indeed, the equation $ab^T = ba^T$ does not even make sense, unless $m = n$; even then, it is not true in general.

Multiplication by identity.   If A is any $m \times n$ matrix, then $AI = A$ and $IA = A$, *i.e.*, when you multiply a matrix by an identity matrix, it has no effect. (Note the different sizes of the identity matrices in the formulas $AI = A$ and $IA = A$.)

Matrix multiplication order matters.   Matrix multiplication is (in general) *not commutative:* We *do not* (in general) have $AB = BA$. In fact, BA may not even make sense, or, if it makes sense, may be a different size than AB. For example, if A is 2×3 and B is 3×4, then AB makes sense (the dimensions are compatible) but BA does not even make sense (the dimensions are incompatible). Even when AB and BA both make sense and are the same size, *i.e.*, when A and B are square, we do not (in general) have $AB = BA$. As a simple example, take the matrices

$$A = \begin{bmatrix} 1 & 6 \\ 9 & 3 \end{bmatrix}, \qquad B = \begin{bmatrix} 0 & -1 \\ -1 & 2 \end{bmatrix}.$$

We have

$$AB = \begin{bmatrix} -6 & 11 \\ -3 & -3 \end{bmatrix}, \qquad BA = \begin{bmatrix} -9 & -3 \\ 17 & 0 \end{bmatrix}.$$

Two matrices A and B that satisfy $AB = BA$ are said to *commute*. (Note that for $AB = BA$ to make sense, A and B must both be square.)

Properties of matrix multiplication. The following properties hold and are easy to verify from the definition of matrix multiplication. We assume that A, B, and C are matrices for which all the operations below are valid, and that γ is a scalar.

- *Associativity:* $(AB)C = A(BC)$. Therefore we can write the product simply as ABC.

- *Associativity with scalar multiplication:* $\gamma(AB) = (\gamma A)B$, where γ is a scalar, and A and B are matrices (that can be multiplied). This is also equal to $A(\gamma B)$. (Note that the products γA and γB are defined as scalar-matrix products, but in general, unless A and B have one row, not as matrix-matrix products.)

- *Distributivity with addition.* Matrix multiplication distributes across matrix addition: $A(B+C) = AB+AC$ and $(A+B)C = AC+BC$. On the right-hand sides of these equations we use the higher precedence of matrix multiplication over addition, so, for example, $AC + BC$ is interpreted as $(AC) + (BC)$.

- *Transpose of product.* The transpose of a product is the product of the transposes, but in the *opposite* order: $(AB)^T = B^T A^T$.

From these properties we can derive others. For example, if A, B, C, and D are square matrices of the same size, we have the identity

$$(A + B)(C + D) = AC + AD + BC + BD.$$

This is the same as the usual formula for expanding a product of sums of scalars; but with matrices, we must be careful to preserve the order of the products.

Inner product and matrix-vector products. As an exercise on matrix-vector products and inner products, one can verify that if A is $m \times n$, x is an n-vector, and y is an m-vector, then

$$y^T(Ax) = (y^T A)x = (A^T y)^T x,$$

i.e., the inner product of y and Ax is equal to the inner product of x and $A^T y$. (Note that when $m \neq n$, these inner products involve vectors with different dimensions.)

Products of block matrices. Suppose A is a block matrix with $m \times p$ block entries A_{ij}, and B is a block matrix with $p \times n$ block entries B_{ij}, and for each $k = 1, \ldots, p$, the matrix product $A_{ik}B_{kj}$ makes sense, *i.e.*, the number of columns of A_{ik} equals the number of rows of B_{kj}. (In this case we say that the block matrices *conform* or are *compatible.*) Then $C = AB$ can be expressed as the $m \times n$ block matrix with entries C_{ij}, given by the formula (10.1). For example, we have

$$\begin{bmatrix} A & B \\ C & D \end{bmatrix} \begin{bmatrix} E & F \\ G & H \end{bmatrix} = \begin{bmatrix} AE + BG & AF + BH \\ CE + DG & CF + DH \end{bmatrix},$$

for any matrices $A, B, \ldots, H$ for which the matrix products above make sense. This formula is the same as the formula for multiplying two 2×2 matrices (*i.e.*, with scalar entries); but when the entries of the matrix are themselves matrices (as in the block matrix above), we must be careful to preserve the multiplication order.

Column interpretation of matrix-matrix product.   We can derive some additional insight into matrix multiplication by interpreting the operation in terms of the columns of the second matrix. Consider the matrix product of an $m \times p$ matrix A and a $p \times n$ matrix B, and denote the columns of B by b_k. Using block-matrix notation, we can write the product AB as

$$AB = A \begin{bmatrix} b_1 & b_2 & \cdots & b_n \end{bmatrix} = \begin{bmatrix} Ab_1 & Ab_2 & \cdots & Ab_n \end{bmatrix}.$$

Thus, the columns of AB are the matrix-vector products of A and the columns of B. The product AB can be interpreted as the matrix obtained by 'applying' A to each of the columns of B.

Multiple sets of linear equations.   We can use the column interpretation of matrix multiplication to express a set of k linear equations with the same $m \times n$ coefficient matrix A,

$$Ax_i = b_i, \quad i = 1, \ldots, k,$$

in the compact form

$$AX = B,$$

where $X = [x_1 \ \cdots \ x_k]$ and $B = [b_1 \ \cdots \ b_k]$. The matrix equation $AX = B$ is sometimes called a *linear equation with matrix right-hand side*, since it looks like $Ax = b$, but X (the variable) and B (the right-hand side) are now $n \times k$ matrices, instead of n-vectors (which are $n \times 1$ matrices).

Row interpretation of matrix-matrix product.   We can give an analogous row interpretation of the product AB, by partitioning A and AB as block matrices with row vector blocks. Let $a_1^T, \ldots, a_m^T$ be the rows of A. Then we have

$$AB = \begin{bmatrix} a_1^T \\ a_2^T \\ \vdots \\ a_m^T \end{bmatrix} B = \begin{bmatrix} a_1^T B \\ a_2^T B \\ \vdots \\ a_m^T B \end{bmatrix} = \begin{bmatrix} (B^T a_1)^T \\ (B^T a_2)^T \\ \vdots \\ (B^T a_m)^T \end{bmatrix}.$$

This shows that the rows of AB are obtained by applying B^T to the transposed row vectors a_k of A, and transposing the result.

Inner product representation. From the definition of the i, j element of AB in (10.1), we also see that the elements of AB are the inner products of the rows of A with the columns of B:

$$AB = \begin{bmatrix} a_1^T b_1 & a_1^T b_2 & \cdots & a_1^T b_n \\ a_2^T b_1 & a_2^T b_2 & \cdots & a_2^T b_n \\ \vdots & \vdots & \ddots & \vdots \\ a_m^T b_1 & a_m^T b_2 & \cdots & a_m^T b_n \end{bmatrix},$$

where a_i^T are the rows of A and b_j are the columns of B. Thus we can interpret the matrix-matrix product as the mn inner products $a_i^T b_j$ arranged in an $m \times n$ matrix.

Gram matrix. For an $m \times n$ matrix A, with columns $a_1, \ldots, a_n$, the matrix product $G = A^T A$ is called the *Gram matrix* associated with the set of m-vectors $a_1, \ldots, a_n$. (It is named after the mathematician Jørgen Pedersen Gram.) From the inner product interpretation above, the Gram matrix can be expressed as

$$G = A^T A = \begin{bmatrix} a_1^T a_1 & a_1^T a_2 & \cdots & a_1^T a_n \\ a_2^T a_1 & a_2^T a_2 & \cdots & a_2^T a_n \\ \vdots & \vdots & \ddots & \vdots \\ a_n^T a_1 & a_n^T a_2 & \cdots & a_n^T a_n \end{bmatrix}.$$

The entries of the Gram matrix G give all inner products of pairs of columns of A. Note that a Gram matrix is symmetric, since $a_i^T a_j = a_j^T a_i$. This can also be seen using the transpose-of-product rule:

$$G^T = (A^T A)^T = (A^T)(A^T)^T = A^T A = G.$$

The Gram matrix will play an important role later in this book.

As an example, suppose the $m \times n$ matrix A gives the membership of m items in n groups, with entries

$$A_{ij} = \begin{cases} 1 & \text{item } i \text{ is in group } j \\ 0 & \text{item } i \text{ is not in group } j. \end{cases}$$

(So the jth column of A gives the membership in the jth group, and the ith row gives the groups that item i is in.) In this case the Gram matrix G has a nice interpretation: G_{ij} is the number of items that are in both groups i and j, and G_{ii} is the number of items in group i.

Outer product representation. If we express the $m \times p$ matrix A in terms of its columns $a_1, \ldots, a_p$ and the $p \times n$ matrix B in terms of its rows $b_1^T, \ldots, b_p^T$,

$$A = \begin{bmatrix} a_1 & \cdots & a_p \end{bmatrix}, \qquad B = \begin{bmatrix} b_1^T \\ \vdots \\ b_p^T \end{bmatrix},$$

then we can express the product matrix AB as a sum of outer products:

$$AB = a_1 b_1^T + \cdots + a_p b_p^T.$$

Complexity of matrix multiplication.   The total number of flops required for a matrix-matrix product $C = AB$ with A of size $m \times p$ and B of size $p \times n$ can be found several ways. The product matrix C has size $m \times n$, so there are mn elements to compute. The i, j element of C is the inner product of row i of A with column j of B. This is an inner product of vectors of length p and requires $2p - 1$ flops. Therefore the total is $mn(2p - 1)$ flops, which we approximate as $2mnp$ flops. The order of computing the matrix-matrix product is mnp, the product of the three dimensions involved.

In some special cases the complexity is less than $2mnp$ flops. As an example, when we compute the $n \times n$ Gram matrix $G = B^T B$ we only need to compute the entries in the upper (or lower) half of G, since G is symmetric. This saves around half the flops, so the complexity is around pn^2 flops. But the order is the same.

Complexity of sparse matrix multiplication.   Multiplying sparse matrices can be done efficiently, since we don't need to carry out any multiplications in which one or the other entry is zero. We start by analyzing the complexity of multiplying a sparse matrix with a non-sparse matrix. Suppose that A is $m \times p$ and sparse, and B is $p \times n$, but not necessarily sparse. The inner product of the ith row a_i^T of A with the jth column of B requires no more than $2\,\mathbf{nnz}(a_i^T)$ flops. Summing over $i = 1, \ldots, m$ and $j = 1, \ldots, n$ we get $2\,\mathbf{nnz}(A)n$ flops. If B is sparse, the total number of flops is no more that $2\,\mathbf{nnz}(B)m$ flops. (Note that these formulas agree with the one given above, $2mnp$, when the sparse matrices have all entries nonzero.)

There is no simple formula for the complexity of multiplying two sparse matrices, but it is certainly no more than $2\min\{\mathbf{nnz}(A)n, \mathbf{nnz}(B)m\}$ flops.

Complexity of matrix triple product.   Consider the product of three matrices,

$$D = ABC$$

with A of size $m \times n$, B of size $n \times p$, and C of size $p \times q$. The matrix D can be computed in two ways, as $(AB)C$ and as $A(BC)$. In the first method we start with AB ($2mnp$ flops) and then form $D = (AB)C$ ($2mpq$ flops), for a total of $2mp(n+q)$ flops. In the second method we compute the product BC ($2npq$ flops) and then form $D = A(BC)$ ($2mnq$ flops), for a total of $2nq(m + p)$ flops.

You might guess that the total number of flops required is the same with the two methods, but it turns out it is not. The first method is less expensive when $2mp(n + q) < 2nq(m + p)$, i.e., when

$$\frac{1}{n} + \frac{1}{q} < \frac{1}{m} + \frac{1}{p}.$$

For example, if $m = p$ and $n = q$, the first method has a complexity proportional to $m^2 n$, while the second method has complexity mn^2, and one would prefer the first method when $m \ll n$.

As a more specific example, consider the product $ab^T c$, where a, b, c are n-vectors. If we first evaluate the outer product ab^T, the cost is n^2 flops, and we need to store n^2 values. We then multiply the vector c by this $n \times n$ matrix, which costs $2n^2$ flops. The total cost is $3n^2$ flops. On the other hand if we first evaluate the inner product $b^T c$, the cost is $2n$ flops, and we only need to store one number (the result). Multiplying the vector a by this number costs n flops, so the total cost is $3n$ flops. For n large, there is a dramatic difference between $3n$ and $3n^2$ flops. (The storage requirements are also dramatically different for the two methods of evaluating $ab^T c$: one number versus n^2 numbers.)

10.2   Composition of linear functions

Matrix-matrix products and composition.   Suppose A is an $m \times p$ matrix and B is $p \times n$. We can associate with these matrices two linear functions $f : \mathbf{R}^p \to \mathbf{R}^m$ and $g : \mathbf{R}^n \to \mathbf{R}^p$, defined as $f(x) = Ax$ and $g(x) = Bx$. The *composition* of the two functions is the function $h : \mathbf{R}^n \to \mathbf{R}^m$ with

$$h(x) = f(g(x)) = A(Bx) = (AB)x.$$

In words: To find $h(x)$, we first apply the function g, to obtain the partial result $g(x)$ (which is a p-vector); then we apply the function f to this result, to obtain $h(x)$ (which is an m-vector). In the formula $h(x) = f(g(x))$, f appears to the left of g; but when we evaluate $h(x)$, we apply g first. The composition h is evidently a linear function, that can be written as $h(x) = Cx$ with $C = AB$.

   Using this interpretation of matrix multiplication as composition of linear functions, it is easy to understand why in general $AB \neq BA$, even when the dimensions are compatible. Evaluating the function $h(x) = ABx$ means we first evaluate $y = Bx$, and then $z = Ay$. Evaluating the function BAx means we first evaluate $y = Ax$, and then $z = By$. In general, the order matters. As an example, take the 2×2 matrices

$$A = \begin{bmatrix} -1 & 0 \\ 0 & 1 \end{bmatrix}, \qquad B = \begin{bmatrix} 0 & 1 \\ 1 & 0 \end{bmatrix},$$

for which

$$AB = \begin{bmatrix} 0 & -1 \\ 1 & 0 \end{bmatrix}, \qquad BA = \begin{bmatrix} 0 & 1 \\ -1 & 0 \end{bmatrix}.$$

The mapping $f(x) = Ax = (-x_1, x_2)$ changes the sign of the first element of the vector x. The mapping $g(x) = Bx = (x_2, x_1)$ reverses the order of two elements of x. If we evaluate $f(g(x)) = ABx = (-x_2, x_1)$, we first reverse the order, and then change the sign of the first element. This result is obviously different from $g(f(x)) = BAx = (x_2, -x_1)$, obtained by changing the sign of the first element, and then reversing the order of the elements.

Second difference matrix.   As a more interesting example of composition of linear functions, consider the $(n-1) \times n$ difference matrix D_n defined in (6.5). (We use the subscript n here to denote size of D.) Let D_{n-1} denote the $(n-2) \times (n-1)$ difference matrix. Their product $D_{n-1}D_n$ is called the *second difference matrix*, and sometimes denoted Δ.

We can interpret Δ in terms of composition of linear functions. Multiplying an n-vector x by D_n yields the $(n-1)$-vector of consecutive differences of the entries:

$$D_n x = (x_2 - x_1, \ldots, x_n - x_{n-1}).$$

Multiplying this vector by D_{n-1} gives the $(n-2)$-vector of consecutive differences of consecutive differences (or second differences) of x:

$$D_{n-1} D_n x = (x_1 - 2x_2 + x_3, \; x_2 - 2x_3 + x_4, \; \ldots, \; x_{n-2} - 2x_{n-1} + x_n).$$

The $(n-2) \times n$ product matrix $\Delta = D_{n-1} D_n$ is the matrix associated with the second difference function.

For the case $n = 5$, $\Delta = D_{n-1} D_n$ has the form

$$
\begin{bmatrix}
1 & -2 & 1 & 0 & 0 \\
0 & 1 & -2 & 1 & 0 \\
0 & 0 & 1 & -2 & 1
\end{bmatrix}
=
\begin{bmatrix}
-1 & 1 & 0 & 0 \\
0 & -1 & 1 & 0 \\
0 & 0 & -1 & 1
\end{bmatrix}
\begin{bmatrix}
-1 & 1 & 0 & 0 & 0 \\
0 & -1 & 1 & 0 & 0 \\
0 & 0 & -1 & 1 & 0 \\
0 & 0 & 0 & -1 & 1
\end{bmatrix}.
$$

The left-hand matrix Δ is associated with the second difference linear function that maps 5-vectors into 3-vectors. The middle matrix D_4 is associated with the difference function that maps 4-vectors into 3-vectors. The right-hand matrix D_5 is associated with the difference function that maps 5-vectors into 4-vectors.

Composition of affine functions. The composition of affine functions is an affine function. Suppose $f : \mathbf{R}^p \to \mathbf{R}^m$ is the affine function given by $f(x) = Ax + b$, and $g : \mathbf{R}^n \to \mathbf{R}^p$ is the affine function given by $g(x) = Cx + d$. The composition h is given by

$$h(x) = f(g(x)) = A(Cx + d) + b = (AC)x + (Ad + b) = \tilde{A}x + \tilde{b},$$

where $\tilde{A} = AC$, $\tilde{b} = Ad + b$.

Chain rule of differentiation. Let $f : \mathbf{R}^p \to \mathbf{R}^m$ and $g : \mathbf{R}^n \to \mathbf{R}^p$ be differentiable functions. The composition of f and g is defined as the function $h : \mathbf{R}^n \to \mathbf{R}^m$ with

$$h(x) = f(g(x)) = f(g_1(x), \ldots, g_p(x)).$$

The function h is differentiable and its partial derivatives follow from those of f and g via the chain rule:

$$\frac{\partial h_i}{\partial x_j}(z) = \frac{\partial f_i}{\partial y_1}(g(z))\frac{\partial g_1}{\partial x_j}(z) + \cdots + \frac{\partial f_i}{\partial y_p}(g(z))\frac{\partial g_p}{\partial x_j}(z)$$

for $i = 1, \ldots, m$ and $j = 1, \ldots, n$. This relation can be expressed concisely as a matrix-matrix product: The derivative matrix of h at z is the product

$$Dh(z) = Df(g(z))Dg(z)$$

of the derivative matrix of f at $g(z)$ and the derivative matrix of g at z. This compact matrix formula generalizes the chain rule for scalar-valued functions of a single variable, *i.e.*, $h'(z) = f'(g(z))g'(z)$.

The first order Taylor approximation of h at z can therefore be written as

$$
\begin{aligned}
\hat{h}(x) &= h(z) + Dh(z)(x - z) \\
&= f(g(z)) + Df(g(z))Dg(z)(x - z).
\end{aligned}
$$

The same result can be interpreted as a composition of two affine functions, the first order Taylor approximation of f at $g(z)$,

$$
\hat{f}(y) = f(g(z)) + Df(g(z))(y - g(z))
$$

and the first order Taylor approximation of g at z,

$$
\hat{g}(x) = g(z) + Dg(z)(x - z).
$$

The composition of these two affine functions is

$$
\begin{aligned}
\hat{f}(\hat{g}(x)) &= \hat{f}(g(z) + Dg(z)(x - z)) \\
&= f(g(z)) + Df(g(z))(g(z) + Dg(z)(x - z) - g(z)) \\
&= f(g(z)) + Df(g(z))Dg(z)(x - z)
\end{aligned}
$$

which is equal to $\hat{h}(x)$.

When f is a scalar-valued function ($m = 1$), the derivative matrices $Dh(z)$ and $Df(g(z))$ are the transposes of the gradients, and we write the chain rule as

$$
\nabla h(z) = Dg(z)^T \nabla f(g(z)).
$$

In particular, if $g(x) = Ax + b$ is affine, then the gradient of $h(x) = f(g(x)) = f(Ax + b)$ is given by $\nabla h(z) = A^T \nabla f(Ax + b)$.

Linear dynamical system with state feedback.   We consider a time-invariant linear dynamical system with n-vector state x_t and m-vector input u_t, with dynamics

$$
x_{t+1} = Ax_t + Bu_t, \quad t = 1, 2, \dots.
$$

Here we think of the input u_t as something we can manipulate, *e.g.*, the control surface deflections for an airplane or the amount of material we order or move in a supply chain. In *state feedback control* the state x_t is measured, and the input u_t is a linear function of the state, expressed as

$$
u_t = Kx_t,
$$

where K is the $m \times n$ *state-feedback gain matrix*. The term *feedback* refers to the idea that the state is measured, and then (after multiplying by K) fed back into the system, via the input. This leads to a loop, where the state affects the input, and the input affects the (next) state. State feedback is very widely used in many applications. (In §17.2.3 we will see methods for choosing or designing an appropriate state feedback matrix.)

With state feedback, we have

$$x_{t+1} = Ax_t + Bu_t = Ax_t + B(Kx_t) = (A + BK)x_t, \quad t = 1, 2, \ldots.$$

This recursion is called the *closed-loop system*. The matrix $A + BK$ is called the *closed-loop dynamics matrix*. (In this context, the recursion $x_{t+1} = Ax_t$ is called the *open-loop system*. It gives the dynamics when $u_t = 0$.)

10.3   Matrix power

It makes sense to multiply a square matrix A by itself to form AA. We refer to this matrix as A^2. Similarly, if k is a positive integer, then k copies of A multiplied together is denoted A^k. If k and l are positive integers, and A is square, then $A^k A^l = A^{k+l}$ and $(A^k)^l = A^{kl}$. By convention we take $A^0 = I$, which makes the formulas above hold for all nonnegative integer values of k and l.

We should mention one ambiguity in matrix power notation that occasionally arises. When A is a square matrix and T is a nonnegative integer, A^T can mean either the transpose of the matrix A or its Tth power. Usually which is meant is clear from the context, or the author explicitly states which meaning is intended. To avoid this ambiguity, some authors use a different symbol for the transpose, such as A^T (with the superscript in roman font) or A', or avoid referring to the Tth power of a matrix. When A is not square there is no ambiguity, since A^T can only be the transpose in this case.

Other matrix powers.   Matrix powers A^k with k a negative integer will be discussed in §11.2. Non-integer powers, such as $A^{1/2}$ (the matrix squareroot), need not make sense, or can be ambiguous, unless certain conditions on A hold. This is an advanced topic in linear algebra that we will not pursue in this book.

Paths in a directed graph.   Suppose A is the $n \times n$ adjacency matrix of a directed graph with n vertices:

$$A_{ij} = \begin{cases} 1 & \text{there is a edge from vertex } j \text{ to vertex } i \\ 0 & \text{otherwise} \end{cases}$$

(see page 112). A *path* of length ℓ is a sequence of $\ell + 1$ vertices, with an edge from the first to the second vertex, an edge from the second to third vertex, and so on. We say the path goes from the first vertex to the last one. An edge can be considered a path of length one. By convention, every vertex has a path of length zero (from the vertex to itself).

The elements of the matrix powers A^ℓ have a simple meaning in terms of paths in the graph. First examine the expression for the i, j element of the square of A:

$$(A^2)_{ij} = \sum_{k=1}^{n} A_{ik} A_{kj}.$$

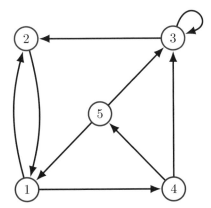

Figure 10.1 Directed graph.

Each term in the sum is 0 or 1, and equal to one only if there is an edge from vertex j to vertex k and an edge from vertex k to vertex i, *i.e.*, a path of length exactly two from vertex j to vertex i via vertex k. By summing over all k, we obtain the total number of paths of length two from j to i.

The adjacency matrix A for the graph in figure 10.1, for example, and its square are given by

$$A = \begin{bmatrix} 0 & 1 & 0 & 0 & 1 \\ 1 & 0 & 1 & 0 & 0 \\ 0 & 0 & 1 & 1 & 1 \\ 1 & 0 & 0 & 0 & 0 \\ 0 & 0 & 0 & 1 & 0 \end{bmatrix}, \qquad A^2 = \begin{bmatrix} 1 & 0 & 1 & 1 & 0 \\ 0 & 1 & 1 & 1 & 2 \\ 1 & 0 & 1 & 2 & 1 \\ 0 & 1 & 0 & 0 & 1 \\ 1 & 0 & 0 & 0 & 0 \end{bmatrix}.$$

We can verify there is exactly one path of length two from vertex 1 to itself, *i.e.*, the path $(1, 2, 1)$), and one path of length two from vertex 3 to vertex 1, *i.e.*, the path $(3, 2, 1)$. There are two paths of length two from vertex 4 to vertex 3, $(4, 3, 3)$ and $(4, 5, 3)$, so $(A^2)_{34} = 2$.

The property extends to higher powers of A. If ℓ is a positive integer, then the i, j element of A^ℓ is the number of paths of length ℓ from vertex j to vertex i. This can be proved by induction on ℓ. We have already shown the result for $\ell = 2$. Assume that it is true that the elements of A^ℓ give the paths of length ℓ between the different vertices. Consider the expression for the i, j element of $A^{\ell+1}$:

$$(A^{\ell+1})_{ij} = \sum_{k=1}^{n} A_{ik}(A^\ell)_{kj}.$$

The kth term in the sum is equal to the number of paths of length ℓ from j to k if there is an edge from k to i, and is equal to zero otherwise. Therefore it is equal to the number of paths of length $\ell + 1$ from j to i that end with the edge (k, i), *i.e.*, of the form $(j, \dots, k, i)$. By summing over all k we obtain the total number of paths of length $\ell + 1$ from vertex j to vertex i.

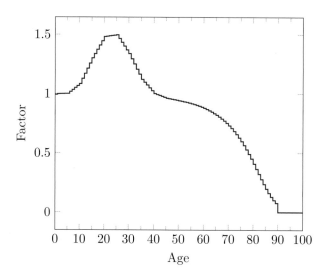

Figure 10.2 Contribution factor per age in 2010 to the total population in 2020. The value for age $i - 1$ is the ith component of the row vector $\mathbf{1}^T A^{10}$.

This can be verified in the example. The third power of A is

$$A^3 = \begin{bmatrix} 1 & 1 & 1 & 1 & 2 \\ 2 & 0 & 2 & 3 & 1 \\ 2 & 1 & 1 & 2 & 2 \\ 1 & 0 & 1 & 1 & 0 \\ 0 & 1 & 0 & 0 & 1 \end{bmatrix}.$$

The $(A^3)_{24} = 3$ paths of length three from vertex 4 to vertex 2 are $(4, 3, 3, 2)$, $(4, 5, 3, 2)$, $(4, 5, 1, 2)$.

Linear dynamical system. Consider a time-invariant linear dynamical system, described by $x_{t+1} = Ax_t$. We have $x_{t+2} = Ax_{t+1} = A(Ax_t) = A^2 x_t$. Continuing this argument, we have

$$x_{t+\ell} = A^\ell x_t,$$

for $\ell = 1, 2, \ldots$. In a linear dynamical system, we can interpret A^ℓ as the matrix that propagates the state forward ℓ time steps.

For example, in a population dynamics model, A^ℓ is the matrix that maps the current population distribution into the population distribution ℓ periods in the future, taking into account births, deaths, and the births and deaths of children, and so on. The total population ℓ periods in the future is given by $\mathbf{1}^T (A^\ell x_t)$, which we can write as $(\mathbf{1}^T A^\ell) x_t$. The row vector $\mathbf{1}^T A^\ell$ has an interesting interpretation: Its ith entry is the contribution to the total population in ℓ periods due to each person with current age $i - 1$. It is plotted in figure 10.2 for the US data given in §9.2.

Matrix powers also come up in the analysis of a time-invariant linear dynamical system with an input. We have

$$x_{t+2} = Ax_{t+1} + Bu_{t+1} = A(Ax_t + Bu_t) = A^2 x_t + ABu_t + Bu_{t+1}.$$

Iterating this over ℓ periods we obtain

$$x_{t+\ell} = A^\ell x_t + A^{\ell-1} Bu_t + A^{\ell-2} Bu_{t+1} + \cdots + Bu_{t+\ell-1}. \tag{10.2}$$

(The first term agrees with the formula for $x_{t+\ell}$ with no input.) The other terms are readily interpreted. The term $A^j Bu_{t+\ell-j}$ is the contribution to the state $x_{t+\ell}$ due to the input at time $t + \ell - j$.

10.4   QR factorization

Matrices with orthonormal columns.   As an application of Gram matrices, we can express the condition that the n-vectors $a_1, \ldots, a_k$ are orthonormal in a simple way using matrix notation:

$$A^T A = I,$$

where A is the $n \times k$ matrix with columns $a_1, \ldots, a_k$. There is no standard term for a matrix whose columns are orthonormal: We refer to a matrix whose columns are orthonormal as 'a matrix whose columns are orthonormal'. But a *square* matrix that satisfies $A^T A = I$ is called *orthogonal*; its columns are an orthonormal basis. Orthogonal matrices have many uses, and arise in many applications.

We have already encountered some orthogonal matrices, including identity matrices, 2-D reflections and rotations (page 129), and permutation matrices (page 132).

Norm, inner product, and angle properties.   Suppose the columns of the $m \times n$ matrix A are orthonormal, and x and y are any n-vectors. We let $f : \mathbf{R}^n \to \mathbf{R}^m$ be the function that maps z to Az. Then we have the following:

- $\|Ax\| = \|x\|$. That is, f is *norm preserving*.

- $(Ax)^T (Ay) = x^T y$. f preserves the inner product between vectors.

- $\angle(Ax, Ay) = \angle(x, y)$. f also preserves angles between vectors.

Note that in each of the three equations above, the vectors appearing in the left- and right-hand sides have different dimensions, m on the left and n on the right.

We can verify these properties using simple matrix properties. We start with the second statement, that multiplication by A preserves the inner product. We have

$$
\begin{aligned}
(Ax)^T (Ay) &= (x^T A^T)(Ay) \\
&= x^T (A^T A) y \\
&= x^T I y \\
&= x^T y.
\end{aligned}
$$

In the first line, we use the transpose-of-product rule; in the second, we re-associate a product of 4 matrices (considering the row vector x^T and column vector x as matrices); in the third line, we use $A^T A = I$; and in the fourth line, we use $Iy = y$.

From the second property we can derive the first one: By taking $y = x$ we get $(Ax)^T(Ax) = x^T x$; taking the squareroot of each side gives $\|Ax\| = \|x\|$. The third property, angle preservation, follows from the first two, since

$$\angle(Ax, Ay) = \arccos\left(\frac{(Ax)^T(Ay)}{\|Ax\|\|Ay\|}\right) = \arccos\left(\frac{x^T y}{\|x\|\|y\|}\right) = \angle(x, y).$$

QR factorization. We can express the result of the Gram–Schmidt algorithm described in §5.4 in a compact form using matrices. Let A be an $n \times k$ matrix with linearly independent columns $a_1, \dots, a_k$. By the independence-dimension inequality, A is tall or square. Let Q be the $n \times k$ matrix with columns $q_1, \dots, q_k$, the orthonormal vectors produced by the Gram–Schmidt algorithm applied to the n-vectors $a_1, \dots, a_k$. Orthonormality of $q_1, \dots, q_k$ is expressed in matrix form as $Q^T Q = I$. We express the equation relating a_i and q_i,

$$a_i = (q_1^T a_i)q_1 + \cdots + (q_{i-1}^T a_i)q_{i-1} + \|\tilde{q}_i\|q_i,$$

where $\tilde{q}_i$ is the vector obtained in the first step of the Gram–Schmidt algorithm, as

$$a_i = R_{1i}q_1 + \cdots + R_{ii}q_i,$$

where $R_{ij} = q_i^T a_j$ for $i < j$ and $R_{ii} = \|\tilde{q}_i\|$. Defining $R_{ij} = 0$ for $i > j$, we can write the equations above in compact matrix form as

$$A = QR.$$

This is called the *QR factorization* of A, since it expresses the matrix A as a product of two matrices, Q and R. The $n \times k$ matrix Q has orthonormal columns, and the $k \times k$ matrix R is upper triangular, with positive diagonal elements. If A is square, with linearly independent columns, then Q is orthogonal and the QR factorization expresses A as a product of two square matrices.

The attributes of the matrices Q and R in the QR factorization come directly from the Gram–Schmidt algorithm. The equation $Q^T Q = I$ follows from the orthonormality of the vectors $q_1, \dots, q_k$. The matrix R is upper triangular because each vector a_i is a linear combination of $q_1, \dots, q_i$.

The Gram–Schmidt algorithm is not the only algorithm for QR factorization. Several other QR factorization algorithms exist, that are more reliable in the presence of round-off errors. (These QR factorization methods may also change the *order* in which the columns of A are processed.)

Sparse QR factorization. There are algorithms for QR factorization that efficiently handle the case when the matrix A is sparse. In this case the matrix Q is stored in a special format that requires much less memory than if it were stored as a generic $n \times k$ matrix, *i.e.*, nk numbers. The flop count for these sparse QR factorizations is also much smaller than $2nk^2$.

Exercises

10.1 *Scalar-row-vector multiplication.* Suppose a is a number and $x = [x_1 \cdots x_n]$ is an n-row-vector. The scalar-row-vector product ax is the n-row-vector $[ax_1 \cdots ax_n]$. Is this a special case of matrix-matrix multiplication? That is, can you interpret scalar-row-vector multiplication as matrix multiplication? (Recall that scalar-vector multiplication, with the scalar on the left, is *not* a special case of matrix-matrix multiplication; see page 177.)

10.2 *Ones matrix.* There is no special notation for an $m \times n$ matrix all of whose entries are one. Give a simple expression for this matrix in terms of matrix multiplication, transpose, and the ones vectors $\mathbf{1}_m$, $\mathbf{1}_n$ (where the subscripts denote the dimension).

10.3 *Matrix sizes.* Suppose A, B, and C are matrices that satisfy $A + BB^T = C$. Determine which of the following statements are necessarily true. (There may be more than one true statement.)

    (a) A is square.

    (b) A and B have the same dimensions.

    (c) A, B, and C have the same number of rows.

    (d) B is a tall matrix.

10.4 *Block matrix notation.* Consider the block matrix

$$A = \begin{bmatrix} I & B & 0 \\ B^T & 0 & 0 \\ 0 & 0 & BB^T \end{bmatrix},$$

where B is 10×5. What are the dimensions of the four zero matrices and the identity matrix in the definition of A? What are the dimensions of A?

10.5 *When is the outer product symmetric?* Let a and b be n-vectors. The inner product is symmetric, *i.e.*, we have $a^T b = b^T a$. The outer product of the two vectors is generally *not* symmetric; that is, we generally have $ab^T \neq ba^T$. What are the conditions on a and b under which $ab = ba^T$? You can assume that all the entries of a and b are nonzero. (The conclusion you come to will hold even when some entries of a or b are zero.) *Hint.* Show that $ab^T = ba^T$ implies that a_i/b_i is a constant (*i.e.*, independent of i).

10.6 *Product of rotation matrices.* Let A be the 2×2 matrix that corresponds to rotation by θ radians, defined in (7.1), and let B be the 2×2 matrix that corresponds to rotation by ω radians. Show that AB is also a rotation matrix, and give the angle by which it rotates vectors. Verify that $AB = BA$ in this case, and give a simple English explanation.

10.7 *Two rotations.* Two 3-vectors x and y are related as follows. First, the vector x is rotated $40°$ around the e_3 axis, counterclockwise (from e_1 toward e_2), to obtain the 3-vector z. Then, z is rotated $20°$ around the e_1 axis, counterclockwise (from e_2 toward e_3), to form y. Find the 3×3 matrix A for which $y = Ax$. Verify that A is an orthogonal matrix. *Hint.* Express A as a product of two matrices, which carry out the two rotations described above.

10.8 *Entries of matrix triple product.* (See page 182.) Suppose A has dimensions $m \times n$, B has dimensions $n \times p$, C has dimensions $p \times q$, and let $D = ABC$. Show that

$$D_{ij} = \sum_{k=1}^{n} \sum_{l=1}^{p} A_{ik} B_{kl} C_{lj}.$$

This is the formula analogous to (10.1) for the product of two matrices.

10.9 *Multiplication by a diagonal matrix.* Suppose that A is an $m \times n$ matrix, D is a diagonal matrix, and $B = DA$. Describe B in terms of A and the entries of D. You can refer to the rows or columns or entries of A.

10.10 *Converting from purchase quantity matrix to purchase dollar matrix.* An $n \times N$ matrix Q gives the purchase history of a set of n products by N customers, over some period, with Q_{ij} being the quantity of product i bought by customer j. The n-vector p gives the product prices. A data analyst needs the $n \times N$ matrix D, where D_{ij} is the total dollar value that customer j spent on product i. Express D in terms of Q and p, using compact matrix/vector notation. You can use any notation or ideas we have encountered, *e.g.*, stacking, slicing, block matrices, transpose, matrix-vector product, matrix-matrix product, inner product, norm, correlation, **diag**(), and so on.

10.11 *Trace of matrix-matrix product.* The sum of the diagonal entries of a square matrix is called the *trace* of the matrix, denoted $\mathbf{tr}(A)$.

   (a) Suppose A and B are $m \times n$ matrices. Show that

$$\mathbf{tr}(A^T B) = \sum_{i=1}^{m} \sum_{j=1}^{n} A_{ij} B_{ij}.$$

   What is the complexity of calculating $\mathbf{tr}(A^T B)$?

   (b) The number $\mathbf{tr}(A^T B)$ is sometimes referred to as the inner product of the matrices A and B. (This allows us to extend concepts like angle to matrices.) Show that $\mathbf{tr}(A^T B) = \mathbf{tr}(B^T A)$.

   (c) Show that $\mathbf{tr}(A^T A) = \|A\|^2$. In other words, the square of the norm of a matrix is the trace of its Gram matrix.

   (d) Show that $\mathbf{tr}(A^T B) = \mathbf{tr}(BA^T)$, even though in general $A^T B$ and BA^T can have different dimensions, and even when they have the same dimensions, they need not be equal.

10.12 *Norm of matrix product.* Suppose A is an $m \times p$ matrix and B is a $p \times n$ matrix. Show that $\|AB\| \le \|A\|\|B\|$, *i.e.*, the (matrix) norm of the matrix product is no more than the product of the norms of the matrices. *Hint.* Let $a_1^T, \ldots, a_m^T$ be the rows of A, and $b_1, \ldots, b_n$ be the columns of B. Then

$$\|AB\|^2 = \sum_{i=1}^{m} \sum_{j=1}^{n} (a_i^T b_j)^2.$$

   Now use the Cauchy–Schwarz inequality.

10.13 *Laplacian matrix of a graph.* Let A be the incidence matrix of a directed graph with n nodes and m edges (see §7.3). The *Laplacian matrix* associated with the graph is defined as $L = AA^T$, which is the Gram matrix of A^T. It is named after the mathematician Pierre-Simon Laplace.

   (a) Show that $\mathcal{D}(v) = v^T L v$, where $\mathcal{D}(v)$ is the Dirichlet energy defined on page 135.

   (b) Describe the entries of L. *Hint.* The following two quantities might be useful: The *degree* of a node, which is the number of edges that connect to the node (in either direction), and the number of edges that connect a pair of distinct nodes (in either direction).

10.14 *Gram matrix.* Let $a_1, \ldots, a_n$ be the columns of the $m \times n$ matrix A. Suppose that the columns all have norm one, and for $i \ne j$, $\angle(a_i, a_j) = 60°$. What can you say about the Gram matrix $G = A^T A$? Be as specific as you can be.

10.15 *Pairwise distances from Gram matrix.* Let A be an $m \times n$ matrix with columns $a_1, \ldots, a_n$, and associated Gram matrix $G = A^T A$. Express $\|a_i - a_j\|$ in terms of G, specifically G_{ii}, G_{ij}, and G_{jj}.

10.16 *Covariance matrix.* Consider a list of k n-vectors $a_1, \ldots, a_k$, and define the $n \times k$ matrix $A = [a_1 \ \cdots \ a_k]$.

(a) Let the k-vector μ give the means of the columns, i.e., $\mu_i = \mathbf{avg}(a_i)$, $i = 1, \ldots, k$. (The symbol μ is a traditional one to denote an average value.) Give an expression for μ in terms of the matrix A.

(b) Let $\tilde{a}_1, \ldots, \tilde{a}_k$ be the de-meaned versions of $a_1, \ldots, a_k$, and define $\tilde{A}$ as the $n \times k$ matrix $\tilde{A} = [\tilde{a}_1 \ \cdots \ \tilde{a}_k]$. Give a matrix expression for $\tilde{A}$ in terms of A and μ.

(c) The *covariance matrix* of the vectors $a_1, \ldots, a_k$ is the $k \times k$ matrix $\Sigma = (1/N)\tilde{A}^T \tilde{A}$, the Gram matrix of $\tilde{A}$ multiplied with $1/N$. Show that

$$\Sigma_{ij} = \begin{cases} \mathbf{std}(a_i)^2 & i = j \\ \mathbf{std}(a_i)\,\mathbf{std}(a_j)\rho_{ij} & i \neq j \end{cases}$$

where ρ_{ij} is the correlation coefficient of a_i and a_j. (The expression for $i \neq j$ assumes that ρ_{ij} is defined, i.e., $\mathbf{std}(a_i)$ and $\mathbf{std}(a_j)$ are nonzero. If not, we interpret the formula as $\Sigma_{ij} = 0$.) Thus the covariance matrix encodes the standard deviations of the vectors, as well as correlations between all pairs. The correlation matrix is widely used in probability and statistics.

(d) Let $z_1, \ldots, z_k$ be the standardized versions of $a_1, \ldots, a_k$. (We assume the de-meaned vectors are nonzero.) Derive a matrix expression for $Z = [z_1 \ \cdots \ z_k]$, the matrix of standardized vectors. Your expression should use A, μ, and the numbers $\mathbf{std}(a_1), \ldots, \mathbf{std}(a_k)$.

10.17 *Patients and symptoms.* Each of a set of N patients can exhibit any number of a set of n symptoms. We express this as an $N \times n$ matrix S, with

$$S_{ij} = \begin{cases} 1 & \text{patient } i \text{ exhibits symptom } j \\ 0 & \text{patient } i \text{ does not exhibit symptom } j. \end{cases}$$

Give simple English descriptions of the following expressions. Include the dimensions, and describe the entries.

(a) $S\mathbf{1}$.

(b) $S^T\mathbf{1}$.

(c) $S^T S$.

(d) SS^T.

10.18 *Students, classes, and majors.* We consider m students, n classes, and p majors. Each student can be in any number of the classes (although we'd expect the number to range from 3 to 6), and can have any number of the majors (although the common values would be 0, 1, or 2). The data about the students' classes and majors are given by an $m \times n$ matrix C and an $m \times p$ matrix M, where

$$C_{ij} = \begin{cases} 1 & \text{student } i \text{ is in class } j \\ 0 & \text{student } i \text{ is not in class } j, \end{cases}$$

and

$$M_{ij} = \begin{cases} 1 & \text{student } i \text{ is in major } j \\ 0 & \text{student } i \text{ is not in major } j. \end{cases}$$

(a) Let E be the n-vector with E_i being the enrollment in class i. Express E using matrix notation, in terms of the matrices C and M.

(b) Define the $n \times p$ matrix S where S_{ij} is the total number of students in class i with major j. Express S using matrix notation, in terms of the matrices C and M.

10.19 *Student group membership.* Let $G \in \mathbf{R}^{m \times n}$ represent a contingency matrix of m students who are members of n groups:

$$G_{ij} = \begin{cases} 1 & \text{student } i \text{ is in group } j \\ 0 & \text{student } i \text{ is not in group } j. \end{cases}$$

(A student can be in any number of the groups.)

(a) What is the meaning of the 3rd column of G?

(b) What is the meaning of the 15th row of G?

(c) Give a simple formula (using matrices, vectors, etc.) for the n-vector M, where M_i is the total membership (*i.e.*, number of students) in group i.

(d) Interpret $(GG^T)_{ij}$ in simple English.

(e) Interpret $(G^T G)_{ij}$ in simple English.

10.20 *Products, materials, and locations.* P different products each require some amounts of M different materials, and are manufactured in L different locations, which have different material costs. We let C_{lm} denote the cost of material m in location l, for $l = 1, \ldots, L$ and $m = 1, \ldots, M$. We let Q_{mp} denote the amount of material m required to manufacture one unit of product p, for $m = 1, \ldots, M$ and $p = 1, \ldots, P$. Let T_{pl} denote the total cost to manufacture product p in location l, for $p = 1, \ldots, P$ and $l = 1, \ldots, L$. Give an expression for the matrix T.

10.21 *Integral of product of polynomials.* Let p and q be two quadratic polynomials, given by

$$p(x) = c_1 + c_2 x + c_3 x^2, \qquad q(x) = d_1 + d_2 x + d_3 x^2.$$

Express the integral $J = \int_0^1 p(x) q(x) \, dx$ in the form $J = c^T G d$, where G is a 3×3 matrix. Give the entries of G (as numbers).

10.22 *Composition of linear dynamical systems.* We consider two time-invariant linear dynamical systems with outputs. The first one is given by

$$x_{t+1} = A x_t + B u_t, \qquad y_t = C x_t, \quad t = 1, 2, \ldots,$$

with state x_t, input u_t, and output y_t. The second is given by

$$\tilde{x}_{t+1} = \tilde{A} \tilde{x}_t + \tilde{B} w_t, \qquad v_t = \tilde{C} \tilde{x}_t, \quad t = 1, 2, \ldots,$$

with state $\tilde{x}_t$, input w_t, and output v_t. We now connect the output of the first linear dynamical system to the input of the second one, which means we take $w_t = y_t$. (This is called the *composition* of the two systems.) Show that this composition can also be expressed as a linear dynamical system with state $z_t = (x_t, \tilde{x}_t)$, input u_t, and output v_t. (Give the state transition matrix, input matrix, and output matrix.)

10.23 Suppose A is an $n \times n$ matrix that satisfies $A^2 = 0$. Does this imply that $A = 0$? (This is the case when $n = 1$.) If this is (always) true, explain why. If it is not, give a specific counterexample, *i.e.*, a matrix A that is nonzero but satisfies $A^2 = 0$.

10.24 *Matrix power identity.* A student says that for any square matrix A,

$$(A + I)^3 = A^3 + 3A^2 + 3A + I.$$

Is she right? If she is, explain why; if she is wrong, give a specific counterexample, *i.e.*, a square matrix A for which it does not hold.

10.25 *Squareroots of the identity.* The number 1 has two squareroots (*i.e.*, numbers who square is 1), 1 and -1. The $n \times n$ identity matrix I_n has many more squareroots.

(a) Find all diagonal squareroots of I_n. How many are there? (For $n = 1$, you should get 2.)

(b) Find a nondiagonal 2×2 matrix A that satisfies $A^2 = I$. This means that in general there are even more squareroots of I_n than you found in part (a).

10.26 *Circular shift matrices.* Let A be the 5×5 matrix

$$
A = \begin{bmatrix}
0 & 0 & 0 & 0 & 1 \\
1 & 0 & 0 & 0 & 0 \\
0 & 1 & 0 & 0 & 0 \\
0 & 0 & 1 & 0 & 0 \\
0 & 0 & 0 & 1 & 0
\end{bmatrix}.
$$

(a) How is Ax related to x? Your answer should be in English. *Hint.* See exercise title.

(b) What is A^5? *Hint.* The answer should make sense, given your answer to part (a).

10.27 *Dynamics of an economy.* Let $x_1, x_2, \ldots$ be n-vectors that give the level of economic activity of a country in years $1, 2, \ldots$, in n different sectors (like energy, defense, manufacturing). Specifically, $(x_t)_i$ is the level of economic activity in economic sector i (say, in billions of dollars) in year t. A common model that connects these economic activity vectors is $x_{t+1} = Bx_t$, where B is an $n \times n$ matrix. (See exercise 9.2.)

Give a matrix expression for the total economic activity across all sectors in year $t = 6$, in terms of the matrix B and the vector of initial activity levels x_1. Suppose you can increase economic activity in year $t = 1$ by some fixed amount (say, one billion dollars) in *one* sector, by government spending. How should you choose which sector to stimulate so as to maximize the total economic output in year $t = 6$?

10.28 *Controllability matrix.* Consider the time-invariant linear dynamical system $x_{t+1} = Ax_t + Bu_t$, with n-vector state x_t and m-vector input u_t. Let $U = (u_1, u_2, \ldots, u_{T-1})$ denote the sequence of inputs, stacked in one vector. Find the matrix C_T for which

$$
x_T = A^{T-1}x_1 + C_T U
$$

holds. The first term is what x_T would be if $u_1 = \cdots = u_{T-1} = 0$; the second term shows how the sequence of inputs $u_1, \ldots, u_{T-1}$ affect x_T. The matrix C_T is called the *controllability matrix* of the linear dynamical system.

10.29 *Linear dynamical system with $2\times$ down-sampling.* We consider a linear dynamical system with n-vector state x_t, m-vector input u_t, and dynamics given by

$$
x_{t+1} = Ax_t + Bu_t, \quad t = 1, 2, \ldots,
$$

where A is $n \times n$ matrix A and B is $n \times m$. Define $z_t = x_{2t-1}$ for $t = 1, 2, \ldots$, i.e.,

$$
z_1 = x_1, \quad z_2 = x_3, \quad z_3 = x_5, \ldots.
$$

(The sequence z_t is the original state sequence x_t 'down-sampled' by $2\times$.) Define the $(2m)$-vectors w_t as $w_t = (u_{2t-1}, u_{2t})$ for $t = 1, 2, \ldots$, i.e.,

$$
w_1 = (u_1, u_2), \quad w_2 = (u_3, u_4), \quad w_3 = (u_5, u_6), \ldots.
$$

(Each entry of the sequence w_t is a stack of two consecutive original inputs.) Show that z_t, w_t satisfy the linear dynamics equation $z_{t+1} = Fz_t + Gw_t$, for $t = 1, 2, \ldots$. Give the matrices F and G in terms of A and B.

10.30 *Cycles in a graph.* A *cycle* of length ℓ in a directed graph is a path of length ℓ that starts and ends at the same vertex. Determine the total number of cycles of length $\ell = 10$ for the directed graph given in the example on page 187. Break this number down into the number of cycles that begin (and end) at vertex 1, vertex 2, $\ldots$, vertex 5. (These should add up to the total.) *Hint.* Do not count the cycles by hand.

10.31 *Diameter of a graph.* A directed graph with n vertices is described by its $n \times n$ adjacency matrix A (see §10.3).

(a) Derive an expression P_{ij} for the total number of paths, with length no more than k, from vertex j to vertex i. (We include in this total the number of paths of length zero, which go from each vertex j to itself.) *Hint.* You can derive an expression for the matrix P, in terms of the matrix A.

(b) The *diameter* D of a graph is the smallest number for which there is a path of length $\leq D$ from node j to node i, for every pair of vertices j and i. Using part (a), explain how to compute the diameter of a graph using matrix operations (such as addition, multiplication).

Remark. Suppose the vertices represent all people on earth, and the graph edges represent acquaintance, *i.e.*, $A_{ij} = 1$ if person j and person i are acquainted. (This graph is symmetric.) Even though n is measured in billions, the diameter of this acquaintance graph is thought to be quite small, perhaps 6 or 7. In other words, any two people on earth can be connected though a set of 6 or 7 (or fewer) acquaintances. This idea, originally conjectured in the 1920s, is sometimes called *six degrees of separation*.

10.32 *Matrix exponential.* You may know that for any real number a, the sequence $(1 + a/k)^k$ converges as $k \to \infty$ to the exponential of a, denoted $\exp a$ or e^a. The *matrix exponential* of a square matrix A is defined as the limit of the matrix sequence $(I + A/k)^k$ as $k \to \infty$. (It can shown that this sequence always converges.) The matrix exponential arises in many applications, and is covered in more advanced courses on linear algebra.

(a) Find $\exp 0$ (the zero matrix) and $\exp I$.

(b) Find $\exp A$, for $A = \begin{bmatrix} 0 & 1 \\ 0 & 0 \end{bmatrix}$.

10.33 *Matrix equations.* Consider two $m \times n$ matrices A and B. Suppose that for $j = 1, \ldots, n$, the jth column of A is a linear combination of the first j columns of B. How do we express this as a matrix equation? Choose one of the matrix equations below and justify your choice.

(a) $A = GB$ for some upper triangular matrix G.

(b) $A = BH$ for some upper triangular matrix H.

(c) $A = FB$ for some lower triangular matrix F.

(d) $A = BJ$ for some lower triangular matrix J.

10.34 Choose one of the responses *always*, *never*, or *sometimes* for each of the statements below. 'Always' means the statement is always true, 'never' means it is never true, and 'Sometimes' means it can be true or false, depending on the particular values of the matrix or matrices. Give a brief justification of each answer.

(a) An upper triangular matrix has linearly independent columns.

(b) The rows of a tall matrix are linearly dependent.

(c) The columns of A are linearly independent, and $AB = 0$ for some nonzero matrix B.

10.35 *Orthogonal matrices.* Let U and V be two orthogonal $n \times n$ matrices. Show that the matrix UV and the $(2n) \times (2n)$ matrix

$$\frac{1}{\sqrt{2}} \begin{bmatrix} U & U \\ V & -V \end{bmatrix}$$

are orthogonal.

10.36 *Quadratic form.* Suppose A is an $n \times n$ matrix and x is an n-vector. The triple product $x^T A x$, a 1×1 matrix which we consider to be a scalar (*i.e.*, number), is called a *quadratic form* of the vector x, with coefficient matrix A. A quadratic form is the vector analog of a quadratic function αu^2, where α and u are both numbers. Quadratic forms arise in many fields and applications.

(a) Show that $x^T A x = \sum_{i,j=1}^{n} A_{ij} x_i x_j$.

(b) Show that $x^T (A^T) x = x^T A x$. In other words, the quadratic form with the transposed coefficient matrix has the same value for any x. *Hint.* Take the transpose of the triple product $x^T A x$.

(c) Show that $x^T ((A + A^T)/2) x = x^T A x$. In other words, the quadratic form with coefficient matrix equal to the symmetric part of a matrix (*i.e.*, $(A + A^T)/2$) has the same value as the original quadratic form.

(d) Express $2x_1^2 - 3x_1 x_2 - x_2^2$ as a quadratic form, with symmetric coefficient matrix A.

10.37 *Orthogonal 2×2 matrices.* In this problem, you will show that every 2×2 orthogonal matrix is either a rotation or a reflection (see §7.1).

(a) Let
$$Q = \begin{bmatrix} a & b \\ c & d \end{bmatrix}$$
be an orthogonal 2×2 matrix. Show that the following equations hold:
$$a^2 + c^2 = 1, \qquad b^2 + d^2 = 1, \qquad ab + cd = 0.$$

(b) Define $s = ad - bc$. Combine the three equalities in part (a) to show that
$$|s| = 1, \qquad b = -sc, \qquad d = sa.$$

(c) Suppose $a = \cos\theta$. Show that there are two possible matrices Q: A rotation (counterclockwise over θ radians), and a reflection (through the line that passes through the origin at an angle of $\theta/2$ radians with respect to horizontal).

10.38 *Orthogonal matrix with nonnegative entries.* Suppose the $n \times n$ matrix A is orthogonal, and all of its entries are nonnegative, *i.e.*, $A_{ij} \geq 0$ for $i, j = 1, \ldots, n$. Show that A must be a permutation matrix, *i.e.*, each entry is either 0 or 1, each row has exactly one entry with value one, and each column has exactly one entry with value one. (See page 132.)

10.39 *Gram matrix and QR factorization.* Suppose the matrix A has linearly independent columns and QR factorization $A = QR$. What is the relationship between the Gram matrix of A and the Gram matrix of R? What can you say about the angles between the columns of A and the angles between the columns of R?

10.40 *QR factorization of first i columns of A.* Suppose the $n \times k$ matrix A has QR factorization $A = QR$. We define the $n \times i$ matrices
$$A_i = \begin{bmatrix} a_1 & \cdots & a_i \end{bmatrix}, \qquad Q_i = \begin{bmatrix} q_1 & \cdots & q_i \end{bmatrix},$$
for $i = 1, \ldots, k$. Define the $i \times i$ matrix R_i as the submatrix of R containing its first i rows and columns, for $i = 1, \ldots, k$. Using index range notation, we have
$$A_i = A_{1:n,1:i}, \quad Q_i = A_{1:n,1:i}, \quad R_i = R_{1:i,1:i}.$$

Show that $A_i = Q_i R_i$ is the QR factorization of A_i. This means that when you compute the QR factorization of A, you are also computing the QR factorization of all submatrices $A_1, \ldots, A_k$.

10.41 *Clustering via k-means as an approximate matrix factorization.* Suppose we run the k-means algorithm on the N n-vectors $x_1, \ldots, x_N$, to obtain the group representatives $z_1, \ldots, z_k$. Define the matrices

$$X = [\ x_1 \quad \cdots \quad x_N\ ], \qquad Z = [\ z_1 \quad \cdots \quad z_N\ ].$$

X has size $n \times N$ and Z has size $n \times k$. We encode the assignment of vectors to groups by the $k \times N$ clustering matrix C, with $C_{ij} = 1$ if x_j is assigned to group i, and $C_{ij} = 0$ otherwise. Each column of C is a unit vector; its transpose is a selector matrix.

(a) Give an interpretation of the columns of the matrix $X - ZC$, and the squared norm (matrix) norm $\|X - ZC\|^2$.

(b) Justify the following statement: The goal of the k-means algorithm is to find an $n \times k$ matrix Z, and a $k \times N$ matrix C, which is the transpose of a selector matrix, so that $\|X - ZC\|$ is small, *i.e.*, $X \approx ZC$.

10.42 A matrix-vector multiplication Ax of an $n \times n$ matrix A and an n-vector x takes $2n^2$ flops in general. Formulate a faster method, with complexity linear in n, for matrix-vector multiplication with the matrix $A = I + ab^T$, where a and b are given n-vectors.

10.43 A particular computer takes about 0.2 seconds to multiply two 1500×1500 matrices. About how long would you guess the computer takes to multiply two 3000×3000 matrices? Give your prediction (*i.e.*, the time in seconds), and your (very brief) reasoning.

10.44 *Complexity of matrix quadruple product.* (See page 182.) We wish to compute the product $E = ABCD$, where A is $m \times n$, B is $n \times p$, C is $p \times q$, and D is $q \times r$.

(a) Find all methods for computing E using three matrix-matrix multiplications. For example, you can compute AB, CD, and then the product $(AB)(CD)$. Give the total number of flops required for each of these methods. *Hint.* There are four other methods.

(b) Which method requires the fewest flops, with dimensions $m = 10$, $n = 1000$, $p = 10$, $q = 1000$, $r = 100$?

Chapter 11

Matrix inverses

In this chapter we introduce the concept of matrix inverse. We show how matrix inverses can be used to solve linear equations, and how they can be computed using the QR factorization.

11.1 Left and right inverses

Recall that for a number a, its (multiplicative) inverse is the number x for which $xa = 1$, which we usually denote as $x = 1/a$ or (less frequently) $x = a^{-1}$. The inverse x exists provided a is nonzero. For matrices the concept of inverse is more complicated than for scalars; in the general case, we need to distinguish between left and right inverses. We start with the left inverse.

Left inverse. A matrix X that satisfies

$$XA = I$$

is called a *left inverse* of A. The matrix A is said to be *left-invertible* if a left inverse exists. Note that if A has size $m \times n$, a left inverse X will have size $n \times m$, the same dimensions as A^T.

Examples.

- If A is a number (*i.e.*, a 1×1 matrix), then a left inverse X is the same as the inverse of the number. In this case, A is left-invertible whenever A is nonzero, and it has only one left inverse.

- Any nonzero n-vector a, considered as an $n \times 1$ matrix, is left-invertible. For any index i with $a_i \neq 0$, the row n-vector $x = (1/a_i)e_i^T$ satisfies $xa = 1$.

- The matrix

$$A = \begin{bmatrix} -3 & -4 \\ 4 & 6 \\ 1 & 1 \end{bmatrix}$$

has two different left inverses:

$$B = \frac{1}{9} \begin{bmatrix} -11 & -10 & 16 \\ 7 & 8 & -11 \end{bmatrix}, \qquad C = \frac{1}{2} \begin{bmatrix} 0 & -1 & 6 \\ 0 & 1 & -4 \end{bmatrix}.$$

This can be verified by checking that $BA = CA = I$. The example illustrates that a left-invertible matrix can have more than one left inverse. (In fact, if it has more than one left inverse, then it has infinitely many; see exercise 11.1.)

- A matrix A with orthonormal columns satisfies $A^T A = I$, so it is left-invertible; its transpose A^T is a left inverse.

Left-invertibility and column independence.   If A has a left inverse C then the columns of A are linearly independent. To see this, suppose that $Ax = 0$. Multiplying on the left by a left inverse C, we get

$$0 = C(Ax) = (CA)x = Ix = x,$$

which shows that the only linear combination of the columns of A that is 0 is the one with all coefficients zero.

We will see below that the converse is also true; a matrix has a left inverse if and only if its columns are linearly independent. So the generalization of 'a number has an inverse if and only if it is nonzero' is 'a matrix has a left inverse if and only if its columns are linearly independent'.

Dimensions of left inverses.   Suppose the $m \times n$ matrix A is wide, *i.e.*, $m < n$. By the independence-dimension inequality, its columns are linearly dependent, and therefore it is not left-invertible. Only square or tall matrices can be left-invertible.

Solving linear equations with a left inverse.   Suppose that $Ax = b$, where A is an $m \times n$ matrix and x is an n-vector. If C is a left inverse of A, we have

$$Cb = C(Ax) = (CA)x = Ix = x,$$

which means that $x = Cb$ is a solution of the set of linear equations. The columns of A are linearly independent (since it has a left inverse), so there is only one solution of the linear equations $Ax = b$; in other words, $x = Cb$ is *the* solution of $Ax = b$.

Now suppose there is no x that satisfies the linear equations $Ax = b$, and let C be a left inverse of A. Then $x = Cb$ does not satisfy $Ax = b$, since no vector satisfies this equation by assumption. This gives a way to check if the linear equations $Ax = b$ have a solution, and to find one when there is one, provided we have a left inverse of A. We simply test whether $A(Cb) = b$. If this holds, then we have found a solution of the linear equations; if it does not, then we can conclude that there is no solution of $Ax = b$.

In summary, a left inverse can be used to determine whether or not a solution of an over-determined set of linear equations exists, and when it does, find the unique solution.

Right inverse. Now we turn to the closely related concept of right inverse. A matrix X that satisfies

$$AX = I$$

is called a *right inverse* of A. The matrix A is *right-invertible* if a right inverse exists. Any right inverse has the same dimensions as A^T.

Left and right inverse of matrix transpose. If A has a right inverse B, then B^T is a left inverse of A^T, since $B^T A^T = (AB)^T = I$. If A has a left inverse C, then C^T is a right inverse of A^T, since $A^T C^T = (CA)^T = I$. This observation allows us to map all the results for left-invertibility given above to similar results for right-invertibility. Some examples are given below.

- A matrix is right-invertible if and only if its rows are linearly independent.

- A tall matrix cannot have a right inverse. Only square or wide matrices can be right-invertible.

Solving linear equations with a right inverse. Consider the set of m linear equations in n variables $Ax = b$. Suppose A is right-invertible, with right inverse B. This implies that A is square or wide, so the linear equations $Ax = b$ are square or under-determined.

Then for *any* m-vector b, the n-vector $x = Bb$ satisfies the equation $Ax = b$. To see this, we note that

$$Ax = A(Bb) = (AB)b = Ib = b.$$

We can conclude that if A is right-invertible, then the linear equations $Ax = b$ can be solved for *any* vector b. Indeed, $x = Bb$ is a solution. (There can be other solutions of $Ax = b$; the solution $x = Bb$ is simply one of them.)

In summary, a right inverse can be used to find *a* solution of a square or under-determined set of linear equations, for any vector b.

Examples. Consider the matrix appearing in the example above on page 199,

$$A = \begin{bmatrix} -3 & -4 \\ 4 & 6 \\ 1 & 1 \end{bmatrix}$$

and the two left inverses

$$B = \frac{1}{9} \begin{bmatrix} -11 & -10 & 16 \\ 7 & 8 & -11 \end{bmatrix}, \qquad C = \frac{1}{2} \begin{bmatrix} 0 & -1 & 6 \\ 0 & 1 & -4 \end{bmatrix}.$$

- The over-determined linear equations $Ax = (1, -2, 0)$ have the unique solution $x = (1, -1)$, which can be obtained from *either* left inverse:

$$x = B(1, -2, 0) = C(1, -2, 0).$$

- The over-determined linear equations $Ax = (1, -1, 0)$ do not have a solution, since $x = C(1, -1, 0) = (1/2, -1/2)$ does not satisfy $Ax = (1, -1, 0)$.

- The under-determined linear equations $A^T y = (1, 2)$ has (different) solutions

$$B^T(1, 2) = (1/3, 2/3, 38/9), \qquad C^T(1, 2) = (0, 1/2, -1).$$

(Recall that B^T and C^T are both right inverses of A^T.) We can find a solution of $A^T y = b$ for any vector b.

Left and right inverse of matrix product. Suppose A and D are compatible for the matrix product AD (*i.e.*, the number of columns in A is equal to the number of rows in D.) If A has a right inverse B and D has a right inverse E, then EB is a right inverse of AD. This follows from

$$(AD)(EB) = A(DE)B = A(IB) = AB = I.$$

If A has a left inverse C and D has a left inverse F, then FC is a left inverse of AD. This follows from

$$(FC)(AD) = F(CA)D = FD = I.$$

11.2 Inverse

If a matrix is left- *and* right-invertible, then the left and right inverses are unique and equal. To see this, suppose that $AX = I$ and $YA = I$, *i.e.*, X is any right inverse and Y is any left inverse of A. Then we have

$$X = (YA)X = Y(AX) = Y,$$

i.e., any left inverse of A is equal to any right inverse of A. This implies that the left inverse is unique: If we have $A\tilde{X} = I$, then the argument above tells us that $\tilde{X} = Y$, so we have $\tilde{X} = X$, *i.e.*, there is only one right inverse of A. A similar argument shows that Y (which is the same as X) is the only left inverse of A.

When a matrix A has both a left inverse Y and a right inverse X, we call the matrix $X = Y$ simply the *inverse* of A, and denote it as A^{-1}. We say that A is *invertible* or *nonsingular*. A square matrix that is not invertible is called *singular*.

Dimensions of invertible matrices. Invertible matrices must be square, since tall matrices are not right-invertible, while wide matrices are not left-invertible. A matrix A and its inverse (if it exists) satisfy

$$AA^{-1} = A^{-1}A = I.$$

If A has inverse A^{-1}, then the inverse of A^{-1} is A; in other words, we have $(A^{-1})^{-1} = A$. For this reason we say that A and A^{-1} are inverses (of each other).

Solving linear equations with the inverse.   Consider the square system of n linear equations with n variables, $Ax = b$. If A is invertible, then for any n-vector b,

$$x = A^{-1}b \qquad (11.1)$$

is a solution of the equations. (This follows since A^{-1} is a right inverse of A.) Moreover, it is the *only* solution of $Ax = b$. (This follows since A^{-1} is a left inverse of A.) We summarize this very important result as

> *The square system of linear equations $Ax = b$, with A invertible, has the unique solution $x = A^{-1}b$, for any n-vector b.*

One immediate conclusion we can draw from the formula (11.1) is that the solution of a square set of linear equations is a linear function of the right-hand side vector b.

Invertibility conditions.   For square matrices, left-invertibility, right-invertibility, and invertibility are equivalent: If a matrix is square and left-invertible, then it is also right-invertible (and therefore invertible) and vice-versa.

To see this, suppose A is an $n \times n$ matrix and left-invertible. This implies that the n columns of A are linearly independent. Therefore they form a basis and so any n-vector can be expressed as a linear combination of the columns of A. In particular, each of the n unit vectors e_i can be expressed as $e_i = Ab_i$ for some n-vector b_i. The matrix $B = \begin{bmatrix} b_1 & b_2 & \cdots & b_n \end{bmatrix}$ satisfies

$$AB = \begin{bmatrix} Ab_1 & Ab_2 & \cdots & Ab_n \end{bmatrix} = \begin{bmatrix} e_1 & e_2 & \cdots & e_n \end{bmatrix} = I.$$

So B is a right inverse of A.

We have just shown that for a square matrix A,

$$\text{left-invertibility} \quad \Longrightarrow \quad \text{column independence} \quad \Longrightarrow \quad \text{right-invertibility.}$$

(The symbol $\Longrightarrow$ means that the left-hand condition implies the right-hand condition.) Applying the same result to the transpose of A allows us to also conclude that

$$\text{right-invertibility} \quad \Longrightarrow \quad \text{row independence} \quad \Longrightarrow \quad \text{left-invertibility.}$$

So all six of these conditions are equivalent; if any one of them holds, so do the other five.

In summary, for a square matrix A, the following are equivalent.

- A is invertible.

- The columns of A are linearly independent.

- The rows of A are linearly independent.

- A has a left inverse.

- A has a right inverse.

Examples.

- The identity matrix I is invertible, with inverse $I^{-1} = I$, since $II = I$.

- A diagonal matrix A is invertible if and only if its diagonal entries are nonzero. The inverse of an $n \times n$ diagonal matrix A with nonzero diagonal entries is

$$A^{-1} = \begin{bmatrix} 1/A_{11} & 0 & \cdots & 0 \\ 0 & 1/A_{22} & \cdots & 0 \\ \vdots & \vdots & \ddots & \vdots \\ 0 & 0 & \cdots & 1/A_{nn} \end{bmatrix},$$

since

$$AA^{-1} = \begin{bmatrix} A_{11}/A_{11} & 0 & \cdots & 0 \\ 0 & A_{22}/A_{22} & \cdots & 0 \\ \vdots & \vdots & \ddots & \vdots \\ 0 & 0 & \cdots & A_{nn}/A_{nn} \end{bmatrix} = I.$$

In compact notation, we have

$$\mathbf{diag}(A_{11}, \ldots, A_{nn})^{-1} = \mathbf{diag}(A_{11}^{-1}, \ldots, A_{nn}^{-1}).$$

Note that the inverse on the left-hand side of this equation is the matrix inverse, while the inverses appearing on the right-hand side are scalar inverses.

- As a non-obvious example, the matrix

$$A = \begin{bmatrix} 1 & -2 & 3 \\ 0 & 2 & 2 \\ -3 & -4 & -4 \end{bmatrix}$$

is invertible, with inverse

$$A^{-1} = \frac{1}{30} \begin{bmatrix} 0 & -20 & -10 \\ -6 & 5 & -2 \\ 6 & 10 & 2 \end{bmatrix}.$$

This can be verified by checking that $AA^{-1} = I$ (or that $A^{-1}A = I$, since either of these implies the other).

- 2×2 *matrices.* A 2×2 matrix A is invertible if and only if $A_{11}A_{22} \neq A_{12}A_{21}$, with inverse

$$A^{-1} = \begin{bmatrix} A_{11} & A_{12} \\ A_{21} & A_{22} \end{bmatrix}^{-1} = \frac{1}{A_{11}A_{22} - A_{12}A_{21}} \begin{bmatrix} A_{22} & -A_{12} \\ -A_{21} & A_{11} \end{bmatrix}.$$

(There are similar formulas for the inverse of a matrix of any size, but they grow very quickly in complexity and so are not very useful in most applications.)

- *Orthogonal matrix.* If A is square with orthonormal columns, we have $A^T A = I$, so A is invertible with inverse $A^{-1} = A^T$.

Inverse of matrix transpose.   If A is invertible, its transpose A^T is also invertible and its inverse is $(A^{-1})^T$:

$$(A^T)^{-1} = (A^{-1})^T.$$

Since the order of the transpose and inverse operations does not matter, this matrix is sometimes written as A^{-T}.

Inverse of matrix product.   If A and B are invertible (hence, square) and of the same size, then AB is invertible, and

$$(AB)^{-1} = B^{-1}A^{-1}. \tag{11.2}$$

The inverse of a product is the product of the inverses, in reverse order.

Dual basis.   Suppose that A is invertible with inverse $B = A^{-1}$. Let $a_1, \ldots, a_n$ be the columns of A, and $b_1^T, \ldots, b_n^T$ denote the *rows* of B, i.e., the columns of B^T:

$$A = \begin{bmatrix} a_1 & \cdots & a_n \end{bmatrix}, \qquad B = \begin{bmatrix} b_1^T \\ \vdots \\ b_n^T \end{bmatrix}.$$

We know that $a_1, \ldots, a_n$ form a basis, since the columns of A are linearly independent. The vectors $b_1, \ldots, b_n$ also form a basis, since the rows of B are linearly independent. They are called the *dual basis* of $a_1, \ldots, a_n$. (The dual basis of $b_1, \ldots, b_n$ is $a_1, \ldots, a_n$, so they called *dual bases*.)

Now suppose that x is any n-vector. It can be expressed as a linear combination of the basis vectors $a_1, \ldots, a_n$:

$$x = \beta_1 a_1 + \cdots + \beta_n a_n.$$

The dual basis gives us a simple way to find the coefficients $\beta_1, \ldots, \beta_n$.

We start with $AB = I$, and multiply by x to get

$$x = ABx = \begin{bmatrix} a_1 & \cdots & a_n \end{bmatrix} \begin{bmatrix} b_1^T \\ \vdots \\ b_n^T \end{bmatrix} x = (b_1^T x)a_1 + \cdots + (b_n^T x)a_n.$$

This means (since the vectors $a_1, \ldots, a_n$ are linearly independent) that $\beta_i = b_i^T x$. In words: The coefficients in the expansion of a vector in a basis are given by the inner products with the dual basis vectors. Using matrix notation, we can say that $\beta = B^T x = (A^{-1})^T x$ is the vector of coefficients of x in the basis given by the columns of A.

As a simple numerical example, consider the basis

$$a_1 = (1, 1), \qquad a_2 = (1, -1).$$

The dual basis consists of the rows of $\begin{bmatrix} a_1 & a_2 \end{bmatrix}^{-1}$, which are

$$b_1^T = \begin{bmatrix} 1/2 & 1/2 \end{bmatrix}, \qquad b_2^T = \begin{bmatrix} 1/2 & -1/2 \end{bmatrix}.$$

To express the vector $x = (-5, 1)$ as a linear combination of a_1 and a_2, we have

$$x = (b_1^T x)a_1 + (b_2^T x)a_2 = (-2)a_1 + (-3)a_2,$$

which can be directly verified.

Negative matrix powers. We can now give a meaning to matrix powers with negative integer exponents. Suppose A is a square invertible matrix and k is a positive integer. Then by repeatedly applying property (11.2), we get

$$(A^k)^{-1} = (A^{-1})^k.$$

We denote this matrix as A^{-k}. For example, if A is square and invertible, then $A^{-2} = A^{-1}A^{-1} = (AA)^{-1}$. With A^0 defined as $A^0 = I$, the identity $A^{k+l} = A^k A^l$ holds for all integers k and l.

Triangular matrix. A triangular matrix with nonzero diagonal elements is invertible. We first discuss this for a lower triangular matrix. Let L be $n \times n$ and lower triangular with nonzero diagonal elements. We show that the columns are linearly independent, *i.e.*, $Lx = 0$ is only possible if $x = 0$. Expanding the matrix-vector product, we can write $Lx = 0$ as

$$
\begin{aligned}
L_{11}x_1 &= 0 \\
L_{21}x_1 + L_{22}x_2 &= 0 \\
L_{31}x_1 + L_{32}x_2 + L_{33}x_3 &= 0 \\
&\vdots \\
L_{n1}x_1 + L_{n2}x_2 + \cdots + L_{n,n-1}x_{n-1} + L_{nn}x_n &= 0.
\end{aligned}
$$

Since $L_{11} \neq 0$, the first equation implies $x_1 = 0$. Using $x_1 = 0$, the second equation reduces to $L_{22}x_2 = 0$. Since $L_{22} \neq 0$, we conclude that $x_2 = 0$. Using $x_1 = x_2 = 0$, the third equation now reduces to $L_{33}x_3 = 0$, and since L_{33} is assumed to be nonzero, we have $x_3 = 0$. Continuing this argument, we find that all entries of x are zero, and this shows that the columns of L are linearly independent. It follows that L is invertible.

A similar argument can be followed to show that an upper triangular matrix with nonzero diagonal elements is invertible. One can also simply note that if R is upper triangular, then $L = R^T$ is lower triangular with the same diagonal, and use the formula $(L^T)^{-1} = (L^{-1})^T$ for the inverse of the transpose.

Inverse via QR factorization. The QR factorization gives a simple expression for the inverse of an invertible matrix. If A is square and invertible, its columns are linearly independent, so it has a QR factorization $A = QR$. The matrix Q is orthogonal and R is upper triangular with positive diagonal entries. Hence Q and R are invertible, and the formula for the inverse product gives

$$A^{-1} = (QR)^{-1} = R^{-1}Q^{-1} = R^{-1}Q^T. \tag{11.3}$$

In the following section we give an algorithm for computing R^{-1}, or more directly, the product $R^{-1}Q^T$. This gives us a method to compute the matrix inverse.

11.3 Solving linear equations

Back substitution. We start with an algorithm for solving a set of linear equations, $Rx = b$, where the $n \times n$ matrix R is upper triangular with nonzero diagonal entries (hence, invertible). We write out the equations as

$$R_{11}x_1 + R_{12}x_2 + \cdots + R_{1,n-1}x_{n-1} + R_{1n}x_n = b_1$$
$$\vdots$$
$$R_{n-2,n-2}x_{n-2} + R_{n-2,n-1}x_{n-1} + R_{n-2,n}x_n = b_{n-2}$$
$$R_{n-1,n-1}x_{n-1} + R_{n-1,n}x_n = b_{n-1}$$
$$R_{nn}x_n = b_n.$$

From the last equation, we find that $x_n = b_n/R_{nn}$. Now that we know x_n, we substitute it into the second to last equation, which gives us

$$x_{n-1} = (b_{n-1} - R_{n-1,n}x_n)/R_{n-1,n-1}.$$

We can continue this way to find $x_{n-2}, x_{n-3}, \ldots, x_1$. This algorithm is known as *back substitution*, since the variables are found one at a time, starting from x_n, and we substitute the ones that are known into the remaining equations.

Algorithm 11.1 BACK SUBSTITUTION

given an $n \times n$ upper triangular matrix R with nonzero diagonal entries, and an n-vector b.

For $i = n, \ldots, 1$,
$$x_i = (b_i - R_{i,i+1}x_{i+1} - \cdots - R_{i,n}x_n)/R_{ii}.$$

(In the first step, with $i = n$, we have $x_n = b_n/R_{nn}$.) The back substitution algorithm computes the solution of $Rx = b$, i.e., $x = R^{-1}b$. It cannot fail since the divisions in each step are by the diagonal entries of R, which are assumed to be nonzero.

Lower triangular matrices with nonzero diagonal elements are also invertible; we can solve equations with lower triangular invertible matrices using *forward substitution*, the obvious analog of the algorithm given above. In forward substitution, we find x_1 first, then x_2, and so on.

Complexity of back substitution. The first step requires 1 flop (division by R_{nn}). The next step requires one multiply, one subtraction, and one division, for a total of 3 flops. The kth step requires $k - 1$ multiplies, $k - 1$ subtractions, and one division, for a total of $2k - 1$ flops. The total number of flops for back substitution is then

$$1 + 3 + 5 + \cdots + (2n - 1) = n^2$$

flops.

This formula can be obtained from the formula (5.7), or directly derived using a similar argument. Here is the argument for the case when n is even; a similar argument works when n is odd. Lump the first entry in the sum together with the last entry, the second entry together with the second-to-last entry, and so on. Each of these pairs add up to $2n$; since there are $n/2$ such pairs, the total is $(n/2)(2n) = n^2$.

Solving linear equations using the QR factorization. The formula (11.3) for the inverse of a matrix in terms of its QR factorization suggests a method for solving a square system of linear equations $Ax = b$ with A invertible. The solution

$$x = A^{-1}b = R^{-1}Q^T b \tag{11.4}$$

can be found by first computing the matrix-vector product $y = Q^T b$, and then solving the triangular equation $Rx = y$ by back substitution.

Algorithm 11.2 SOLVING LINEAR EQUATIONS VIA QR FACTORIZATION

given an $n \times n$ invertible matrix A and an n-vector b.

1. *QR factorization.* Compute the QR factorization $A = QR$.

2. Compute $Q^T b$.

3. *Back substitution.* Solve the triangular equation $Rx = Q^T b$ using back substitution.

The first step requires $2n^3$ flops (see §5.4), the second step requires $2n^2$ flops, and the third step requires n^2 flops. The total number of flops is then

$$2n^3 + 3n^2 \approx 2n^3,$$

so the order is n^3, cubic in the number of variables, which is the same as the number of equations.

In the complexity analysis above, we found that the first step, the QR factorization, dominates the other two; that is, the cost of the other two is negligible in comparison to the cost of the first step. This has some interesting practical implications, which we discuss below.

Factor-solve methods. Algorithm 11.2 is similar to many methods for solving a set of linear equations and is sometimes referred to as a *factor-solve* scheme. A factor-solve scheme consists of two steps. In the first (factor) step the coefficient matrix is factored as a product of matrices with special properties. In the second (solve) step one or more linear equations that involve the factors in the factorization are solved. (In algorithm 11.2, the solve step consists of steps 2 and 3.) The complexity of the solve step is smaller than the complexity of the factor step, and in many cases, it is negligible by comparison. This is the case in algorithm 11.2, where the factor step has order n^3 and the solve step has order n^2.

Factor-solve methods with multiple right-hand sides. Now suppose that we must solve several sets of linear equations,

$$Ax_1 = b_1, \quad \dots, \quad Ax_k = b_k,$$

all with the same coefficient matrix A, but different right-hand sides. We can express this as the matrix equation $AX = B$, where X is the $n \times k$ matrix with columns $x_1, \ldots, x_k$, and B is the $n \times k$ matrix with columns $b_1, \ldots, b_k$ (see page 180). Assuming A is invertible, the solution of $AX = B$ is $X = A^{-1}B$.

A naïve way to solve the k problems $Ax_i = b_i$ (or in matrix notation, compute $X = A^{-1}B$) is to apply algorithm 11.2 k times, which costs $2kn^3$ flops. A more efficient method exploits the fact that A is the same matrix in each problem, so we can re-use the matrix factorization in step 1 and only need to repeat steps 2 and 3 to compute $\hat{x}_k = R^{-1}Q^T b_k$ for $l = 1, \ldots, k$. (This is sometimes referred to as *factorization caching*, since we save or cache the factorization after carrying it out, for later use.) The cost of this method is $2n^3 + 3kn^2$ flops, or approximately $2n^3$ flops if $k \ll n$. The (surprising) conclusion is that we can solve *multiple* sets of linear equations, with the same coefficient matrix A, at essentially the same cost as solving *one* set of linear equations.

Backslash notation.   In several software packages for manipulating matrices, $A \backslash b$ is taken to mean the solution of $Ax = b$, *i.e.*, $A^{-1}b$, when A is invertible. This *backslash* notation is extended to matrix right-hand sides: $A \backslash B$, with B an $n \times k$ matrix, denotes $A^{-1}B$, the solution of the matrix equation $AX = B$. (The computation is implemented as described above, by factoring A just once, and carrying out k back substitutions.) This backslash notation is not standard mathematical notation, however, so we will not use it in this book.

Computing the matrix inverse.   We can now describe a method to compute the inverse $B = A^{-1}$ of an (invertible) $n \times n$ matrix A. We first compute the QR factorization of A, so $A^{-1} = R^{-1}Q^T$. We can write this as $RB = Q^T$, which, written out by columns is

$$Rb_i = \tilde{q}_i, \quad i = 1, \ldots, n,$$

where b_i is the ith column of B and $\tilde{q}_i$ is the ith column of Q^T. We can solve these equations using back substitution, to get the columns of the inverse B.

Algorithm 11.3 COMPUTING THE INVERSE VIA QR FACTORIZATION

given an $n \times n$ invertible matrix A.

1. *QR factorization.* Compute the QR factorization $A = QR$.

2. For $i = 1, \ldots, n$,
   Solve the triangular equation $Rb_i = \tilde{q}_i$ using back substitution.

The complexity of this method is $2n^3$ flops (for the QR factorization) and n^3 for n back substitutions, each of which costs n^2 flops. So we can compute the matrix inverse in around $3n^3$ flops.

This gives an alternative method for solving the square set of linear equations $Ax = b$: We first compute the inverse matrix A^{-1}, and then the matrix-vector product $x = (A^{-1})b$. This method has a higher flop count than directly solving the equations using algorithm 11.2 ($3n^3$ versus $2n^3$), so algorithm 11.2 is the usual method of choice. While the matrix inverse appears in many formulas (such as the solution of a set of linear equations), it is *computed* far less often.

Sparse linear equations. Systems of linear equations with sparse coefficient matrix arise in many applications. By exploiting the sparsity of the coefficient matrix, these linear equations can be solved far more efficiently than by using the generic algorithm 11.2. One method is to use the same basic algorithm 11.2, replacing the QR factorization with a variant that handles sparse matrices (see page 190). The memory usage and complexity of these methods depends in a complicated way on the sparsity pattern of the coefficient matrix. In order, the memory usage is typically a modest multiple of $\mathbf{nnz}(A) + n$, the number of scalars required to specify the problem data A and b, which is typically much smaller than $n^2 + n$, the number of scalars required to store A and b if they are not sparse. The flop count for solving sparse linear equations is also typically closer in order to $\mathbf{nnz}(A)$ than n^3, the order when the matrix A is not sparse.

11.4   Examples

Polynomial interpolation. The 4-vector c gives the coefficients of a cubic polynomial,

$$p(x) = c_1 + c_2 x + c_3 x^2 + c_4 x^3$$

(see pages 154 and 120). We seek the coefficients that satisfy

$$p(-1.1) = b_1, \qquad p(-0.4) = b_2, \qquad p(0.2) = b_3, \qquad p(0.8) = b_4.$$

We can express this as the system of 4 equations in 4 variables $Ac = b$, where

$$A = \begin{bmatrix} 1 & -1.1 & (-1.1)^2 & (-1.1)^3 \\ 1 & -0.4 & (-0.4)^2 & (-0.4)^3 \\ 1 & 0.2 & (0.2)^2 & (0.2)^3 \\ 1 & 0.8 & (0.8)^2 & (0.8)^3 \end{bmatrix},$$

which is a specific Vandermonde matrix (see (6.7)). The unique solution is $c = A^{-1}b$, where

$$A^{-1} = \begin{bmatrix} -0.5784 & 1.9841 & -2.1368 & 0.7310 \\ 0.3470 & 0.1984 & -1.4957 & 0.9503 \\ 0.1388 & -1.8651 & 1.6239 & 0.1023 \\ -0.0370 & 0.3492 & 0.7521 & -0.0643 \end{bmatrix}$$

(to 4 decimal places). This is illustrated in figure 11.1, which shows the two cubic polynomials that interpolate the two sets of points shown as filled circles and squares, respectively.

The columns of A^{-1} are interesting: They give the coefficients of a polynomial that evaluates to 0 at three of the points, and 1 at the other point. For example, the

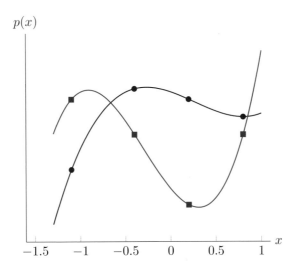

Figure 11.1 Cubic interpolants through two sets of points, shown as circles and squares.

first column of A^{-1}, which is $A^{-1}e_1$, gives the coefficients of the polynomial that has value 1 at -1.1, and value 0 at -0.4, 0.2, and 0.8. The four polynomials with coefficients given by the columns of A^{-1} are called the *Lagrange polynomials* associated with the points -1.1, -0.4, 0.2, 0.8. These are plotted in figure 11.2. (The Lagrange polynomials are named after the mathematician Joseph-Louis Lagrange, whose name will re-appear in several other contexts.)

The rows of A^{-1} are also interesting: The ith row shows how the values b_1, ..., b_4, the polynomial values at the points -1.1, -0.4, 0.2, 0.8, map into the ith coefficient of the polynomial, c_i. For example, we see that the coefficient c_4 is not very sensitive to the value of b_1 (since $(A^{-1})_{41}$ is small). We can also see that for each increase of one in b_4, the coefficient c_2 increases by around 0.95.

Balancing chemical reactions.   (See page 154 for background.) We consider the problem of balancing the chemical reaction

$$a_1 Cr_2 O_7^{2-} + a_2 Fe^{2+} + a_3 H^+ \longrightarrow b_1 Cr^{3+} + b_2 Fe^{3+} + b_3 H_2O,$$

where the superscript gives the charge of each reactant and product. There are 4 atoms (Cr, O, Fe, H) and charge to balance. The reactant and product matrices are (using the order just listed)

$$R = \begin{bmatrix} 2 & 0 & 0 \\ 7 & 0 & 0 \\ 0 & 1 & 0 \\ 0 & 0 & 1 \\ -2 & 2 & 1 \end{bmatrix}, \qquad P = \begin{bmatrix} 1 & 0 & 0 \\ 0 & 0 & 1 \\ 0 & 1 & 0 \\ 0 & 0 & 2 \\ 3 & 3 & 0 \end{bmatrix}.$$

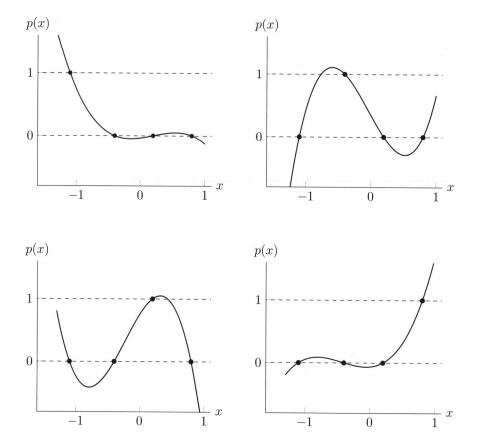

Figure 11.2 Lagrange polynomials associated with the points -1.1, -0.4, 0.2, 0.8.

Imposing the condition that $a_1 = 1$ we obtain a square set of 6 linear equations,

$$
\begin{bmatrix}
2 & 0 & 0 & -1 & 0 & 0 \\
7 & 0 & 0 & 0 & 0 & -1 \\
0 & 1 & 0 & 0 & -1 & 0 \\
0 & 0 & 1 & 0 & 0 & -2 \\
-2 & 2 & 1 & -3 & -3 & 0 \\
1 & 0 & 0 & 0 & 0 & 0
\end{bmatrix}
\begin{bmatrix}
a_1 \\ a_2 \\ a_3 \\ b_1 \\ b_2 \\ b_3
\end{bmatrix}
=
\begin{bmatrix}
0 \\ 0 \\ 0 \\ 0 \\ 0 \\ 1
\end{bmatrix}.
$$

Solving these equations we obtain

$$
a_1 = 1, \quad a_2 = 6, \quad a_3 = 14, \quad b_1 = 2, \quad b_2 = 6, \quad b_3 = 7.
$$

(Setting $a_1 = 1$ could have yielded fractional values for the other coefficients, but in this case, it did not.) The balanced reaction is

$$
\mathrm{Cr_2O_7^{2-}} + 6\mathrm{Fe^{2+}} + 14\mathrm{H^+} \longrightarrow 2\mathrm{Cr^{3+}} + 6\mathrm{Fe^{3+}} + 7\mathrm{H_2O}.
$$

Heat diffusion.   We consider a diffusion system as described on page 155. Some of the nodes have fixed potential, *i.e.*, e_i is given; for the other nodes, the associated external source s_i is zero. This would model a thermal system in which some nodes are in contact with the outside world or a heat source, which maintains their temperatures (via external heat flows) at constant values; the other nodes are internal, and have no heat sources. This gives us a set of n additional equations:

$$
e_i = e_i^{\text{fix}}, \quad i \in \mathcal{P}, \qquad s_i = 0, \quad i \notin \mathcal{P},
$$

where $\mathcal{P}$ is the set of indices of nodes with fixed potential. We can write these n equations in matrix-vector form as

$$
Bs + Ce = d,
$$

where B and C are the $n \times n$ diagonal matrices, and d is the n-vector given by

$$
B_{ii} = \begin{cases} 0 & i \in \mathcal{P} \\ 1 & i \notin \mathcal{P}, \end{cases} \qquad
C_{ii} = \begin{cases} 1 & i \in \mathcal{P} \\ 0 & i \notin \mathcal{P}, \end{cases} \qquad
d_i = \begin{cases} e_i^{\text{fix}} & i \in \mathcal{P} \\ 0 & i \notin \mathcal{P}. \end{cases}
$$

We assemble the flow conservation, edge flow, and the boundary conditions into one set of $m + 2n$ equations in $m + 2n$ variables (f, s, e):

$$
\begin{bmatrix}
A & I & 0 \\
R & 0 & A^T \\
0 & B & C
\end{bmatrix}
\begin{bmatrix}
f \\ s \\ e
\end{bmatrix}
=
\begin{bmatrix}
0 \\ 0 \\ d
\end{bmatrix}.
$$

(The matrix A is the incidence matrix of the graph, and R is the resistance matrix; see page 155.) Assuming the coefficient matrix is invertible, we have

$$
\begin{bmatrix}
f \\ s \\ e
\end{bmatrix}
=
\begin{bmatrix}
A & I & 0 \\
R & 0 & A^T \\
0 & B & C
\end{bmatrix}^{-1}
\begin{bmatrix}
0 \\ 0 \\ d
\end{bmatrix}.
$$

This is illustrated with an example in figure 11.3. The graph is a 100×100 grid, with 10000 nodes, and edges connecting each node to its horizontal and vertical neighbors. The resistance on each edge is the same. The nodes at the top and bottom are held at zero temperature, and the three sets of nodes with rectilinear shapes are held at temperature one. All other nodes have zero source value.

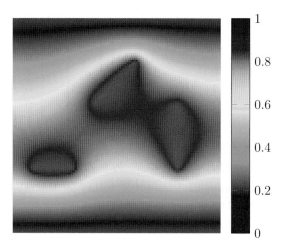

Figure 11.3 Temperature distribution on a 100×100 grid of nodes. Nodes in the top and bottom rows are held at zero temperature. The three sets of nodes with rectilinear shapes are held at temperature one.

11.5   Pseudo-inverse

Linearly independent columns and Gram invertibility.   We first show that an $m \times n$ matrix A has linearly independent columns if and only if its $n \times n$ Gram matrix $A^T A$ is invertible.

First suppose that the columns of A are linearly independent. Let x be an n-vector which satisfies $(A^T A)x = 0$. Multiplying on the left by x^T we get

$$0 = x^T 0 = x^T (A^T A x) = x^T A^T A x = \|Ax\|^2,$$

which implies that $Ax = 0$. Since the columns of A are linearly independent, we conclude that $x = 0$. Since the only solution of $(A^T A)x = 0$ is $x = 0$, we conclude that $A^T A$ is invertible.

Now let's show the converse. Suppose the columns of A are linearly dependent, which means there is a nonzero n-vector x which satisfies $Ax = 0$. Multiply on the left by A^T to get $(A^T A)x = 0$. This shows that the Gram matrix $A^T A$ is singular.

Pseudo-inverse of square or tall matrix.   We show here that if A has linearly independent columns (and therefore, is square or tall) then it has a left inverse. (We already have observed the converse, that a matrix with a left inverse has linearly independent columns.) Assuming A has linearly independent columns, we know that $A^T A$ is invertible. We now observe that the matrix $(A^T A)^{-1} A^T$ is a left inverse of A:

$$\left((A^T A)^{-1} A^T \right) A = (A^T A)^{-1} (A^T A) = I.$$

This particular left-inverse of A will come up in the sequel, and has a name, the *pseudo-inverse* of A. It is denoted $A^\dagger$ (or A^+):

$$A^\dagger = (A^T A)^{-1} A^T. \tag{11.5}$$

The pseudo-inverse is also called the *Moore–Penrose inverse*, after the mathematicians Eliakim Moore and Roger Penrose.

When A is square, the pseudo-inverse $A^\dagger$ reduces to the ordinary inverse:

$$A^\dagger = (A^T A)^{-1} A^T = A^{-1} A^{-T} A^T = A^{-1} I = A^{-1}.$$

Note that this equation does not make sense (and certainly is not correct) when A is not square.

Pseudo-inverse of a square or wide matrix. Transposing all the equations, we can show that a (square or wide) matrix A has a right inverse if and only if its rows are linearly independent. Indeed, one right inverse is given by

$$A^T (A A^T)^{-1}. \tag{11.6}$$

(The matrix $A A^T$ is invertible if and only if the rows of A are linearly independent.)

The matrix in (11.6) is also referred to as the pseudo-inverse of A, and denoted $A^\dagger$. The only possible confusion in defining the pseudo-inverse using the two different formulas (11.5) and (11.6) occurs when the matrix A is square. In this case, however, they both reduce to the ordinary inverse:

$$A^T (A A^T)^{-1} = A^T A^{-T} A^{-1} = A^{-1}.$$

Pseudo-inverse in other cases. The pseudo-inverse $A^\dagger$ is defined for any matrix, including the case when A is tall but its columns are linearly dependent, the case when A is wide but its rows are linearly dependent, and the case when A is square but not invertible. In these cases, however, it is not a left inverse, right inverse, or inverse, respectively. We mention it here since the reader may encounter it. (We will see what $A^\dagger$ means in these cases in exercise 15.11.)

Pseudo-inverse via QR factorization. The QR factorization gives a simple formula for the pseudo-inverse. If A is left-invertible, its columns are linearly independent and the QR factorization $A = QR$ exists. We have

$$A^T A = (QR)^T (QR) = R^T Q^T Q R = R^T R,$$

so

$$A^\dagger = (A^T A)^{-1} A^T = (R^T R)^{-1} (QR)^T = R^{-1} R^{-T} R^T Q^T = R^{-1} Q^T.$$

We can compute the pseudo-inverse using the QR factorization, followed by back substitution on the columns of Q^T. (This is exactly the same as algorithm 11.3 when A is square and invertible.) The complexity of this method is $2n^2 m$ flops (for the QR factorization), and mn^2 flops for the m back substitutions. So the total is $3mn^2$ flops.

Similarly, if A is right-invertible, the QR factorization $A^T = QR$ of its transpose exists. We have $AA^T = (QR)^T(QR) = R^TQ^TQR = R^TR$ and

$$A^\dagger = A^T(AA^T)^{-1} = QR(R^TR)^{-1} = QRR^{-1}R^{-T} = QR^{-T}.$$

We can compute it using the method described above, using the formula

$$(A^T)^\dagger = (A^\dagger)^T.$$

Solving over- and under-determined systems of linear equations.   The pseudo-inverse gives us a method for solving over-determined and under-determined systems of linear equations, provided the columns of the coefficient matrix are linearly independent (in the over-determined case), or the rows are linearly independent (in the under-determined case). If the columns of A are linearly independent, and the over-determined equations $Ax = b$ have a solution, then $x = A^\dagger b$ is it. If the rows of A are linearly independent, the under-determined equations $Ax = b$ have a solution for any vector b, and $x = A^\dagger b$ is a solution.

Numerical example.   We illustrate these ideas with a simple numerical example, using the 3×2 matrix A used in earlier examples on pages 199 and 201,

$$A = \begin{bmatrix} -3 & -4 \\ 4 & 6 \\ 1 & 1 \end{bmatrix}.$$

This matrix has linearly independent columns, and QR factorization with (to 4 digits)

$$Q = \begin{bmatrix} -0.5883 & 0.4576 \\ 0.7845 & 0.5230 \\ 0.1961 & -0.7191 \end{bmatrix}, \qquad R = \begin{bmatrix} 5.0990 & 7.2563 \\ 0 & 0.5883 \end{bmatrix}.$$

It has pseudo-inverse (to 4 digits)

$$A^\dagger = R^{-1}Q^T = \begin{bmatrix} -1.2222 & -1.1111 & 1.7778 \\ 0.7778 & 0.8889 & -1.2222 \end{bmatrix}.$$

We can use the pseudo-inverse to check if the over-determined systems of equations $Ax = b$, with $b = (1, -2, 0)$, has a solution, and to find a solution if it does. We compute $x = A^\dagger(1, -2, 0) = (1, -1)$ and check whether $Ax = b$ holds. It does, so we have found the unique solution of $Ax = b$.

Exercises

11.1 *Affine combinations of left inverses.* Let Z be a tall $m \times n$ matrix with linearly independent columns, and let X and Y be left inverses of Z. Show that for any scalars α and β satisfying $\alpha + \beta = 1$, $\alpha X + \beta Y$ is also a left inverse of Z. It follows that if a matrix has two different left inverses, it has an infinite number of different left inverses.

11.2 *Left and right inverses of a vector.* Suppose that x is a nonzero n-vector with $n > 1$.

(a) Does x have a left inverse?

(b) Does x have a right inverse?

In each case, if the answer is yes, give a left or right inverse; if the answer is no, give a specific nonzero vector and show that it is not left- or right-invertible.

11.3 *Matrix cancellation.* Suppose the scalars a, x, and y satisfy $ax = ay$. When $a \neq 0$ we can conclude that $x = y$; that is, we can cancel the a on the left of the equation. In this exercise we explore the matrix analog of cancellation, specifically, what properties of A are needed to conclude $X = Y$ from $AX = AY$, for matrices A, X, and Y?

(a) Give an example showing that $A \neq 0$ is not enough to conclude that $X = Y$.

(b) Show that if A is left-invertible, we can conclude from $AX = AY$ that $X = Y$.

(c) Show that if A is not left-invertible, there are matrices X and Y with $X \neq Y$, and $AX = AY$.

Remark. Parts (b) and (c) show that you can cancel a matrix on the left when, and only when, the matrix is left-invertible.

11.4 *Transpose of orthogonal matrix.* Let U be an orthogonal $n \times n$ matrix. Show that its transpose U^T is also orthogonal.

11.5 *Inverse of a block matrix.* Consider the $(n+1) \times (n+1)$ matrix

$$A = \begin{bmatrix} I & a \\ a^T & 0 \end{bmatrix},$$

where a is an n-vector.

(a) When is A invertible? Give your answer in terms of a. Justify your answer.

(b) Assuming the condition you found in part (a) holds, give an expression for the inverse matrix A^{-1}.

11.6 *Inverse of a block upper triangular matrix.* Let B and D be invertible matrices of sizes $m \times m$ and $n \times n$, respectively, and let C be any $m \times n$ matrix. Find the inverse of

$$A = \begin{bmatrix} B & C \\ 0 & D \end{bmatrix}$$

in terms of B^{-1}, C, and D^{-1}. (The matrix A is called *block upper triangular*.)

Hints. First get an idea of what the solution should look like by considering the case when B, C, and D are scalars. For the matrix case, your goal is to find matrices W, X, Y, Z (in terms of B^{-1}, C, and D^{-1}) that satisfy

$$A \begin{bmatrix} W & X \\ Y & Z \end{bmatrix} = I.$$

Use block matrix multiplication to express this as a set of four matrix equations that you can then solve. The method you will find is sometimes called *block back substitution*.

11.7 *Inverse of an upper triangular matrix.* Suppose the $n \times n$ matrix R is upper triangular and invertible, *i.e.*, its diagonal entries are all nonzero. Show that R^{-1} is also upper triangular. *Hint.* Use back substitution to solve $Rs_k = e_k$, for $k = 1, \ldots, n$, and argue that $(s_k)_i = 0$ for $i > k$.

11.8 *If a matrix is small, its inverse is large.* If a number a is small, its inverse $1/a$ (assuming $a \neq 0$) is large. In this exercise you will explore a matrix analog of this idea. Suppose the $n \times n$ matrix A is invertible. Show that $\|A^{-1}\| \geq \sqrt{n}/\|A\|$. This implies that if a matrix is small, its inverse is large. *Hint.* You can use the inequality $\|AB\| \leq \|A\|\|B\|$, which holds for any matrices for which the product makes sense. (See exercise 10.12.)

11.9 *Push-through identity.* Suppose A is $m \times n$, B is $n \times m$, and the $m \times m$ matrix $I + AB$ is invertible.

    (a) Show that the $n \times n$ matrix $I + BA$ is invertible. *Hint.* Show that $(I + BA)x = 0$ implies $(I + AB)y = 0$, where $y = Ax$.

    (b) Establish the identity

$$B(I + AB)^{-1} = (I + BA)^{-1}B.$$

This is sometimes called the *push-through identity* since the matrix B appearing on the left 'moves' into the inverse, and 'pushes' the B in the inverse out to the right side. *Hint.* Start with the identity

$$B(I + AB) = (I + BA)B,$$

and multiply on the right by $(I + AB)^{-1}$, and on the left by $(I + BA)^{-1}$.

11.10 *Reverse-time linear dynamical system.* A linear dynamical system has the form

$$x_{t+1} = Ax_t,$$

where x_t in the (n-vector) state in period t, and A is the $n \times n$ dynamics matrix. This formula gives the state in the next period as a function of the current state.
We want to derive a recursion of the form

$$x_{t-1} = A^{\mathrm{rev}}x_t,$$

which gives the previous state as a function of the current state. We call this the *reverse time linear dynamical system.*

    (a) When is this possible? When it is possible, what is A^{rev}?

    (b) For the specific linear dynamical system with dynamics matrix

$$A = \begin{bmatrix} 3 & 2 \\ -1 & 4 \end{bmatrix},$$

    find A^{rev}, or explain why the reverse time linear dynamical system doesn't exist.

11.11 *Interpolation of rational functions.* (Continuation of exercise 8.8.) Find a rational function

$$f(t) = \frac{c_1 + c_2 t + c_3 t^2}{1 + d_1 t + d_2 t^2}$$

that satisfies the following interpolation conditions:

$$f(1) = 2, \qquad f(2) = 5, \qquad f(3) = 9, \qquad f(4) = -1, \qquad f(5) = -4.$$

In exercise 8.8 these conditions were expressed as a set of linear equations in the coefficients c_1, c_2, c_3, d_1 and d_2; here we are asking you to form and (numerically) solve the system of equations. Plot the rational function you find over the range $x = 0$ to $x = 6$. Your plot should include markers at the interpolation points $(1, 2), \ldots, (5, -4)$. (Your rational function graph should pass through these points.)

11.12 *Combinations of invertible matrices.* Suppose the $n \times n$ matrices A and B are both invertible. Determine whether each of the matrices given below is invertible, without any further assumptions about A and B.

(a) $A + B$.

(b) $\begin{bmatrix} A & 0 \\ 0 & B \end{bmatrix}$.

(c) $\begin{bmatrix} A & A+B \\ 0 & B \end{bmatrix}$.

(d) ABA.

11.13 *Another left inverse.* Suppose the $m \times n$ matrix A is tall and has linearly independent columns. One left inverse of A is the pseudo-inverse $A^\dagger$. In this problem we explore another one. Write A as the block matrix

$$A = \begin{bmatrix} A_1 \\ A_2 \end{bmatrix},$$

where A_1 is $n \times n$. We assume that A_1 is invertible (which need not happen in general). Show that the following matrix is a left inverse of A:

$$\tilde{A} = \begin{bmatrix} A_1^{-1} & 0_{n \times (m-n)} \end{bmatrix}.$$

11.14 *Middle inverse.* Suppose A is an $n \times p$ matrix and B is a $q \times n$ matrix. If a $p \times q$ matrix X exists that satisfies $AXB = I$, we call it a *middle inverse* of the pair A, B. (This is not a standard concept.) Note that when A or B is an identity matrix, the middle inverse reduces to the right or left inverse, respectively.

(a) Describe the conditions on A and B under which a middle inverse X exists. Give your answer using only the following four concepts: Linear independence of the rows or columns of A, and linear independence of the rows or columns of B. You must justify your answer.

(b) Give an expression for a middle inverse, assuming the conditions in part (a) hold.

11.15 *Invertibility of population dynamics matrix.* Consider the population dynamics matrix

$$A = \begin{bmatrix} b_1 & b_2 & \cdots & b_{99} & b_{100} \\ 1-d_1 & 0 & \cdots & 0 & 0 \\ 0 & 1-d_2 & \cdots & 0 & 0 \\ \vdots & \vdots & \ddots & \vdots & \vdots \\ 0 & 0 & \cdots & 1-d_{99} & 0 \end{bmatrix},$$

where $b_i \geq 0$ are the birth rates and $0 \leq d_i \leq 1$ are death rates. What are the conditions on b_i and d_i under which A is invertible? (If the matrix is never invertible or always invertible, say so.) Justify your answer.

11.16 *Inverse of running sum matrix.* Find the inverse of the $n \times n$ running sum matrix,

$$S = \begin{bmatrix} 1 & 0 & \cdots & 0 & 0 \\ 1 & 1 & \cdots & 0 & 0 \\ \vdots & \vdots & \ddots & \vdots & \vdots \\ 1 & 1 & \cdots & 1 & 0 \\ 1 & 1 & \cdots & 1 & 1 \end{bmatrix}.$$

Does your answer make sense?

11.17 *A matrix identity.* Suppose A is a square matrix that satisfies $A^k = 0$ for some integer k. (Such a matrix is called *nilpotent*.) A student guesses that $(I - A)^{-1} = I + A + \cdots + A^{k-1}$, based on the infinite series $1/(1 - a) = 1 + a + a^2 + \cdots$, which holds for numbers a that satisfy $|a| < 1$.

Is the student right or wrong? If right, show that her assertion holds with no further assumptions about A. If she is wrong, give a counterexample, *i.e.*, a matrix A that satisfies $A^k = 0$, but $I + A + \cdots + A^{k-1}$ is not the inverse of $I - A$.

11.18 *Tall-wide product.* Suppose A is an $n \times p$ matrix and B is a $p \times n$ matrix, so $C = AB$ makes sense. Explain why C cannot be invertible if A is tall and B is wide, *i.e.*, if $p < n$. *Hint.* First argue that the columns of B must be linearly dependent.

11.19 *Control restricted to one time period.* A linear dynamical system has the form $x_{t+1} = Ax_t + u_t$, where the n-vector x_t is the state and u_t is the input at time t. Our goal is to choose the input sequence $u_1, \ldots, u_{N-1}$ so as to achieve $x_N = x^{\text{des}}$, where x^{des} is a given n-vector, and N is given. The input sequence must satisfy $u_t = 0$ unless $t = K$, where $K < N$ is given. In other words, the input can only act at time $t = K$. Give a formula for u_K that achieves this goal. Your formula can involve A, N, K, x_1, and x^{des}. You can assume that A is invertible. *Hint.* First derive an expression for x_K, then use the dynamics equation to find x_{K+1}. From x_{K+1} you can find x_N.

11.20 *Immigration.* The population dynamics of a country is given by $x_{t+1} = Ax_t + u$, $t = 1, \ldots, T - 1$, where the 100-vector x_t gives the population age distribution in year t, and u gives the immigration age distribution (with negative entries meaning emigration), which we assume is constant (*i.e.*, does not vary with t). You are given A, x_1, and x^{des}, a 100-vector that represents a desired population distribution in year T. We seek a constant level of immigration u that achieves $x_T = x^{\text{des}}$.

Give a matrix formula for u. If your formula only makes sense when some conditions hold (for example invertibility of one or more matrices), say so.

11.21 *Quadrature weights.* Consider a quadrature problem (see exercise 8.12) with $n = 4$, with points $t = (-0.6, -0.2, 0.2, 0.6)$. We require that the quadrature rule be exact for all polynomials of degree up to $d = 3$.

Set this up as a square system of linear equations in the weight vector. Numerically solve this system to get the weights. Compute the true value and the quadrature estimate,

$$\alpha = \int_{-1}^{1} f(x)\, dx, \qquad \hat{\alpha} = w_1 f(-0.6) + w_2 f(-0.2) + w_3 f(0.2) + w_4 f(0.6),$$

for the specific function $f(x) = e^x$.

11.22 *Properties of pseudo-inverses.* For an $m \times n$ matrix A and its pseudo-inverse $A^\dagger$, show that $A = AA^\dagger A$ and $A^\dagger = A^\dagger AA^\dagger$ in each of the following cases.

 (a) A is tall with linearly independent columns.

 (b) A is wide with linearly independent rows.

 (c) A is square and invertible.

11.23 *Product of pseudo-inverses.* Suppose A and D are right-invertible matrices and the product AD exists. We have seen that if B is a right inverse of A and E is a right inverse of D, then EB is a right inverse of AD. Now suppose B is the pseudo-inverse of A and E is the pseudo-inverse of D. Is EB the pseudo-inverse of AD? Prove that this is always true or give an example for which it is false.

11.24 *Simultaneous left inverse.* The two matrices

$$A = \begin{bmatrix} 1 & 2 \\ 3 & 1 \\ 2 & 1 \\ 2 & 2 \end{bmatrix}, \qquad B = \begin{bmatrix} 3 & 2 \\ 1 & 0 \\ 2 & 1 \\ 1 & 3 \end{bmatrix}$$

and both left-invertible, and have multiple left inverses. Do they have a common left inverse? Explain how to find a 2×4 matrix C that satisfies $CA = CB = I$, or determine that no such matrix exists. (You can use numerical computing to find C.) *Hint.* Set up a set of linear equations for the entries of C. *Remark.* There is nothing special about the particular entries of the two matrices A and B.

11.25 *Checking the computed solution of linear equations.* One of your colleagues says that whenever you compute the solution x of a square set of n equations $Ax = b$ (say, using QR factorization), you should compute the number $\|Ax - b\|$ and check that it is small. (It is not exactly zero due to the small rounding errors made in floating point computations.) Another colleague says that this would be nice to do, but the additional cost of computing $\|Ax - b\|$ is too high. Briefly comment on your colleagues' advice. Who is right?

11.26 *Sensitivity of solution of linear equations.* Let A be an invertible $n \times n$ matrix, and b and x be n-vectors satisfying $Ax = b$. Suppose we perturb the jth entry of b by $\epsilon \neq 0$ (which is a traditional symbol for a small quantity), so b becomes $\tilde{b} = b + \epsilon e_j$. Let $\tilde{x}$ be the n-vector that satisfies $A\tilde{x} = \tilde{b}$, *i.e.*, the solution of the linear equations using the perturbed right-hand side. We are interested in $\|x - \tilde{x}\|$, which is how much the solution changes due to the change in the right-hand side. The ratio $\|x - \tilde{x}\|/|\epsilon|$ gives the sensitivity of the solution to changes (perturbations) of the jth entry of b.

(a) Show that $\|x - \tilde{x}\|$ does not depend on b; it only depends on the matrix A, ϵ, and j.

(b) How would you find the index j that maximizes the value of $\|x - \tilde{x}\|$? By part (a), your answer should be in terms of A (or quantities derived from A) and ϵ only.

Remark. If a small change in the right-hand side vector b can lead to a large change in the solution, we say that the linear equations $Ax = b$ are *poorly conditioned* or *ill-conditioned*. As a practical matter it means that unless you are very confident in what the entries of b are, the solution $A^{-1}b$ may not be useful in practice.

11.27 *Timing test.* Generate a random $n \times n$ matrix A and an n-vector b, for $n = 500$, $n = 1000$, and $n = 2000$. For each of these, compute the solution $x = A^{-1}b$ (for example using the backslash operator, if the software you are using supports it), and verify that $Ax - b$ is (very) small. Report the time it takes to solve each of these three sets of linear equations, and for each one work out the implied speed of your processor in Gflop/s, based on the $2n^3$ complexity of solving equations using the QR factorization.

11.28 *Solving multiple linear equations efficiently.* Suppose the $n \times n$ matrix A is invertible. We can solve the system of linear equations $Ax = b$ in around $2n^3$ flops using algorithm 11.2. Once we have done that (specifically, computed the QR factorization of A), we can solve an additional set of linear equations with same matrix but different right-hand side, $Ay = c$, in around $3n^2$ additional flops. Assuming we have solved both of these sets of equations, suppose we want to solve $Az = d$, where $d = \alpha b + \beta c$ is a linear combination of b and c. (We are given the coefficients α and β.) Suggest a method for doing this that is even faster than re-using the QR factorization of A. Your method should have a complexity that is *linear* in n. Give rough estimates for the time needed to solve $Ax = b$, $Ay = c$, and $Az = d$ (using your method) for $n = 3000$ on a computer capable of carrying out 1 Gflop/s.

Part III

Least squares

Chapter 12

Least squares

In this chapter we look at the powerful idea of finding approximate solutions of over-determined systems of linear equations by minimizing the sum of the squares of the errors in the equations. The method, and some extensions we describe in later chapters, are widely used in many application areas. It was discovered independently by the mathematicians Carl Friedrich Gauss and Adrien-Marie Legendre around the beginning of the 19th century.

12.1 Least squares problem

Suppose that the $m \times n$ matrix A is tall, so the system of linear equations $Ax = b$, where b is an m-vector, is over-determined, *i.e.*, there are more equations (m) than variables to choose (n). These equations have a solution only if b is a linear combination of the columns of A.

For most choices of b, however, there is no n-vector x for which $Ax = b$. As a compromise, we seek an x for which $r = Ax - b$, which we call the *residual* (for the equations $Ax = b$), is as small as possible. This suggests that we should choose x so as to minimize the norm of the residual, $\|Ax - b\|$. If we find an x for which the residual vector is small, we have $Ax \approx b$, *i.e.*, x almost satisfies the linear equations $Ax = b$. (Some authors define the residual as $b - Ax$, which will not affect us since $\|Ax - b\| = \|b - Ax\|$.)

Minimizing the norm of the residual and its square are the same, so we can just as well minimize

$$\|Ax - b\|^2 = \|r\|^2 = r_1^2 + \cdots + r_m^2,$$

the sum of squares of the residuals. The problem of finding an n-vector $\hat{x}$ that minimizes $\|Ax - b\|^2$, over all possible choices of x, is called the *least squares problem*. It is denoted using the notation

$$\text{minimize} \quad \|Ax - b\|^2, \tag{12.1}$$

where we should specify that the *variable* is x (meaning that we should choose x). The matrix A and the vector b are called the *data* for the problem (12.1), which means that they are given to us when we are asked to choose x. The quantity to be minimized, $\|Ax - b\|^2$, is called the *objective function* (or just objective) of the least squares problem (12.1).

The problem (12.1) is sometimes called *linear* least squares to emphasize that the residual r (whose norm squared we are to minimize) is an affine function of x, and to distinguish it from the *nonlinear* least squares problem, in which we allow the residual r to be an arbitrary function of x. We will study the nonlinear least squares problem in chapter 18.

Any vector $\hat{x}$ that satisfies $\|A\hat{x} - b\|^2 \le \|Ax - b\|^2$ for all x is a *solution* of the least squares problem (12.1). Such a vector is called a *least squares approximate solution* of $Ax = b$. It is very important to understand that a least squares approximate solution $\hat{x}$ of $Ax = b$ need not satisfy the equations $A\hat{x} = b$; it simply makes the norm of the residual as small as it can be. Some authors use the confusing phrase '$\hat{x}$ solves $Ax = b$ in the least squares sense', but we emphasize that a least squares approximate solution $\hat{x}$ does not, in general, solve the equation $Ax = b$.

If $\|A\hat{x} - b\|$ (which we call the *optimal residual norm*) is small, then we can say that $\hat{x}$ *approximately* solves $Ax = b$. On the other hand, if there is an n-vector x that satisfies $Ax = b$, then it is a solution of the least squares problem, since its associated residual norm is zero.

Another name for the least squares problem (12.1), typically used in data fitting applications (the topic of the next chapter), is *regression*. We say that $\hat{x}$, a solution of the least squares problem, is the result of *regressing* the vector b onto the columns of A.

Column interpretation. If the columns of A are the m-vectors $a_1, \ldots, a_n$, then the least squares problem (12.1) is the problem of finding a linear combination of the columns that is closest to the m-vector b; the vector x gives the coefficients:

$$\|Ax - b\|^2 = \|(x_1 a_1 + \cdots + x_n a_n) - b\|^2.$$

If $\hat{x}$ is a solution of the least squares problem, then the vector

$$A\hat{x} = \hat{x}_1 a_1 + \cdots + \hat{x}_n a_n$$

is closest to the vector b, among all linear combinations of the vectors $a_1, \ldots, a_n$.

Row interpretation. Suppose the rows of A are the n-row-vectors $\tilde{a}_1^T, \ldots, \tilde{a}_m^T$, so the residual components are given by

$$r_i = \tilde{a}_i^T x - b_i, \quad i = 1, \ldots, m.$$

The least squares objective is then

$$\|Ax - b\|^2 = (\tilde{a}_1^T x - b_1)^2 + \cdots + (\tilde{a}_m^T x - b_m)^2,$$

the sum of the squares of the residuals in m scalar linear equations. Minimizing this sum of squares of the residuals is a reasonable compromise if our goal is to choose x so that all of them are small.

Example.   We consider the least squares problem with data

$$A = \begin{bmatrix} 2 & 0 \\ -1 & 1 \\ 0 & 2 \end{bmatrix}, \qquad b = \begin{bmatrix} 1 \\ 0 \\ -1 \end{bmatrix}.$$

The over-determined set of three equations in two variables $Ax = b$,

$$2x_1 = 1, \qquad -x_1 + x_2 = 0, \qquad 2x_2 = -1,$$

has no solution. (From the first equation we have $x_1 = 1/2$, and from the last equation we have $x_2 = -1/2$; but then the second equation does not hold.) The corresponding least squares problem is

$$\text{minimize} \quad (2x_1 - 1)^2 + (-x_1 + x_2)^2 + (2x_2 + 1)^2.$$

This least squares problem can be solved using the methods described in the next section (or simple calculus). Its unique solution is $\hat{x} = (1/3, -1/3)$. The least squares approximate solution $\hat{x}$ does not satisfy the equations $Ax = b$; the corresponding residuals are

$$\hat{r} = A\hat{x} - b = (-1/3, -2/3, 1/3),$$

with sum of squares value $\|A\hat{x} - b\|^2 = 2/3$. Let us compare this to another choice of x, $\tilde{x} = (1/2, -1/2)$, which corresponds to (exactly) solving the first and last of the three equations in $Ax = b$. It gives the residual

$$\tilde{r} = A\tilde{x} - b = (0, -1, 0),$$

with sum of squares value $\|A\tilde{x} - b\|^2 = 1$.

The column interpretation tells us that

$$(1/3) \begin{bmatrix} 2 \\ -1 \\ 0 \end{bmatrix} + (-1/3) \begin{bmatrix} 0 \\ 1 \\ 2 \end{bmatrix} = \begin{bmatrix} 2/3 \\ -2/3 \\ -2/3 \end{bmatrix}$$

is the linear combination of the columns of A that is closest to b.

Figure 12.1 shows the values of the least squares objective $\|Ax - b\|^2$ versus $x = (x_1, x_2)$, with the least squares solution $\hat{x}$ shown as the dark point, with objective value $\|A\hat{x} - b\|^2 = 2/3$. The curves show the points x that have objective value $\|A\hat{x} - b\|^2 + 1$, $\|A\hat{x} - b\|^2 + 2$, and so on.

12.2   Solution

In this section we derive several expressions for the solution of the least squares problem (12.1), under one assumption on the data matrix A:

$$\textit{The columns of } A \textit{ are linearly independent.} \tag{12.2}$$

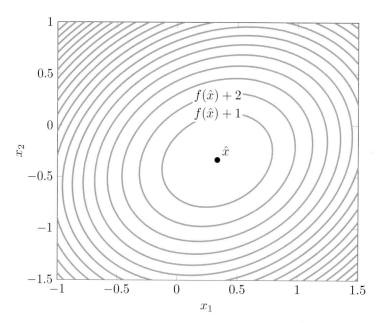

Figure 12.1 Level curves of the function $\|Ax - b\|^2 = (2x_1 - 1)^2 + (-x_1 + x_2)^2 + (2x_2 + 1)^2$. The point $\hat{x}$ minimizes the function.

Solution via calculus.   In this section we find the solution of the least squares problem using some basic results from calculus, reviewed in §C.2. (We will also give an independent verification of the result, that does not rely on calculus, below.) We know that any minimizer $\hat{x}$ of the function $f(x) = \|Ax - b\|^2$ must satisfy

$$\frac{\partial f}{\partial x_i}(\hat{x}) = 0, \quad i = 1, \dots, n,$$

which we can express as the vector equation

$$\nabla f(\hat{x}) = 0,$$

where $\nabla f(\hat{x})$ is the gradient of f evaluated at $\hat{x}$. The gradient can be expressed in matrix form as

$$\nabla f(x) = 2A^T(Ax - b). \tag{12.3}$$

This formula can be derived from the chain rule given on page 184, and the gradient of the sum of squares function, given in §C.1. For completeness, we will derive the formula (12.3) from scratch here. Writing the least squares objective out as a sum, we get

$$f(x) = \|Ax - b\|^2 = \sum_{i=1}^{m}\left(\sum_{j=1}^{n} A_{ij}x_j - b_i\right)^2.$$

To find $\nabla f(x)_k$ we take the partial derivative of f with respect to x_k. Differentiating the sum term by term, we get

$$
\begin{aligned}
\nabla f(x)_k &= \frac{\partial f}{\partial x_k}(x) \\
&= \sum_{i=1}^{m} 2 \left(\sum_{j=1}^{n} A_{ij}x_j - b_i \right) (A_{ik}) \\
&= \sum_{i=1}^{m} 2(A^T)_{ki}(Ax - b)_i \\
&= \left(2A^T(Ax - b) \right)_k.
\end{aligned}
$$

This is our formula (12.3), written out in terms of its components.

Now we continue the derivation of the solution of the least squares problem. Any minimizer $\hat{x}$ of $\|Ax - b\|^2$ must satisfy

$$
\nabla f(\hat{x}) = 2A^T(A\hat{x} - b) = 0,
$$

which can be written as

$$
A^T A\hat{x} = A^T b. \tag{12.4}
$$

These equations are called the *normal equations*. The coefficient matrix $A^T A$ is the Gram matrix associated with A; its entries are inner products of columns of A.

Our assumption (12.2) that the columns of A are linearly independent implies that the Gram matrix $A^T A$ is invertible (§11.5, page 214). This implies that

$$
\hat{x} = (A^T A)^{-1} A^T b \tag{12.5}
$$

is the only solution of the normal equations (12.4). So this must be the unique solution of the least squares problem (12.1).

We have already encountered the matrix $(A^T A)^{-1} A^T$ that appears in (12.5): It is the pseudo-inverse of the matrix A, given in (11.5). So we can write the solution of the least squares problem in the simple form

$$
\hat{x} = A^\dagger b. \tag{12.6}
$$

We observed in §11.5 that $A^\dagger$ is a left inverse of A, which means that $\hat{x} = A^\dagger b$ solves $Ax = b$ if this set of over-determined equations has a solution. But now we see that $\hat{x} = A^\dagger b$ is the least squares approximate solution, *i.e.*, it minimizes $\|Ax - b\|^2$. (And if there is a solution of $Ax = b$, then $\hat{x} = A^\dagger b$ is it.)

The equation (12.6) looks very much like the formula for solution of the linear equations $Ax = b$, when A is square and invertible, *i.e.*, $x = A^{-1}b$. It is very important to understand the difference between the formula (12.6) for the least squares approximate solution, and the formula for the solution of a square set of linear equations, $x = A^{-1}b$. In the case of linear equations and the inverse, $x = A^{-1}b$ actually satisfies $Ax = b$. In the case of the least squares approximate solution, $\hat{x} = A^\dagger b$ generally *does not* satisfy $A\hat{x} = b$.

The formula (12.6) shows us that the solution $\hat{x}$ of the least squares problem is a *linear* function of b. This generalizes the fact that the solution of a square invertible set of linear equations is a linear function of its right-hand side.

Direct verification of least squares solution.   In this section we directly show that $\hat{x} = (A^T A)^{-1} A^T b$ is the solution of the least squares problem (12.1), without relying on calculus. We will show that for any $x \neq \hat{x}$, we have

$$\|A\hat{x} - b\|^2 < \|Ax - b\|^2,$$

establishing that $\hat{x}$ is the unique vector that minimizes $\|Ax - b\|^2$.

We start by writing

$$
\begin{aligned}
\|Ax - b\|^2 &= \|(Ax - A\hat{x}) + (A\hat{x} - b)\|^2 \\
&= \|Ax - A\hat{x}\|^2 + \|A\hat{x} - b\|^2 + 2(Ax - A\hat{x})^T(A\hat{x} - b), \quad (12.7)
\end{aligned}
$$

where we use the identity

$$\|u + v\|^2 = (u + v)^T(u + v) = \|u\|^2 + \|v\|^2 + 2u^T v.$$

The third term in (12.7) is zero:

$$
\begin{aligned}
(Ax - A\hat{x})^T(A\hat{x} - b) &= (x - \hat{x})^T A^T(A\hat{x} - b) \\
&= (x - \hat{x})^T(A^T A\hat{x} - A^T b) \\
&= (x - \hat{x})^T 0 \\
&= 0,
\end{aligned}
$$

where we use $(A^T A)\hat{x} = A^T b$ (the normal equations) in the third line. With this simplification, (12.7) reduces to

$$\|Ax - b\|^2 = \|A(x - \hat{x})\|^2 + \|A\hat{x} - b\|^2.$$

The first term on the right-hand side is nonnegative and therefore

$$\|Ax - b\|^2 \geq \|A\hat{x} - b\|^2.$$

This shows that $\hat{x}$ minimizes $\|Ax - b\|^2$; we now show that it is the unique minimizer. Suppose equality holds above, that is, $\|Ax - b\|^2 = \|A\hat{x} - b\|^2$. Then we have $\|A(x - \hat{x})\|^2 = 0$, which implies $A(x - \hat{x}) = 0$. Since A has linearly independent columns, we conclude that $x - \hat{x} = 0$, i.e., $x = \hat{x}$. So the only x with $\|Ax - b\|^2 = \|A\hat{x} - b\|^2$ is $x = \hat{x}$; for all $x \neq \hat{x}$, we have $\|Ax - b\|^2 > \|A\hat{x} - b\|^2$.

Row form.   The formula for the least squares approximate solution can be expressed in a useful form in terms of the rows $\tilde{a}_i^T$ of the matrix A.

$$\hat{x} = (A^T A)^{-1} A^T b = \left(\sum_{i=1}^{m} \tilde{a}_i \tilde{a}_i^T \right)^{-1} \left(\sum_{i=1}^{m} b_i \tilde{a}_i \right). \quad (12.8)$$

In this formula we express the $n \times n$ Gram matrix $A^T A$ as a sum of m outer products, and the n-vector $A^T b$ as a sum of m n-vectors.

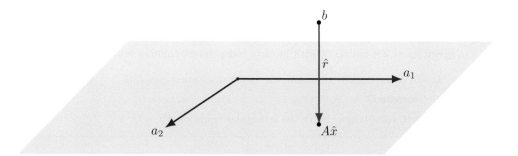

Figure 12.2 Illustration of orthogonality principle for a least squares problem of size $m = 3$, $n = 2$. The optimal residual $\hat{r}$ is orthogonal to any linear combination of a_1 and a_2, the two columns of A.

Orthogonality principle.   The point $A\hat{x}$ is the linear combination of the columns of A that is closest to b. The optimal residual is $\hat{r} = A\hat{x} - b$. The optimal residual satisfies a property that is sometimes called the *orthogonality principle*: It is orthogonal to the columns of A, and therefore, it is orthogonal to any linear combination of the columns of A. In other words, for any n-vector z, we have

$$(Az) \perp \hat{r}. \tag{12.9}$$

We can derive the orthogonality principle from the normal equations, which can be expressed as $A^T(A\hat{x} - b) = 0$. For any n-vector z, we have

$$(Az)^T \hat{r} = (Az)^T(A\hat{x} - b) = z^T A^T(A\hat{x} - b) = 0.$$

The orthogonality principle is illustrated in figure 12.2, for a least squares problem with $m = 3$ and $n = 2$. The shaded plane is the set of all linear combinations $z_1 a_1 + z_2 a_2$ of a_1 and a_2, the two columns of A. The point $A\hat{x}$ is the closest point in the plane to b. The optimal residual $\hat{r}$ is shown as the vector from b to $A\hat{x}$. This vector is orthogonal to any point in the shaded plane.

12.3   Solving least squares problems

We can use the QR factorization to compute the least squares approximate solution (12.5). Let $A = QR$ be the QR factorization of A (which exists by our assumption (12.2) that its columns are linearly independent). We have already seen that the pseudo-inverse $A^\dagger$ can be expressed as $A^\dagger = R^{-1}Q^T$, so we have

$$\hat{x} = R^{-1}Q^T b. \tag{12.10}$$

To compute $\hat{x}$ we first multiply b by Q^T; then we compute $R^{-1}(Q^T b)$ using back substitution. This is summarized in the following algorithm, which computes the least squares approximate solution $\hat{x}$, given A and b.

Algorithm 12.1 LEAST SQUARES VIA QR FACTORIZATION

given an $m \times n$ matrix A with linearly independent columns and an m-vector b.

1. *QR factorization.* Compute the QR factorization $A = QR$.
2. Compute $Q^T b$.
3. *Back substitution.* Solve the triangular equation $R\hat{x} = Q^T b$.

Comparison to solving a square system of linear equations. Recall that the solution of the square invertible system of linear equations $Ax = b$ is $x = A^{-1}b$. We can express x using the QR factorization of A as $x = R^{-1}Q^T b$ (see (11.4)). This equation is formally identical to (12.10). The only difference is that in (12.10), A and Q need not be square, and $R^{-1}Q^T b$ is the least squares approximate solution, which is not (in general) a solution of $Ax = b$.

Indeed, algorithm 12.1 is formally the same as algorithm 11.2, the QR factorization method for solving linear equations. (The only difference is that in algorithm 12.1, A and Q can be tall.)

When A is square, solving the linear equations $Ax = b$ and the least squares problem of minimizing $\|Ax - b\|^2$ are the same, and algorithm 11.2 and algorithm 12.1 are the same. So we can think of algorithm 12.1 as a generalization of algorithm 11.2, which solves the equation $Ax = b$ when A is square, and computes the least squares approximate solution when A is tall.

Backslash notation. Several software packages for manipulating matrices extend the backslash operator (see page 209) to mean the least squares approximate solution of an over-determined set of linear equations. In these packages $A \backslash b$ is taken to mean the solution $A^{-1}b$ of $Ax = b$ when A is square and invertible, and the least squares approximate solution $A^{\dagger}b$ when A is tall and has linearly independent columns. (We remind the reader that this backslash notation is not standard mathematical notation.)

Complexity. The complexity of the first step of algorithm 12.1 is $2mn^2$ flops. The second step involves a matrix-vector multiplication, which takes $2mn$ flops. The third step requires n^2 flops. The total number of flops is

$$2mn^2 + 2mn + n^2 \approx 2mn^2,$$

neglecting the second and third terms, which are smaller than the first by factors of n and $2m$, respectively. The order of the algorithm is mn^2. The complexity is linear in the row dimension of A and quadratic in the number of variables.

Sparse least squares. Least squares problems with sparse A arise in several applications and can be solved more efficiently, for example by using a QR factorization tailored for sparse matrices (see page 190) in the generic algorithm 12.1.

Another simple approach for exploiting sparsity of A is to solve the normal equations $A^T A \hat{x} = A^T b$ by solving a larger (but sparse) system of equations,

$$\begin{bmatrix} 0 & A^T \\ A & I \end{bmatrix} \begin{bmatrix} \hat{x} \\ \hat{y} \end{bmatrix} = \begin{bmatrix} 0 \\ b \end{bmatrix}. \tag{12.11}$$

This is a square set of $m+n$ linear equations. Its coefficient matrix is sparse when A is sparse. If $(\hat{x}, \hat{y})$ satisfies these equations, it is easy to see that $\hat{x}$ satisfies (12.11); conversely, if $\hat{x}$ satisfies the normal equations, $(\hat{x}, \hat{y})$ satisfies (12.11) with $\hat{y} = b - A\hat{x}$. Any method for solving a sparse system of linear equations can be used to solve (12.11).

Matrix least squares.   A simple extension of the least squares problem is to choose the $n \times k$ *matrix* X so as to minimize $\|AX - B\|^2$. Here A is an $m \times n$ matrix and B is an $m \times k$ matrix, and the norm is the matrix norm. This is sometimes called the *matrix least squares problem*. When $k = 1$, x and b are vectors, and the matrix least squares problem reduces to the usual least squares problem.

The matrix least squares problem is in fact nothing but a set of k ordinary least squares problems. To see this, we note that

$$\|AX - B\|^2 = \|Ax_1 - b_1\|^2 + \cdots + \|Ax_k - b_k\|^2,$$

where x_j is the jth column of X and b_j is the jth column of B. (Here we use the property that the square of the matrix norm is the sum of the squared norms of the columns of the matrix.) So the objective is a sum of k terms, with each term depending on only one column of X. It follows that we can choose the columns x_j independently, each one by minimizing its associated term $\|Ax_j - b_j\|^2$. Assuming that A has linearly independent columns, the solution is $\hat{x}_j = A^\dagger b_j$. The solution of the matrix least squares problem is therefore

$$\begin{aligned} \hat{X} &= \begin{bmatrix} \hat{x}_1 & \cdots & \hat{x}_k \end{bmatrix} \\ &= \begin{bmatrix} A^\dagger b_1 & \cdots & A^\dagger b_k \end{bmatrix} \\ &= A^\dagger \begin{bmatrix} b_1 & \cdots & b_k \end{bmatrix} \\ &= A^\dagger B. \end{aligned} \tag{12.12}$$

The very simple solution $\hat{X} = A^\dagger B$ of the matrix least squares problem agrees with the solution of the ordinary least squares problem when $k = 1$ (as it must). Many software packages for linear algebra use the backslash operator $A \backslash B$ to denote $A^\dagger B$, but this is not standard mathematical notation.

The matrix least squares problem can be solved efficiently by exploiting the fact that algorithm 12.1 is another example of a factor-solve algorithm. To compute $\hat{X} = A^\dagger B$ we carry out the QR factorization of A once; we carry out steps 2 and 3 of algorithm 12.1 for each of the k columns of B. The total cost is $2mn^2 + k(2mn + n^2)$ flops. When k is small compared to n this is roughly $2mn^2$ flops, the same cost as solving a single least squares problem (*i.e.*, one with a vector right-hand side).

12.4   Examples

Advertising purchases.   We have m demographic groups or audiences that we want to advertise to, with a target number of impressions or views for each group, which we give as a vector v^{des}. (The entries are positive.) To reach these audiences, we purchase advertising in n different channels (say, different web publishers, radio, print, ...), in amounts that we give as an n-vector s. (The entries of s are non-negative, which we ignore.) The $m \times n$ matrix R gives the number of impressions in each group per dollar spending in the channels: R_{ij} is the number of impressions in group i per dollar spent on advertising in channel j. (These entries are estimated, and are nonnegative.) The jth column of R gives the effectiveness or reach (in impressions per dollar) for channel j. The ith row of R shows which media demographic group i is exposed to. The total number of impressions in each demographic group is the m-vector v, which is given by $v = Rs$. The goal is to find s so that $v = Rs \approx v^{\text{des}}$. We can do this using least squares, by choosing s to minimize $\|Rs - v^{\text{des}}\|^2$. (We are not guaranteed that the resulting channel spend vector will be nonnegative.) This least squares formulation does not take into account the total cost of the advertising; we will see in chapter 16 how this can be done.

We consider a simple numerical example, with $n = 3$ channels and $m = 10$ demographic groups, and matrix

$$
R = \begin{bmatrix}
0.97 & 1.86 & 0.41 \\
1.23 & 2.18 & 0.53 \\
0.80 & 1.24 & 0.62 \\
1.29 & 0.98 & 0.51 \\
1.10 & 1.23 & 0.69 \\
0.67 & 0.34 & 0.54 \\
0.87 & 0.26 & 0.62 \\
1.10 & 0.16 & 0.48 \\
1.92 & 0.22 & 0.71 \\
1.29 & 0.12 & 0.62
\end{bmatrix},
$$

with units of 1000 views per dollar. The entries of R range over an 18:1 range, so the 3 channels are quite different in terms of their audience reach; see figure 12.3.

We take $v^{\text{des}} = (10^3)\mathbf{1}$, *i.e.*, our goal is to reach one million customers in each of the 10 demographic groups. Least squares gives the advertising budget allocation

$$
\hat{s} = (62, 100, 1443),
$$

which achieves a views vector with RMS error 132, or 13.2% of the target values. The views vector is shown in figure 12.4.

Illumination.   A set of n lamps illuminates an area that we divide into m regions or pixels. We let l_i denote the lighting level in region i, so the m-vector l gives the illumination levels across all regions. We let p_i denote the power at which lamp i operates, so the n-vector p gives the set of lamp powers. (The lamp powers are nonnegative and also must not exceed a maximum allowed power, but we ignore these issues here.)

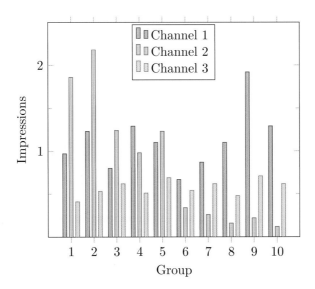

Figure 12.3 Number of impressions in ten demographic groups, per dollar spent on advertising in three channels. The units are 1000 views per dollar.

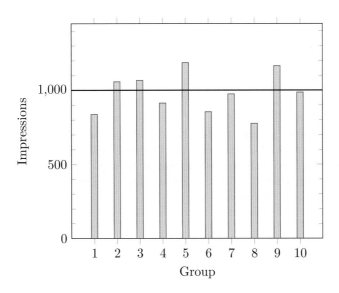

Figure 12.4 Views vector that best approximates the target of one million impressions in each group.

The vector of illumination levels is a linear function of the lamp powers, so we have $l = Ap$ for some $m \times n$ matrix A. The jth column of A gives the illumination pattern for lamp j, *i.e.*, the illumination when lamp j has power 1 and all other lamps are off. We will assume that A has linearly independent columns (and therefore is tall). The ith row of A gives the sensitivity of pixel i to the n lamp powers.

The goal is to find lamp powers that result in a desired illumination pattern l^{des}, such as $l^{\mathrm{des}} = \alpha \mathbf{1}$, which is uniform illumination with value α across the area. In other words, we seek p so that $Ap \approx l^{\mathrm{des}}$. We can use least squares to find $\hat{p}$ that minimizes the sum square deviation from the desired illumination, $\|Ap - l^{\mathrm{des}}\|^2$. This gives the lamp power levels

$$\hat{p} = A^{\dagger} l^{\mathrm{des}} = (A^T A)^{-1} A^T l^{\mathrm{des}}.$$

(We are not guaranteed that these powers are nonnegative, or less than the maximum allowed power level.)

An example is shown in figure 12.5. The area is a 25×25 grid with $m = 625$ pixels, each (say) 1m square. The lamps are at various heights ranging from 3m to 6m, and at the positions shown in the figure. The illumination decays with an inverse square law, so A_{ij} is proportional to d_{ij}^{-2}, where d_{ij} is the (3-D) distance between the center of the pixel and the lamp position. The matrix A is scaled so that when all lamps have power one, the average illumination level is one. The desired illumination pattern is $\mathbf{1}$, *i.e.*, uniform with value 1.

With $p = \mathbf{1}$, the resulting illumination pattern is shown in the top part of figure 12.5. The RMS illumination error is 0.24. We can see that the corners are quite a bit darker than the center, and there are pronounced bright spots directly beneath each lamp. Using least squares we find the lamp powers

$$\hat{p} = (1.46, 0.79, 2.97, 0.74, 0.08, 0.21, 0.21, 2.05, 0.91, 1.47).$$

The resulting illumination pattern has an RMS error of 0.14, about half of the RMS error with all lamp powers set to one. The illumination pattern is shown in the bottom plot of figure 12.5; we can see that the illumination is more uniform than when all lamps have power 1. Most illumination values are near the target level 1, with the corners a bit darker and the illumination a bit brighter directly below each lamp, but less so than when all lamps have power one. This is clear from figure 12.6, which shows the histogram of patch illumination values for all lamp powers one, and for lamp powers $\hat{p}$.

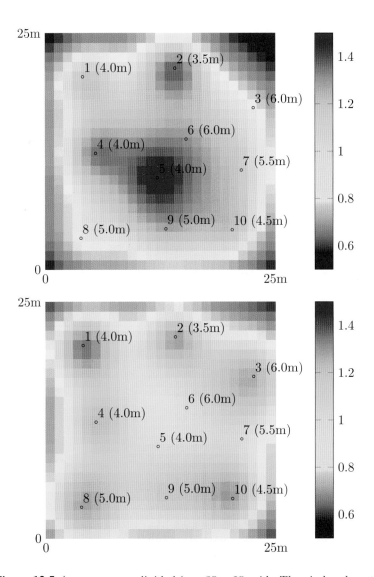

Figure 12.5 A square area divided in a 25×25 grid. The circles show the positions of 10 lamps, the number in parentheses next to each circle is the height of the lamp. The top plot shows the illumination pattern with lamps set to power one. The bottom plot shows the illumination pattern for the lamp powers that minimize the sum square deviation with a desired uniform illumination of one.

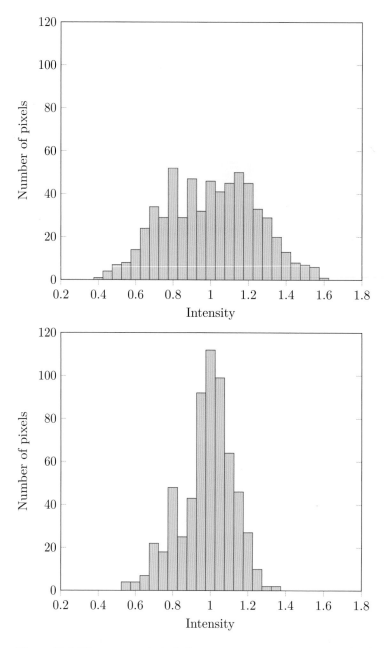

Figure 12.6 Histograms of pixel illumination values using $p = \mathbf{1}$ (top) and $\hat{p}$ (bottom). The target intensity value is one.

Exercises

12.1 *Approximating a vector as a multiple of another one.* In the special case $n = 1$, the general least squares problem (12.1) reduces to finding a scalar x that minimizes $\|ax - b\|^2$, where a and b are m-vectors. (We write the matrix A here in lower case, since it is an m-vector.) Assuming a and b are nonzero, show that $\|a\hat{x} - b\|^2 = \|b\|^2(\sin\theta)^2$, where $\theta = \angle(a, b)$. This shows that the optimal relative error in approximating one vector by a multiple of another one depends on their angle.

12.2 *Least squares with orthonormal columns.* Suppose the $m \times n$ matrix Q has orthonormal columns and b is an m-vector. Show that $\hat{x} = Q^T b$ is the vector that minimizes $\|Qx - b\|^2$. What is the complexity of computing $\hat{x}$, given Q and b, and how does it compare to the complexity of a general least squares problem with an $m \times n$ coefficient matrix?

12.3 *Least angle property of least squares.* Suppose the $m \times n$ matrix A has linearly independent columns, and b is an m-vector. Let $\hat{x} = A^\dagger b$ denote the least squares approximate solution of $Ax = b$.

(a) Show that for any n-vector x, $(Ax)^T b = (Ax)^T (A\hat{x})$, *i.e.*, the inner product of Ax and b is the same as the inner product of Ax and $A\hat{x}$. *Hint.* Use $(Ax)^T b = x^T (A^T b)$ and $(A^T A)\hat{x} = A^T b$.

(b) Show that when $A\hat{x}$ and b are both nonzero, we have

$$\frac{(A\hat{x})^T b}{\|A\hat{x}\|\|b\|} = \frac{\|A\hat{x}\|}{\|b\|}.$$

The left-hand side is the cosine of the angle between $A\hat{x}$ and b. *Hint.* Apply part (a) with $x = \hat{x}$.

(c) *Least angle property of least squares.* The choice $x = \hat{x}$ minimizes the distance between Ax and b. Show that $x = \hat{x}$ also minimizes the angle between Ax and b. (You can assume that Ax and b are nonzero.) *Remark.* For any positive scalar α, $x = \alpha\hat{x}$ also minimizes the angle between Ax and b.

12.4 *Weighted least squares.* In least squares, the objective (to be minimized) is

$$\|Ax - b\|^2 = \sum_{i=1}^m (\tilde{a}_i^T x - b_i)^2,$$

where $\tilde{a}_i^T$ are the rows of A, and the n-vector x is to chosen. In the *weighted least squares problem*, we minimize the objective

$$\sum_{i=1}^m w_i (\tilde{a}_i^T x - b_i)^2,$$

where w_i are given positive weights. The weights allow us to assign different weights to the different components of the residual vector. (The objective of the weighted least squares problem is the square of the weighted norm, $\|Ax - b\|_w^2$, as defined in exercise 3.28.)

(a) Show that the weighted least squares objective can be expressed as $\|D(Ax - b)\|^2$ for an appropriate diagonal matrix D. This allows us to solve the weighted least squares problem as a standard least squares problem, by minimizing $\|Bx - d\|^2$, where $B = DA$ and $d = Db$.

(b) Show that when A has linearly independent columns, so does the matrix B.

(c) The least squares approximate solution is given by $\hat{x} = (A^T A)^{-1} A^T b$. Give a similar formula for the solution of the weighted least squares problem. You might want to use the matrix $W = \mathbf{diag}(w)$ in your formula.

12.5 *Approximate right inverse.* Suppose the tall $m \times n$ matrix A has linearly independent columns. It does not have a right inverse, *i.e.*, there is no $n \times m$ matrix X for which $AX = I$. So instead we seek the $n \times m$ matrix X for which the residual matrix $R = AX - I$ has the smallest possible matrix norm. We call this matrix the *least squares approximate right inverse* of A. Show that the least squares right inverse of A is given by $X = A^\dagger$. *Hint.* This is a matrix least squares problem; see page 233.

12.6 *Least squares equalizer design.* (See exercise 7.15.) You are given a channel impulse response, the n-vector c. Your job is to find an equalizer impulse response, the n-vector h, that minimizes $\|h * c - e_1\|^2$. You can assume that $c_1 \neq 0$. *Remark.* h is called an equalizer since it approximately inverts, or undoes, convolution by c.

Explain how to find h. Apply your method to find the equalizer h for the channel $c = (1.0, 0.7, -0.3, -0.1, 0.05)$. Plot c, h, and $h * c$.

12.7 *Network tomography.* A network consists of n links, labeled $1, \ldots, n$. A *path* through the network is a subset of the links. (The order of the links on a path does not matter here.) Each link has a (positive) *delay*, which is the time it takes to traverse it. We let d denote the n-vector that gives the link delays. The total travel time of a path is the sum of the delays of the links on the path. Our goal is to estimate the link delays (*i.e.*, the vector d), from a large number of (noisy) measurements of the travel times along different paths. This data is given to you as an $N \times n$ matrix P, where

$$
P_{ij} = \begin{cases} 1 & \text{link } j \text{ is on path } i \\ 0 & \text{otherwise,} \end{cases}
$$

and an N-vector t whose entries are the (noisy) travel times along the N paths. You can assume that $N > n$. You will choose your estimate $\hat{d}$ by minimizing the RMS deviation between the measured travel times (t) and the travel times predicted by the sum of the link delays. Explain how to do this, and give a matrix expression for $\hat{d}$. If your expression requires assumptions about the data P or t, state them explicitly.

Remark. This problem arises in several contexts. The network could be a computer network, and a path gives the sequence of communication links data packets traverse. The network could be a transportation system, with the links representing road segments.

12.8 *Least squares and QR factorization.* Suppose A is an $m \times n$ matrix with linearly independent columns and QR factorization $A = QR$, and b is an m-vector. The vector $A\hat{x}$ is the linear combination of the columns of A that is closest to the vector b, *i.e.*, it is the projection of b onto the set of linear combinations of the columns of A.

(a) Show that $A\hat{x} = QQ^T b$. (The matrix QQ^T is called the *projection matrix*.)

(b) Show that $\|A\hat{x} - b\|^2 = \|b\|^2 - \|Q^T b\|^2$. (This is the square of the distance between b and the closest linear combination of the columns of A.)

12.9 *Invertibility of matrix in sparse least squares formulation.* Show that the $(m+n) \times (m+n)$ coefficient matrix appearing in equation (12.11) is invertible if and only if the columns of A are linearly independent.

12.10 *Numerical check of the least squares approximate solution.* Generate a random 30×10 matrix A and a random 30-vector b. Compute the least squares approximate solution $\hat{x} = A^\dagger b$ and the associated residual norm squared $\|A\hat{x} - b\|^2$. (There may be several ways to do this, depending on the software package you use.) Generate three different random 10-vectors d_1, d_2, d_3, and verify that $\|A(\hat{x} + d_i) - b\|^2 > \|A\hat{x} - b\|^2$ holds. (This shows that $x = \hat{x}$ has a smaller associated residual than the choices $x = \hat{x} + d_i$, $i = 1, 2, 3$.)

12.11 *Complexity of matrix least squares problem.* Explain how to compute the matrix least squares approximate solution of $AX = B$, given by $\hat{X} = A^\dagger B$ (see (12.12)), in no more than $2mn^2 + 3mnk$ flops. (In contrast, solving k vector least squares problems to obtain the columns of $\hat{X}$, in a naïve way, requires $2mn^2 k$ flops.)

12.12 *Least squares placement.* The 2-vectors $p_1, \ldots, p_N$ represent the locations or positions of N objects, for example, factories, warehouses, and stores. The last K of these locations are fixed and given; the goal in a *placement problem* to choose the locations of the first $N - K$ objects. Our choice of the locations is guided by an undirected graph; an edge between two objects means we would like them to be close to each other. In *least squares placement*, we choose the locations $p_1, \ldots, p_{N-K}$ so as to minimize the sum of the squares of the distances between objects connected by an edge,

$$\|p_{i_1} - p_{j_1}\|^2 + \cdots + \|p_{i_L} - p_{j_L}\|^2,$$

where the L edges of the graph are given by $(i_1, j_1), \ldots, (i_L, j_L)$.

(a) Let $\mathcal{D}$ be the Dirichlet energy of the graph, as defined on page 135. Show that the sum of the squared distances between the N objects can be expressed as $\mathcal{D}(u) + \mathcal{D}(v)$, where $u = ((p_1)_1, \ldots, (p_N)_1)$ and $v = ((p_1)_2, \ldots, (p_N)_2)$ are N-vectors containing the first and second coordinates of the objects, respectively.

(b) Express the least squares placement problem as a least squares problem, with variable $x = (u_{1:(N-K)}, v_{1:(N-K)})$. In other words, express the objective above (the sum of squares of the distances across edges) as $\|Ax - b\|^2$, for an appropriate $m \times n$ matrix A and m-vector b. You will find that $m = 2L$. *Hint.* Recall that $\mathcal{D}(y) = \|B^T y\|^2$, where B is the incidence matrix of the graph.

(c) Solve the least squares placement problem for the specific problem with $N = 10$, $K = 4$, $L = 13$, fixed locations

$$p_7 = (0, 0), \quad p_8 = (0, 1), \quad p_8 = (1, 1), \quad p_{10} = (1, 0),$$

and edges

$$(1, 3), \quad (1, 4), \quad (1, 7), \quad (2, 3), \quad (2, 5), \quad (2, 8), \quad (2, 9),$$

$$(3, 4), \quad (3, 5), \quad (4, 6), \quad (5, 6), \quad (6, 9), \quad (6, 10).$$

Plot the locations, showing the graph edges as lines connecting the locations.

12.13 *Iterative method for least squares problem.* Suppose that A has linearly independent columns, so $\hat{x} = A^\dagger b$ minimizes $\|Ax - b\|^2$. In this exercise we explore an iterative method, due to the mathematician Lewis Richardson, that can be used to compute $\hat{x}$. We define $x^{(1)} = 0$ and for $k = 1, 2, \ldots,$

$$x^{(k+1)} = x^{(k)} - \mu A^T (Ax^{(k)} - b),$$

where μ is a positive parameter, and the superscripts denote the iteration number. This defines a sequence of vectors that converge to $\hat{x}$ provided μ is not too large; the choice $\mu = 1/\|A\|^2$, for example, always works. The iteration is terminated when $A^T (Ax^{(k)} - b)$ is small enough, which means the least squares optimality conditions are almost satisfied. To implement the method we only need to multiply vectors by A and by A^T. If we have efficient methods for carrying out these two matrix-vector multiplications, this iterative method can be faster than algorithm 12.1 (although it does not give the exact solution). Iterative methods are often used for very large scale least squares problems.

(a) Show that if $x^{(k+1)} = x^{(k)}$, we have $x^{(k)} = \hat{x}$.

(b) Express the vector sequence $x^{(k)}$ as a linear dynamical system with constant dynamics matrix and offset, *i.e.*, in the form $x^{(k+1)} = Fx^{(k)} + g$.

(c) Generate a random 20×10 matrix A and 20-vector b, and compute $\hat{x} = A^\dagger b$. Run the Richardson algorithm with $\mu = 1/\|A\|^2$ for 500 iterations, and plot $\|x^{(k)} - \hat{x}\|$ to verify that $x^{(k)}$ appears to be converging to $\hat{x}$.

12.14 *Recursive least squares.* In some applications of least squares the rows of the coefficient matrix A become available (or are added) sequentially, and we wish to solve the resulting family of growing least squares problems. Define the $k \times n$ matrices and k-vectors

$$A^{(k)} = \begin{bmatrix} a_1^T \\ \vdots \\ a_k^T \end{bmatrix} = A_{1:k,1:n}, \qquad b^{(k)} = \begin{bmatrix} b_1 \\ \vdots \\ b_k \end{bmatrix} = b_{1:k},$$

for $k = 1, \ldots, m$. We wish to compute $\hat{x}^{(k)} = A^{(k)\dagger} b^{(k)}$, for $k = n, n+1, \ldots, m$. We will assume that the columns of $A^{(n)}$ are linearly independent, which implies that the columns of $A^{(k)}$ are linearly independent for $k = n, \ldots, m$. We will also assume that m is much larger than n. The naïve method for computing $x^{(k)}$ requires $2kn^2$ flops, so the total cost for $k = n, \ldots, m$ is

$$\sum_{k=n}^{m} 2kn^2 = \left(\sum_{k=n}^{m} k \right) (2n^2) = \left(\frac{m^2 - n^2 + m + n}{2} \right) (2n^2) \approx m^2 n^2 \text{ flops.}$$

A simple trick allows us to compute $x^{(k)}$ for $k = n \ldots, m$ much more efficiently, with a cost that grows linearly with m. The trick also requires memory storage order n^2, which does not depend on m. for $k = 1, \ldots, m$, define

$$G^{(k)} = (A^{(k)})^T A^{(k)}, \qquad h^{(k)} = (A^{(k)})^T b^{(k)}.$$

(a) Show that $\hat{x}^{(k)} = (G^{(k)})^{-1} h^{(k)}$ for $k = n, \ldots, m$. *Hint.* See (12.8).

(b) Show that $G^{(k+1)} = G^{(k)} + a_k a_k^T$ and $h^{(k+1)} = h^{(k)} + b_k a_k$, for $k = 1, \ldots, m-1$.

(c) *Recursive least squares* is the following algorithm. For $k = n, \ldots, m$, compute $G^{(k+1)}$ and $h^{(k+1)}$ using (b); then compute $\hat{x}^{(k)}$ using (a). Work out the total flop count for this method, keeping only dominant terms. (You can include the cost of computing $G^{(n)}$ and $h^{(n)}$, which should be negligible in the total.) Compare to the flop count for the naïve method.

Remark. A further trick called the matrix inversion lemma (which is beyond the scope of this book) can be used to reduce the complexity of recursive least squares to order mn^2.

12.15 *Minimizing a squared norm plus an affine function.* A generalization of the least squares problem (12.1) adds an affine function to the least squares objective,

$$\text{minimize} \quad \|Ax - b\|^2 + c^T x + d,$$

where the n-vector x is the variable to be chosen, and the (given) data are the $m \times n$ matrix A, the m-vector b, the n-vector c, and the number d. We will use the same assumption we use in least squares: The columns of A are linearly independent. This generalized problem can be solved by reducing it to a standard least squares problem, using a trick called *completing the square*.

Show that the objective of the problem above can be expressed in the form

$$\|Ax - b\|^2 + c^T x + d = \|Ax - b + f\|^2 + g,$$

for some m-vector f and some constant g. It follows that we can solve the generalized least squares problem by minimizing $\|Ax - (b - f)\|$, an ordinary least squares problem with solution $\hat{x} = A^{\dagger}(b - f)$.

Hints. Express the norm squared term on the right-hand side as $\|(Ax - b) + f\|^2$ and expand it. Then argue that the equality above holds provided $2A^T f = c$. One possible choice is $f = (1/2)(A^{\dagger})^T c$. (You must justify these statements.)

12.16 *Gram method for computing least squares approximate solution.* Algorithm 12.1 uses the QR factorization to compute the least squares approximate solution $\hat{x} = A^\dagger b$, where the $m \times n$ matrix A has linearly independent columns. It has a complexity of $2mn^2$ flops. In this exercise we consider an alternative method: First, form the Gram matrix $G = A^T A$ and the vector $h = A^T b$; and then compute $\hat{x} = G^{-1}h$ (using algorithm 11.2). What is the complexity of this method? Compare it to algorithm 12.1. *Remark.* You might find that the Gram algorithm appears to be a bit faster than the QR method, but the factor is not large enough to have any practical significance. The idea is useful in situations where G is partially available and can be computed more efficiently than by multiplying A and its transpose. An example is exercise 13.21.

Chapter 13

Least squares data fitting

In this chapter we introduce one of the most important applications of least squares methods, to the problem of data fitting. The goal is to find a mathematical model, or an approximate model, of some relation, given some observed data.

13.1 Least squares data fitting

Least squares is widely used as a method to construct a mathematical model from some data, experiments, or observations. Suppose we have an n-vector x, and a scalar y, and we believe that they are related, perhaps approximately, by some function $f : \mathbf{R}^n \to \mathbf{R}$:

$$y \approx f(x).$$

The vector x might represent a set of n feature values, and is called the *feature vector* or the vector of *independent variables*, depending on the context. The scalar y represents some *outcome* (also called *response variable*) that we are interested in. Or x might represent the previous n values of a time series, and y represents the next value.

Data. We don't know f, although we might have some idea about its general form. But we do have some *data*, given by

$$x^{(1)}, \ldots, x^{(N)}, \qquad y^{(1)}, \ldots, y^{(N)},$$

where the n-vector $x^{(i)}$ is the feature vector and the scalar $y^{(i)}$ is the associated value of the outcome for data sample i. We sometimes refer to the pair $x^{(i)}, y^{(i)}$ as the ith *data pair*. These data are also called *observations, examples, samples,* or *measurements*, depending on the context. Here we use the superscript (i) to denote the ith data point: $x^{(i)}$ is an n-vector, the ith independent variable; the number $x_j^{(i)}$ is the value of jth feature for example i.

Model. We will form a *model* of the relationship between x and y, given by

$$y \approx \hat{f}(x),$$

where $\hat{f} : \mathbf{R}^n \to \mathbf{R}$. We write $\hat{y} = \hat{f}(x)$, where $\hat{y}$ is the (scalar) *prediction* (of the outcome y), given the independent variable (vector) x. The hat appearing over f is traditional notation that suggests that the function $\hat{f}$ is an approximation of the function f. The function $\hat{f}$ is called the model, prediction function, or predictor. For a specific value of the feature vector x, $\hat{y} = \hat{f}(x)$ is the prediction of the outcome.

Linear in the parameters model. We will focus on a specific form for the model, which has the form

$$\hat{f}(x) = \theta_1 f_1(x) + \cdots + \theta_p f_p(x),$$

where $f_i : \mathbf{R}^n \to \mathbf{R}$ are *basis functions* or *feature mappings* that we choose, and θ_i are the *model parameters* that we choose. This form of model is called *linear in the parameters*, since for each x, $\hat{f}(x)$ is a linear function of the model parameter p-vector θ. The basis functions are usually chosen based on our idea of what f looks like. (We will see many examples of this below.) Once the basis functions have been chosen, there is the question of how to choose the model parameters, given our set of data.

Prediction error. Our goal is to choose the model $\hat{f}$ so that it is consistent with the data, *i.e.*, we have $y^{(i)} \approx \hat{f}(x^{(i)})$, for $i = 1, \ldots, N$. (There is another goal in choosing $\hat{f}$, that we will discuss in §13.2.) For data sample i, our model predicts the value $\hat{y}^{(i)} = \hat{f}(x^{(i)})$, so the *prediction error* or *residual* for this data point is

$$r^{(i)} = y^{(i)} - \hat{y}^{(i)}.$$

(Some authors define the prediction error in the opposite way, as $\hat{y}^{(i)} - y^{(i)}$. We will see that this does not affect the methods developed in this chapter.)

Vector notation for outcomes, predictions, and residuals. For our data set and model, we have the observed response $y^{(i)}$, the prediction $\hat{y}^{(i)}$, and the residual or prediction error $r^{(i)}$, for each example $i = 1, \ldots, N$. We will now use vector notation to express these as N-vectors,

$$y^{\mathrm{d}} = (y^{(1)}, \ldots, y^{(N)}), \qquad \hat{y}^{\mathrm{d}} = (\hat{y}^{(1)}, \ldots, \hat{y}^{(N)}), \qquad r^{\mathrm{d}} = (r^{(1)}, \ldots, r^{(N)}),$$

respectively. (In the notation used above to describe the approximate relation between the feature vector and the outcome, $y \approx f(x)$, and the prediction function $\hat{y} = \hat{f}(x)$, the symbols y and $\hat{y}$ refer to generic scalar values. With the superscript d (for 'data'), y^{d}, $\hat{y}^{\mathrm{d}}$, and r^{d} refer to the N-vectors of observed data values, predicted values, and associated residuals.)

Using this vector notation we can express the (vector of) residuals as $r^{\mathrm{d}} = y^{\mathrm{d}} - \hat{y}^{\mathrm{d}}$. A natural measure of how well the model predicts the observed data, or how consistent it is with the observed data, is the RMS prediction error $\mathbf{rms}(r^{\mathrm{d}})$. The ratio $\mathbf{rms}(r^{\mathrm{d}}) / \mathbf{rms}(y^{\mathrm{d}})$ gives a relative prediction error. For example, if the relative prediction error is 0.1, we might say that the model predicts the outcomes, or fits the data, within 10%.

Least squares model fitting.   A very common method for choosing the model parameters $\theta_1, \ldots, \theta_p$ is to minimize the RMS prediction error on the given data set, which is the same as minimizing the sum of squares of the prediction errors, $\|r^{\mathrm{d}}\|^2$. We now show that this is a least squares problem.

Expressing $\hat{y}^{(i)} = \hat{f}(x^{(i)})$ in terms of the model parameters, we have

$$\hat{y}^{(i)} = A_{i1}\theta_1 + \cdots + A_{ip}\theta_p, \quad i = 1, \ldots, N,$$

where we define the $N \times p$ matrix A as

$$A_{ij} = \hat{f}_j(x^{(i)}), \quad i = 1, \ldots, N, \quad j = 1, \ldots, p, \tag{13.1}$$

and the p-vector θ as $\theta = (\theta_1, \ldots, \theta_p)$. The jth column of A is the jth basis function, evaluated at each of the data points $x^{(1)}, \ldots, x^{(N)}$. Its ith row gives the values of the p basis functions on the ith data point $x^{(i)}$. In matrix-vector notation we have

$$\hat{y}^{\mathrm{d}} = A\theta.$$

This simple equation shows how our choice of model parameters maps into the vector of predicted values of the outcomes in the N different experiments. We know the matrix A from the given data points, and choice of basis functions; our goal is to choose the p-vector of model coefficients θ.

The sum of squares of the residuals is then

$$\|r^{\mathrm{d}}\|^2 = \|y^{\mathrm{d}} - \hat{y}^{\mathrm{d}}\|^2 = \|y^{\mathrm{d}} - A\theta\|^2 = \|A\theta - y^{\mathrm{d}}\|^2.$$

(In the last step we use the fact that the norm of a vector is the same as the norm of its negative.) Choosing θ to minimize this is evidently a least squares problem, of the same form as (12.1). Provided the columns of A are linearly independent, we can solve this least squares problem to find $\hat{\theta}$, the model parameter values that minimize the norm of the prediction error on our data set, as

$$\hat{\theta} = (A^T A)^{-1} A^T y^{\mathrm{d}} = A^{\dagger} y^{\mathrm{d}}. \tag{13.2}$$

We say that the model parameter values $\hat{\theta}$ are obtained by *least squares fitting on the data set*.

We can interpret each term in $\|y^{\mathrm{d}} - A\theta\|^2$. The term $\hat{y}^{\mathrm{d}} = A\theta$ is the N-vector of measurements or outcomes that is predicted by our model, with the parameter vector θ. The term y^{d} is the N-vector of actual observed or measured outcomes. The difference $y^{\mathrm{d}} - A\theta$ is the N-vector of prediction errors. Finally, $\|y^{\mathrm{d}} - A\theta\|^2$ is the sum of squares of the prediction errors, also called the residual sum of squares (RSS). This is minimized by the least squares fit $\theta = \hat{\theta}$.

The number $\|y^{\mathrm{d}} - A\hat{\theta}\|^2$ is called the minimum sum square error (for the given model basis and data set). The number $\|y^{\mathrm{d}} - A\hat{\theta}\|^2/N$ is called the *minimum mean square error* (MMSE) (of our model, on the data set). Its squareroot is the minimum RMS fitting error. The model performance on the data set can be visualized by plotting $\hat{y}^{(i)}$ versus $y^{(i)}$ on a scatter plot, with a dashed line showing $\hat{y} = y$ for reference.

Since $\|y^{\mathrm{d}} - A\theta\|^2 = \|A\theta - y^{\mathrm{d}}\|^2$, the same least squares model parameter is obtained when the residual or prediction error is defined as $\hat{y}^{\mathrm{d}} - y^{\mathrm{d}}$ instead of (our definition) $y^{\mathrm{d}} - \hat{y}^{\mathrm{d}}$. The residual sum of squares, minimum mean square error, and RMS fitting error also agree using this alternate definition of prediction error.

Some notation differences from chapter 12. Before proceeding we note some differences in the meanings of symbols used in chapter 12 (on least squares) and in this chapter on data fitting, that the reader will need to keep in mind. In chapter 12, the symbol x denotes a generic variable, the vector that we would like to find, and b refers to the so-called right-hand side, the vector we seek to approximate. In this chapter, in the context of fitting a model to data, the symbol x generically refers to a feature vector; we want to find θ, the vector of coefficients in our model, and the vector we seek to approximate is y^{d}, a vector of (observed) data outcomes. When we use least squares in this chapter, we will need to transcribe the results or formulas from chapter 12 to the current context, as in the formula (13.2).

Least squares fit with a constant. We start with the simplest possible fit: We take $p = 1$, with $f_1(x) = 1$ for all x. In this case the model $\hat{f}$ is a constant function, with $\hat{f}(x) = \theta_1$ for all x. Least squares fitting in this case is the same as choosing the best constant value θ_1 to approximate the data $y^{(1)}, \ldots, y^{(N)}$.

In this simple case, the matrix A in (13.1) is the $N \times 1$ matrix $\mathbf{1}$, which always has linearly independent columns (since it has one column, which is nonzero). The formula (13.2) is then

$$\hat{\theta}_1 = (A^T A)^{-1} A^T y^{\mathrm{d}} = N^{-1} \mathbf{1}^T y^{\mathrm{d}} = \mathbf{avg}(y^{\mathrm{d}}),$$

where we use $\mathbf{1}^T \mathbf{1} = N$. So the best constant fit to the data is simply its mean,

$$\hat{f}(x) = \mathbf{avg}(y^{\mathrm{d}}).$$

The RMS fit to the data (*i.e.*, the RMS value of the optimal residual) is

$$\mathbf{rms}(y^{\mathrm{d}} - \mathbf{avg}(y^{\mathrm{d}})\mathbf{1}) = \mathbf{std}(y^{\mathrm{d}}),$$

the standard deviation of the data. This gives a nice interpretation of the average value and the standard deviation of the outcomes, as the best constant fit and the associated RMS error, respectively. It is common to compare the RMS fitting error for a more sophisticated model with the standard deviation of the outcomes, which is the optimal RMS fitting error for a constant model.

A simple example of a constant fit is shown in figure 13.1. In this example we have $n = 1$, so the data points $x^{(i)}$ are scalars. The green circles in the left-hand plot show the data points; the blue line shows the prediction function $\hat{f}(x)$ (which has constant value). The right-hand plot shows a scatter plot of the data outcomes $y^{(i)}$ versus the predicted values $\hat{y}^{(i)}$ (all of which are the same), with a dashed line showing $y = \hat{y}$.

Independent column assumption. To use least squares fitting we assume that the columns of the matrix A in (13.1) are linearly independent. We can give an interesting interpretation of what it means when this assumption fails. If the columns of A are linearly dependent, it means that one of the columns can be expressed as a linear combination of the others. Suppose, for example, that the last column can be expressed as a linear combination of the first $p - 1$ columns. Using $A_{ij} = f_j(x^{(i)})$, this means

$$f_p(x^{(i)}) = \beta_1 f_1(x^{(i)}) + \cdots + \beta_{p-1} f_{p-1}(x^{(i)}), \quad i = 1, \ldots, N.$$

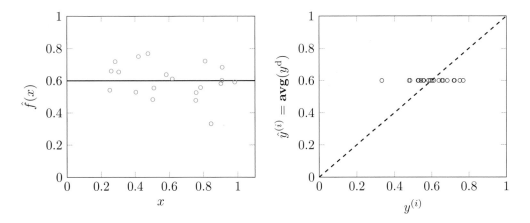

Figure 13.1 The constant fit $\hat{f}(x) = \mathbf{avg}(y^{\mathrm{d}})$ to $N = 20$ data points and a scatter plot of $\hat{y}^{(i)}$ versus $y^{(i)}$.

This says that the value of the pth basis function can be expressed as a linear combination of the values of the first $p - 1$ basis functions *on the given data set.* Evidently, then, the pth basis function is redundant (on the given data set).

13.1.1   Fitting univariate functions

Suppose that $n = 1$, so the feature vector x is a scalar (as is the outcome y). The relationship $y \approx f(x)$ says that y is approximately a (univariate) function f of x. We can plot the data $(x^{(i)}, y^{(i)})$ as points in the (x, y) plane, and we can plot the model $\hat{f}$ as a curve in the (x, y)-plane. This allows us to visualize the fit of our model to the data.

Straight-line fit.   We take basis functions $f_1(x) = 1$ and $f_2(x) = x$. Our model has the form
$$\hat{f}(x) = \theta_1 + \theta_2 x,$$
which is a straight line when plotted. (This is perhaps why $\hat{f}$ is sometimes called a linear model, even though it is in general an affine, and not linear, function of x.) Figure 13.2 shows an example. The matrix A in (13.1) is given by
$$A = \begin{bmatrix} 1 & x^{(1)} \\ 1 & x^{(2)} \\ \vdots & \vdots \\ 1 & x^{(N)} \end{bmatrix} = \begin{bmatrix} \mathbf{1} & x^{\mathrm{d}} \end{bmatrix},$$
where in the right-hand side we use x^{d} to denote the N-vector of values $x^{\mathrm{d}} = (x^{(1)}, \ldots, x^{(N)})$. Provided that there are at least two different values appearing in

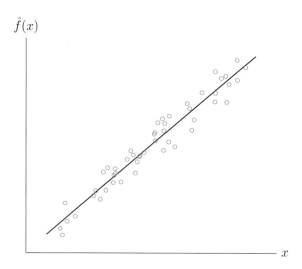

Figure 13.2 Straight-line fit to 50 points $(x^{(i)}, y^{(i)})$ in a plane.

$x^{(1)}, \ldots, x^{(N)}$, this matrix has linearly independent columns. The parameters in the optimal straight-line fit to the data are given by

$$\begin{bmatrix} \theta_1 \\ \theta_2 \end{bmatrix} = (A^T A)^{-1} A^T y^{\mathrm{d}}.$$

This expression is simple enough for us to work it out explicitly, although there is no computational advantage to doing so. The Gram matrix is

$$A^T A = \begin{bmatrix} N & \mathbf{1}^T x^{\mathrm{d}} \\ \mathbf{1}^T x^{\mathrm{d}} & (x^{\mathrm{d}})^T x^{\mathrm{d}} \end{bmatrix}.$$

The 2-vector $A^T y^{\mathrm{d}}$ is

$$A^T y = \begin{bmatrix} \mathbf{1}^T y^{\mathrm{d}} \\ (x^{\mathrm{d}})^T y^{\mathrm{d}} \end{bmatrix},$$

so we have (using the formula for the inverse of a 2×2 matrix)

$$\begin{bmatrix} \hat{\theta}_1 \\ \hat{\theta}_2 \end{bmatrix} = \frac{1}{N(x^{\mathrm{d}})^T x^{\mathrm{d}} - (\mathbf{1}^T x^{\mathrm{d}})^2} \begin{bmatrix} (x^{\mathrm{d}})^T x^{\mathrm{d}} & -\mathbf{1}^T x^{\mathrm{d}} \\ -\mathbf{1}^T x^{\mathrm{d}} & N \end{bmatrix} \begin{bmatrix} \mathbf{1}^T y^{\mathrm{d}} \\ (x^{\mathrm{d}})^T y^{\mathrm{d}} \end{bmatrix}.$$

Multiplying the scalar term by N^2, and dividing the matrix and vector terms by N, we can express this as

$$\begin{bmatrix} \hat{\theta}_1 \\ \hat{\theta}_2 \end{bmatrix} = \frac{1}{\mathbf{rms}(x^{\mathrm{d}})^2 - \mathbf{avg}(x^{\mathrm{d}})^2} \begin{bmatrix} \mathbf{rms}(x^{\mathrm{d}})^2 & -\mathbf{avg}(x^{\mathrm{d}}) \\ -\mathbf{avg}(x^{\mathrm{d}}) & 1 \end{bmatrix} \begin{bmatrix} \mathbf{avg}(y^{\mathrm{d}}) \\ (x^{\mathrm{d}})^T y^{\mathrm{d}}/N \end{bmatrix}.$$

The optimal slope $\hat{\theta}_2$ of the straight line fit can be expressed more simply in terms of the correlation coefficient ρ between the data vectors x^{d} and y^{d}, and their standard deviations. We have

$$
\begin{aligned}
\hat{\theta}_2 &= \frac{N(x^{\mathrm{d}})^T y^{\mathrm{d}} - (\mathbf{1}^T x^{\mathrm{d}})(\mathbf{1}^T y^{\mathrm{d}})}{N(x^{\mathrm{d}})^T x^{\mathrm{d}} - (\mathbf{1}^T x^{\mathrm{d}})^2} \\
&= \frac{(x^{\mathrm{d}} - \mathbf{avg}(x^{\mathrm{d}})\mathbf{1})^T (y^{\mathrm{d}} - \mathbf{avg}(y^{\mathrm{d}})\mathbf{1})}{\|x^{\mathrm{d}} - \mathbf{avg}(x^{\mathrm{d}})\mathbf{1}\|^2} \\
&= \frac{\mathbf{std}(y^{\mathrm{d}})}{\mathbf{std}(x^{\mathrm{d}})}\rho.
\end{aligned}
$$

In the last step we used the definitions

$$
\rho = \frac{(x^{\mathrm{d}} - \mathbf{avg}(x^{\mathrm{d}})\mathbf{1})^T (y^{\mathrm{d}} - \mathbf{avg}(y^{\mathrm{d}})\mathbf{1})}{N\,\mathbf{std}(x^{\mathrm{d}})\,\mathbf{std}(y^{\mathrm{d}})}, \qquad \mathbf{std}(x^{\mathrm{d}}) = \frac{\|x^{\mathrm{d}} - \mathbf{avg}(x^{\mathrm{d}})\mathbf{1}\|}{\sqrt{N}}
$$

from chapter 3. From the first of the two normal equations, $N\theta_1 + (\mathbf{1}^T x^{\mathrm{d}})\theta_2 = \mathbf{1}^T y^{\mathrm{d}}$, we also obtain a simple expression for $\hat{\theta}_1$:

$$
\hat{\theta}_1 = \mathbf{avg}(y^{\mathrm{d}}) - \hat{\theta}_2\,\mathbf{avg}(x^{\mathrm{d}}).
$$

Putting these results together, we can write the least squares fit as

$$
\hat{f}(x) = \mathbf{avg}(y^{\mathrm{d}}) + \rho\frac{\mathbf{std}(y^{\mathrm{d}})}{\mathbf{std}(x^{\mathrm{d}})}(x - \mathbf{avg}(x^{\mathrm{d}})). \tag{13.3}
$$

(Note that x and y are generic scalar values, while x^{d} and y^{d} are vectors of the observed data values.) When $\mathbf{std}(y^{\mathrm{d}}) \neq 0$, this can be expressed in the more symmetric form

$$
\frac{\hat{y} - \mathbf{avg}(y^{\mathrm{d}})}{\mathbf{std}(y^{\mathrm{d}})} = \rho\frac{x - \mathbf{avg}(x^{\mathrm{d}})}{\mathbf{std}(x^{\mathrm{d}})},
$$

which has a nice interpretation. The left-hand side is the difference between the predicted response value and the mean response value, divided by its standard deviation. The right-hand side is the correlation coefficient ρ times the same quantity, computed for the dependent variable.

The least squares straight-line fit is used in many application areas.

Asset α and β in finance.   In finance, the straight-line fit is used to predict the return of an individual asset from the return of the whole market. (The return of the whole market is typically taken to be a sum of the individual asset returns, weighted by their capitalizations.) The straight-line model $\hat{f}(x) = \theta_1 + \theta_2 x$ predicts the asset return from the market return x. The least squares straight-line fit is computed from observed market returns $r_1^{\mathrm{mkt}}, \ldots, r_T^{\mathrm{mkt}}$ and individual asset returns $r_1^{\mathrm{ind}}, \ldots, r_T^{\mathrm{ind}}$ over some period of length T. We therefore take

$$
x^{\mathrm{d}} = (r_1^{\mathrm{mkt}}, \ldots, r_T^{\mathrm{mkt}}), \qquad y^{\mathrm{d}} = (r_1^{\mathrm{ind}}, \ldots, r_T^{\mathrm{ind}})
$$

in (13.3). The model is typically written in the form

$$
\hat{f}(x) = (r^{\mathrm{rf}} + \alpha) + \beta(x - \mu^{\mathrm{mkt}}),
$$

where r^{rf} is the risk-free interest rate over the period and $\mu^{\mathrm{mkt}} = \mathbf{avg}(x^{\mathrm{d}})$ is the average market return. Comparing this formula to the straight-line model $\hat{f}(x) = \theta_1 + \theta_2 x$, we find that $\theta_2 = \beta$, and $\theta_1 = r^{\mathrm{rf}} + \alpha - \beta \mu^{\mathrm{mkt}}$.

The prediction of asset return $\hat{f}(x)$ has two components: A constant $r^{\mathrm{rf}} + \alpha$, and one that is proportional to the de-meaned market performance, $\beta(x - \mu^{\mathrm{mkt}})$. The second component, which has average value zero, relates market return fluctuations to the asset return fluctuations, and is related to the correlation of the asset and market returns; see exercise 13.4. The parameter α is the average asset return, over and above the risk-free interest rate. This model of asset return in terms of the market return is so common that the terms 'Alpha' and 'Beta' are widely used in finance. (Though not always with exactly the same meaning, since there are a few variations on how the parameters are defined.)

Time series trend. Suppose the data represents a series of samples of a quantity y at time (epoch) $x^{(i)} = i$. The straight-line fit to the time series data,

$$\hat{y}^{(i)} = \theta_1 + \theta_2 i, \quad i = 1, \ldots, N,$$

is called the *trend line*. Its slope, which is θ_2, is interpreted as the *trend* in the quantity over time. Subtracting the trend line from the original time series we get the *de-trended time series*, $y^{\mathrm{d}} - \hat{y}^{\mathrm{d}}$. The de-trended time series shows how the time series compares with its straight-line fit: When it is positive, it means the time series is above its straight-line fit, and when it is negative, it is below the straight-line fit.

An example is shown in figures 13.3 and 13.4. Figure 13.3 shows world petroleum consumption versus year, along with the straight-line fit. Figure 13.4 shows the de-trended world petroleum consumption.

Estimation of trend and seasonal component. In the previous example, we used least squares to approximate a time series $y^{\mathrm{d}} = (y^{(1)}, \ldots, y^{(N)})$ of length N by a sum of two components: $y^{\mathrm{d}} \approx \hat{y}^{\mathrm{d}} = \hat{y}^{\mathrm{const}} + \hat{y}^{\mathrm{lin}}$ where

$$\hat{y}^{\mathrm{const}} = \theta_1 \mathbf{1}, \qquad \hat{y}^{\mathrm{lin}} = \theta_2 \begin{bmatrix} 1 \\ 2 \\ \vdots \\ N \end{bmatrix}.$$

In many applications, the de-trended time series has a clear periodic component, *i.e.*, a component that repeats itself periodically. As an example, figure 13.5 shows an estimate of the road traffic (total number of miles traveled in vehicles) in the US, for each month between January 2000 and December 2014. The most striking aspect of the time series is the pattern that is (approximately) repeated every year, with a peak in the summer and a minimum in the winter. In addition there is a slowly increasing long term trend. The bottom figure shows the least squares fit of a sum of two components

$$y^{\mathrm{d}} \approx \hat{y}^{\mathrm{d}} = \hat{y}^{\mathrm{lin}} + \hat{y}^{\mathrm{seas}},$$

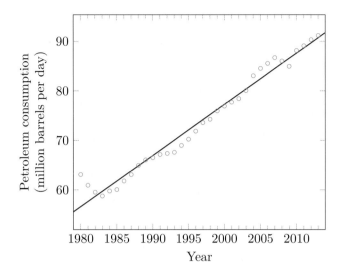

Figure 13.3 World petroleum consumption between 1980 and 2013 (dots) and least squares straight-line fit (data from www.eia.gov).

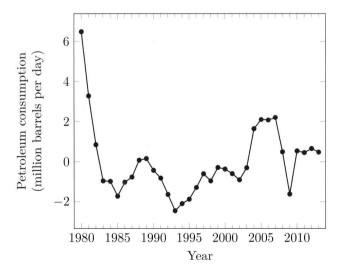

Figure 13.4 De-trended world petroleum consumption.

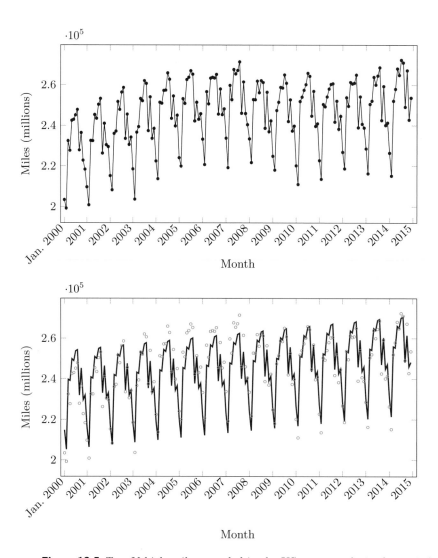

Figure 13.5 *Top.* Vehicle miles traveled in the US, per month, in the period January 2000 – December 2014 (U.S. Department of Transportation, Bureau of Transportation Statistics, www.transtats.bts.gov). *Bottom.* Least squares fit of a sum of two time series: A linear trend and a seasonal component with a 12-month period.

where $\hat{y}^{\mathrm{lin}}$ and $\hat{y}^{\mathrm{seas}}$ are defined as

$$\hat{y}^{\mathrm{lin}} = \theta_1 \begin{bmatrix} 1 \\ 2 \\ \vdots \\ N \end{bmatrix}, \qquad \hat{y}^{\mathrm{seas}} = \begin{bmatrix} \theta_{2:(P+1)} \\ \theta_{2:(P+1)} \\ \vdots \\ \theta_{2:(P+1)} \end{bmatrix}.$$

The second component is periodic or *seasonal*, with period $P = 12$, and consists of the pattern $(\theta_2, \dots, \theta_{P+1})$, repeated N/P times (we assume N is a multiple of P). The constant term is omitted in the model because it would be redundant: It has the same effect as adding a constant to the parameters $\theta_2, \dots, \theta_{P+1}$.

The least squares fit is computed by minimizing $\|A\theta - y^{\mathrm{d}}\|^2$ where θ is a $(P+1)$-vector and the matrix A in (13.1) is given by

$$A = \begin{bmatrix} 1 & 1 & 0 & \cdots & 0 \\ 2 & 0 & 1 & \cdots & 0 \\ \vdots & \vdots & \vdots & \ddots & \vdots \\ P & 0 & 0 & \cdots & 1 \\ P+1 & 1 & 0 & \cdots & 0 \\ P+2 & 0 & 1 & \cdots & 0 \\ \vdots & \vdots & \vdots & \ddots & \vdots \\ 2P & 0 & 0 & \cdots & 1 \\ \vdots & \vdots & \vdots & & \vdots \\ N-P+1 & 1 & 0 & \cdots & 0 \\ N-P+2 & 0 & 1 & \cdots & 0 \\ \vdots & \vdots & \vdots & \ddots & \vdots \\ N & 0 & 0 & \cdots & 1 \end{bmatrix}.$$

In this example, $N = 15P = 180$. The residual or prediction error in this case is called the de-trended, seasonally-adjusted series.

Polynomial fit.   A simple extension beyond the straight-line fit is a *polynomial fit*, with

$$f_i(x) = x^{i-1}, \quad i = 1, \dots, p,$$

so $\hat{f}$ is a polynomial of degree at most $p - 1$,

$$\hat{f}(x) = \theta_1 + \theta_2 x + \cdots + \theta_p x^{p-1}.$$

(Note that here, x^i means the generic scalar value x raised to the ith power; $x^{(i)}$ means the ith observed scalar data value.) In this case the matrix A in (13.1) has the form

$$A = \begin{bmatrix} 1 & x^{(1)} & \cdots & (x^{(1)})^{p-1} \\ 1 & x^{(2)} & \cdots & (x^{(2)})^{p-1} \\ \vdots & \vdots & & \vdots \\ 1 & x^{(N)} & \cdots & (x^{(N)})^{p-1} \end{bmatrix},$$

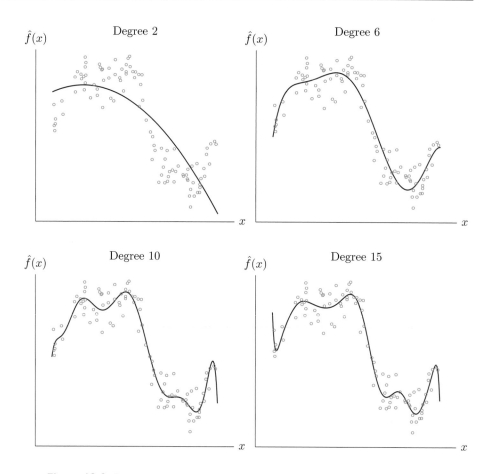

Figure 13.6 Least squares polynomial fits of degree 2, 6, 10, and 15 to 100 points.

i.e., it is a Vandermonde matrix (see (6.7)). Its columns are linearly independent provided the numbers $x^{(1)}, \ldots, x^{(N)}$ include at least p different values. Figure 13.6 shows an example of the least squares fit of polynomials of degree 2, 6, 10, and 15 to a set of 100 data points. Since any polynomial of degree less than r is also a polynomial of degree less than s, for $r \leq s$, it follows that the RMS fit attained by a polynomial with a larger degree is smaller (or at least, no larger) than that obtained by a fit with a smaller degree polynomial. This suggests that we should use the largest degree polynomial that we can, since this results in the smallest residual and the best RMS fit. But we will see in §13.2 that this is not true, and explore rational methods for choosing a model from among several candidates.

Piecewise-linear fit. A *piecewise-linear* function, with *knot points* or *kink points* $a_1 < a_2 < \cdots < a_k$, is a continuous function that is affine in between the knot points. (Such functions should be called piecewise-affine.) We can describe any

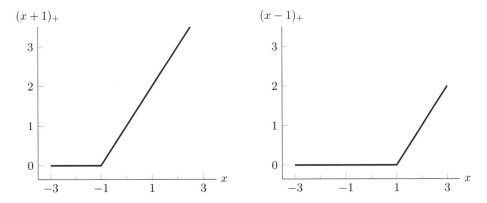

Figure 13.7 The piecewise-linear functions $(x + 1)_+ = \max\{x + 1, 0\}$ and $(x - 1)_+ = \max\{x - 1, 0\}$.

piecewise-linear function with k knot points using the $p = k + 2$ basis functions

$$f_1(x) = 1, \qquad f_2(x) = x, \qquad f_{i+2}(x) = (x - a_i)_+, \quad i = 1, \ldots, k,$$

where $(u)_+ = \max\{u, 0\}$. These basis functions are shown in figure 13.7 for $k = 2$ knot points at $a_1 = -1$, $a_2 = 1$. An example of a piecewise-linear fit with these knot points is shown in figure 13.8.

13.1.2   Regression

We now return to the general case when x is an n-vector. Recall that the regression model has the form

$$\hat{y} = x^T \beta + v,$$

where β is the weight vector and v is the offset. We can put this model in our general data fitting form using the basis functions $f_1(x) = 1$, and

$$f_i(x) = x_{i-1}, \quad i = 2, \ldots, n + 1,$$

so $p = n + 1$. The regression model can then be expressed as

$$\hat{y} = x^T \theta_{2:(n+1)} + \theta_1,$$

and we see that $\beta = \theta_{2:n+1}$ and $v = \theta_1$.

The $N \times (n + 1)$ matrix A in our general data fitting form is given by

$$A = \begin{bmatrix} \mathbf{1} & X^T \end{bmatrix},$$

where X is the feature matrix with columns $x^{(1)}, \ldots, x^{(N)}$. So the regression model is a special case of our general linear in the parameters model.

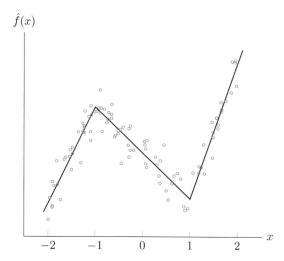

Figure 13.8 Piecewise-linear fit to 100 points.

General fitting model as regression. The regression model is a special case of our general data fitting model. Conversely, we can think of our linear in the parameters model as regression, with a different set of feature vectors of dimension $p - 1$. Assuming the first basis element f_1 is the constant function with value one, we consider the new or generated feature vectors $\tilde{x}$ given by

$$\tilde{x} = \begin{bmatrix} f_2(x) \\ \vdots \\ f_p(x) \end{bmatrix}.$$

A regression model for the outcome y and the new generated or mapped features $\tilde{x}$ has the form

$$\hat{y} = \tilde{x}^T \beta + v,$$

where β has dimension $p - 1$, and v is a number. Comparing this to our linear in the parameters model

$$\hat{y} = \theta_1 f_1(x) + \cdots + \theta_p f_p(x),$$

we see that they are the same, with $v = \theta_1$, $\beta = \theta_{2:p}$. So we can think of the general linear in the parameters model as nothing more than simple regression, but applied to the transformed, mapped, or generated features $f_1(x), \ldots, f_p(x)$. (This idea is discussed more in §13.3.)

House price regression. In §2.3 we described a simple regression model for the selling price of a house based on two attributes, area and number of bedrooms. The values of the parameters β and the offset v given in (2.9) were computed by least squares fitting on a set of data consisting of 774 house sales in Sacramento over a 5 day period. The RMS fitting error for the model is 74.8 (in thousands of dollars). For comparison, the standard deviation of the prices in the data set is 112.8. So this very basic regression model predicts the prices substantially better than a constant model (*i.e.*, the mean price of the houses in the data set).

Auto-regressive time series model.   Suppose that $z_1, z_2, \ldots$ is a time series. An *auto-regressive model* (also called *AR model*) for the time series has the form

$$\hat{z}_{t+1} = \theta_1 z_t + \cdots + \theta_M z_{t-M+1}, \quad t = M, M+1, \ldots$$

where M is the *memory* or *lag* of the model. Here $\hat{z}_{t+1}$ is the prediction of z_{t+1} made at time t (when $z_t, \ldots, z_{t-M+1}$ are known). This prediction is a linear function of the previous M values of the time series. With good choice of model parameters, the AR model can be used to predict the next value in a time series, given the current and previous M values. This has many practical uses.

We can use least squares (or regression) to choose the AR model parameters, based on the observed data $z_1, \ldots, z_T$, by minimizing the sum of squares of the *prediction errors* $z_t - \hat{z}_t$ over $t = M+1, \ldots, T$, i.e.,

$$(z_{M+1} - \hat{z}_{M+1})^2 + \cdots + (z_T - \hat{z}_T)^2.$$

(We must start the predictions at $t = M+1$, since each prediction involves the previous M time series values, and we do not know $z_0, z_{-1}, \ldots$)

The AR model can be put into the general linear in the parameters model form by taking

$$y^{(i)} = z_{M+i}, \quad x^{(i)} = (z_{M+i-1}, \ldots, z_i), \quad i = 1, \ldots, T - M.$$

We have $N = T - M$ examples, and $n = M$ features.

As an example, consider the time series of hourly temperature at Los Angeles International Airport, May 1–31, 2016, with length $31 \cdot 24 = 744$. The simple constant prediction $\hat{z}_{t+1} = 61.76°F$ (the average temperature) has RMS prediction error $3.05°F$ (the standard deviation). The very simple predictor $\hat{z}_{t+1} = z_t$, i.e., guessing that the temperature next hour is the same as the current temperature, has RMS error $1.16°F$. The predictor $\hat{z}_{t+1} = z_{t-23}$, i.e., guessing that the temperature next hour is what is was yesterday at the same time, has RMS error $1.73°F$.

We fit an AR model with memory $M = 8$ using least squares, with $N = 31 \cdot 24 - 8 = 736$ samples. The RMS error of this predictor is $0.98°F$, smaller than the RMS errors for the simple predictors described above. Figure 13.9 shows the temperature and the predictions for the first five days.

13.1.3   Log transform of dependent variable

When the dependent variable y is positive and varies over a large range, it is common to replace it with its logarithm $w = \log y$, and then use least squares to develop a model for w, $\hat{w} = \hat{g}(x)$. We then form our estimate of y using $\hat{y} = e^{\hat{g}(x)}$. When we fit a model $\hat{w} = \hat{g}(x)$ to the logarithm $w = \log y$, the fitting error for w can be interpreted in terms of the *percentage* or *relative* error between $\hat{y}$ and y, defined as

$$\eta = \max\{\hat{y}/y, y/\hat{y}\} - 1.$$

So $\eta = 0.1$ means either $\hat{y} = 1.1y$ (i.e., we over-estimate by 10%) or $\hat{y} = (1/1.1)y$ (i.e., we under-estimate by 10%). The connection between the relative error between $\hat{y}$ and y, and the residual r in predicting w, is

$$\eta = e^{|r|} - 1$$

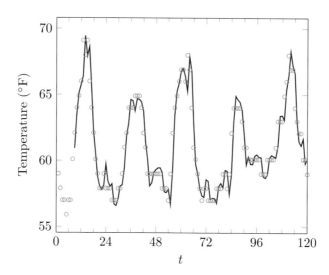

Figure 13.9 Hourly temperature at Los Angeles International Airport between 12:53AM on May 1, 2016, and 11:53PM on May 5, 2016, shown as circles. The solid line is the prediction of an auto-regressive model with eight coefficients.

(see exercise 13.16). For example, a residual with $|r| = 0.05$ corresponds (approximately) to a relative error in our prediction $\hat{y}$ of y of 5%. (Here we use the approximation $e^{|r|} - 1 \approx |r|$ for r smaller than, say, 0.15.) So if our RMS error in predicting $w = \log y$ across our examples is 0.05, our predictions of y are typically within $\pm 5\%$.

As an example suppose we wish to model house sale prices over an area and period that includes sale prices varying from under \$100k to over \$10M. Fitting the prices directly means that we care equally about absolute errors in price predictions, so, for example, a prediction error of \$20k should bother us equally for a house that sells for \$70k and one that sells for \$6.5M. But (at least for some purposes) in the first case we have made a very poor estimate, and the second case, a remarkably good one. When we fit the logarithm of the house sale price, we are seeking low percentage prediction errors, not necessarily low absolute errors.

Whether or not to use a logarithmic transform on the dependent variable (when it is positive) is a judgment call that depends on whether we seek small absolute prediction errors or small relative or percentage prediction errors.

13.2   Validation

Generalization ability. In this section we address a key point in model fitting: The goal of model fitting is typically *not* to just achieve a good fit on the given data set, but rather to achieve a good fit on *new data that we have not yet seen.*

This leads us to a basic question: How well can we expect a model to predict y for future or other unknown values of x? Without some assumptions about the future data, there is no good way to answer this question.

One very common assumption is that the data are described by a formal probability model. With this assumption, techniques from probability and statistics can be used to predict how well a model will work on new, unseen data. This approach has been very successful in many applications, and we hope that you will learn about these methods in another course. In this book, however, we will take a simple intuitive approach to this issue.

If a model predicts the outcomes for new unseen data values as well, or nearly as well, as it predicts the outcomes on the data used to form the model, it is said to have *good generalization ability*. In the opposite case, when the model makes predictions on new unseen data that are much worse than the predictions on the data used to form the model, the model is said to have *poor generalization ability*. So our question is: How can we assess the generalization ability of a model?

Validation on a test set.   A simple but effective method for assessing the generalization ability of a model is called *out-of-sample validation*. We divide the data we have into two sets: A *training set* and a *test set* (also called a *validation set*). This is often done randomly, with 80% of the data put into the training set and 20% put into the test set. A common way to describe this is to say that '20% of the data were reserved for validation'. Another common choice for the split ratio between the training set and the test set is 90%–10%.

To fit our model, we use *only the data in the training set*. The model that we come up with is based only on the data in the training set; the data in the test set has never been 'seen' by the model. Then we judge the model by its RMS fit *on the test set*. Since the model was developed without any knowledge of the test set data, the test data are effectively data that are new and unseen, and the performance of our model on this data gives us at least an idea of how our model will perform on new, unseen data. If the RMS prediction error on the test set is much larger than the RMS prediction error on the training set, we conclude that our model has poor generalization ability. Assuming that the test data are 'typical' of future data, the RMS prediction error on the test set is what we might guess our RMS prediction error will be on new data.

If the RMS prediction error of the model on the training set is similar to the RMS prediction error on the test set, we have increased confidence that our model has reasonable generalization ability. (A more sophisticated validation method called *cross-validation*, described below, can be used to gain even more confidence.)

For example, if our model achieves an RMS prediction error of 10% (compared to $\mathbf{rms}(y)$) on the training set and 11% on the test set, we can *guess* that it will have a similar RMS prediction error on other unseen data. But there is no guarantee of this, without further assumptions about the data. The basic assumption we are making here is that the future data will 'look like' the test data, or that the test data were 'typical'. Ideas from statistics can make this idea more precise, but we will leave this idea informal and intuitive.

Over-fitting. When the RMS prediction error on the training set is much smaller than the RMS prediction error on the test set, we say that the model is *over-fit*. It tells us that, for the purposes of making predictions on new, unseen data, the model is much less valuable than its performance on the training data suggests. Roughly speaking, an over-fit model trusts the data it has seen (*i.e.*, the training set) too much; it is too sensitive to the changes in the data that will likely be seen in the future data. One method for avoiding over-fit is to keep the model simple; another technique, called regularization, is discussed in chapter 15. Over-fit can be detected and (one hopes) avoided by validating a model on a test set.

Model prediction quality and generalization ability. Model generalization ability and training set prediction quality are not the same. A model can perform poorly and yet have good generalization ability. As an example, consider the (very simple) model that always makes the prediction $\hat{y} = 0$. This model will (likely) perform poorly on the training set and the test set data, with similar RMS errors, assuming the two data sets are 'similar'. So this model has good generalization ability, but has poor prediction quality. In general, we seek a model that *makes good predictions on the training data set* and also *makes good predictions on the test data set*. In other words, we seek a model with good performance and generalization ability. We care much more about a model's performance on the test data set than the training data set, since its performance on the test data set is much more likely to predict how the model will do on (other) unseen data.

Choosing among different models. We can use least squares fitting to fit multiple models to the same data. For example, in univariate fitting, we can fit a constant, an affine function, a quadratic, or a higher order polynomial. Which is the best model among these? Assuming that the goal is to make good predictions on new, unseen data, *we should choose the model with the smallest RMS prediction error on the test set*. Since the RMS prediction error on the test set is only a guess about what we might expect for performance on new, unseen data, we can soften this advice to *we should choose a model that has test set RMS error that is near the minimum over the candidates*. If multiple candidates achieve test set performance near the minimum, we should choose the 'simplest' one among these candidates.

We observed earlier that when we add basis functions to a model, our fitting error on the training data can only decrease (or stay the same). But this is not true for the test error. The test error need not decrease when we add more basis functions. Indeed, when we have too many basis functions, we can expect over-fit, *i.e.*, larger error on the test set.

If we have a sequence of basis functions $f_1, f_2, \ldots$ we can consider models based on using just f_1 (which is often the constant function 1), then f_1 and f_2, and so on. As we increase p, the number of basis functions, our training error will go down (or stay the same). But the test error typically decreases at first and then starts to increase for larger p. The intuition for this typical behavior is that for p too small, our model is 'too simple' to fit the data well, and so cannot make good predictions; when p is too large, our model is 'too complex' and suffers from over-fit, and so makes poor predictions. Somewhere in the middle, where the model achieves near minimum test set performance, is a good choice (or several good choices) of p.

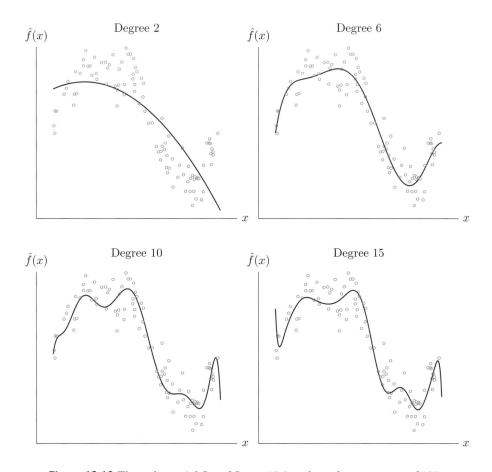

Figure 13.10 The polynomial fits of figure 13.6 evaluated on a test set of 100 points.

Example. To illustrate these ideas, we consider the example shown in figure 13.6. Using a training set of 100 points, we find least squares fits of polynomials of degrees $0, 1, \ldots, 20$. (The polynomial fits of degrees 2, 6, 10, and 15 are shown in the figure.) We now obtain a new set of data for validation, also with 100 points. These test data are plotted along with the polynomial fits obtained from the training data in figure 13.10. This is a real check of our models, since these data points were not used to develop the models. Figure 13.11 shows the RMS training and test errors for polynomial fits of different degrees. We can see that the RMS training error decreases with every increase in degree. The RMS test error decreases until degree 6 and starts to increase for degrees larger than 6. This plot suggests that a polynomial fit of degree 6 is a reasonable choice. (Degree 4 is another reasonable choice, considering that it achieves nearly minimum test error, and is 'simpler' than the model with degree 6.) Note also that the models with degrees 0, 1, and 2 have good generalization ability (*i.e.*, similar performance on the training and test sets), but worse prediction performance than models with higher degrees.

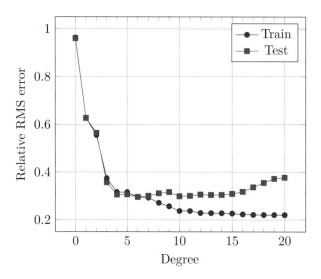

Figure 13.11 RMS error versus polynomial degree for the fitting example in figures 13.6 and 13.10. Circles indicate RMS errors on the training set. Squares show RMS errors on the test set.

With a 6th degree polynomial, the relative RMS test error for both training and test sets is around 0.3. It is a good sign, in terms of generalization ability, that the training and test errors are similar. While there are no guarantees, we can guess that the 6th degree polynomial model will have a relative RMS error around 0.3 on new, unseen data, provided the new, unseen data is sufficiently similar to the test set data.

Cross-validation. Cross-validation is an extension of out-of-sample validation that can be used to get even more confidence in the generalization ability of a model, or more accurately, a choice of basis functions used to construct a model. We divide the original data set into 10 sets, called *folds*. We then fit the model using folds $1, 2, \ldots, 9$ as training data, and fold 10 as test data. (So far, this is the same as out-of-sample validation.) We then fit the model using folds $1, 2, \ldots, 8, 10$ as training data and fold 9 as the test data. We continue, fitting a model for each choice of one of the folds as the test set. We end up with 10 (presumably different) models, and 10 assessments of these models using the fold that was not used to fit the model. (We have described 10-fold cross-validation here; 5-fold cross-validation is also commonly used.) If the test fit performance of these 10 models is similar, we can expect the same, or at least similar, performance on new unseen data. In cross-validation we can also check for *stability* of the model coefficients. This means that the model coefficients found in the different folds are similar to each other. Stability of the model coefficients further enhances our confidence in the model.

To obtain a single number that is our guess of the prediction RMS error we can expect on new, unseen data, it is common practice to compute the RMS test set error across all 10 folds. For example, if $\epsilon_1, \ldots, \epsilon_{10}$ are the RMS prediction errors

	Model parameters			RMS error	
Fold	v	β_1	β_2	Train	Test
1	60.65	143.36	-18.00	74.00	78.44
2	54.00	151.11	-20.30	75.11	73.89
3	49.06	157.75	-21.10	76.22	69.93
4	47.96	142.65	-14.35	71.16	88.35
5	60.24	150.13	-21.11	77.28	64.20

Table 13.1 Five-fold cross-validation for the simple regression model of the house sales data set. The RMS cross-validation error is 75.41.

obtained by our models on the test folds, we take

$$\sqrt{(\epsilon_1^2 + \cdots + \epsilon_{10}^2)/10} \tag{13.4}$$

as our guess of the RMS error our models might make on new data. In a plot like that in figure 13.11, the RMS test error over all folds is plotted, instead of the RMS test error on the single data or validation set, as in that plot. The single number (13.4) is called the *RMS cross-validation error*, or simply the *RMS test error* (when cross-validation is used).

Note that cross-validation does not check a particular model, since it creates 10 different (but hopefully not very different) models. Cross-validation checks a selection of basis functions. Once cross-validation is used to verify that a choice of basis functions produces models that predict and generalize well, there is the question of which of the 10 models one should use. The models should be not too different, so the choice really should not matter much. One reasonable choice is to use the parameters obtained by fitting a model over all the data; another option is to use the average of the model parameters from the different folds.

House price regression model.   As an example, we apply cross-validation to assess the generalization ability of the simple regression model of the house sales data discussed in §2.3 and on page 258. The simple regression model described there, based on house area and number of bedrooms, has an RMS fitting error of 74.8 thousand dollars. Cross-validation will help us answer the question of how the model might do on different, unseen houses.

We randomly partition the data set of 774 sales records into five folds, four of size 155 and one of size 154. Then we fit five regression models, each of the form

$$\hat{y} = v + \beta_1 x_1 + \beta_2 x_2$$

to the data set after removing one of the folds. Table 13.1 summarizes the results. The model parameters for the 5 different regression models are not exactly the same, but quite similar. The training and test RMS errors are reasonably similar, which suggests that our model does not suffer from over-fit. Scanning the RMS error on the test sets, we can expect that our prediction error on new houses will

Fold	v	RMS error (train)	RMS error (test)
1	230.11	110.93	119.91
2	230.25	113.49	109.96
3	228.04	114.47	105.79
4	225.23	110.35	122.27
5	230.23	114.51	105.59

Table 13.2 Five-fold cross-validation for the constant model of the house sales data set. The RMS cross-validation error is 119.93.

be around 70–80 (thousand dollars) RMS. The RMS cross-validation error (13.4) is 75.41. We can also see that the model parameters change a bit, but not drastically, in each of the folds. This gives us more confidence that, for example, β_2 being negative is not a fluke of the data.

For comparison, table 13.2 shows the RMS errors for the constant model $\hat{y} = v$, where v is the mean price of the training set. The results suggest that the constant model can predict house prices with a prediction error around 105–120 (thousand dollars). The RMS cross-validation error for the constant model is 119.93.

Figure 13.12 shows the scatter plots of actual and regression model predicted prices for each of the five training and test sets. The results for training and test sets are reasonably similar in each case, which gives us confidence that the regression model will have similar performance on new, unseen houses.

Validating time series predictions.   When the original data are unordered, for example, patient records or customer purchase histories, the division of the data into training and test sets is typically done randomly. This same method can be used to validate a time series prediction model, such as an AR model, but it does not give the best emulation of how the model will ultimately be used. In practice, the model will be trained on past data and then used to make predictions on future data. When the training data in a time series prediction model are randomly chosen, the model is being built with some knowledge of the future, a phenomenon called *look-ahead* or *peek-ahead*. Look-ahead can make a model look better than it really is at making predictions.

To avoid look-ahead, the training set for a time series prediction model is typically taken to be the data examples up to some point in time, and the test data are chosen as points that are past that time (and sometimes, at least M samples past that time, taking into account the memory of the predictor). In this way we can say that the model is being tested by making predictions on data it has never seen. As an example, we might train an AR model for some daily quantity using data from the years 2006 through 2008, and then test the resulting AR model on the data from year 2009.

As an example, we return to the AR model of hourly temperatures at Los Angeles International Airport described on page 259. We divide the one month of data into a training set (May 1–24) and a test set (May 25–31). The coefficients in

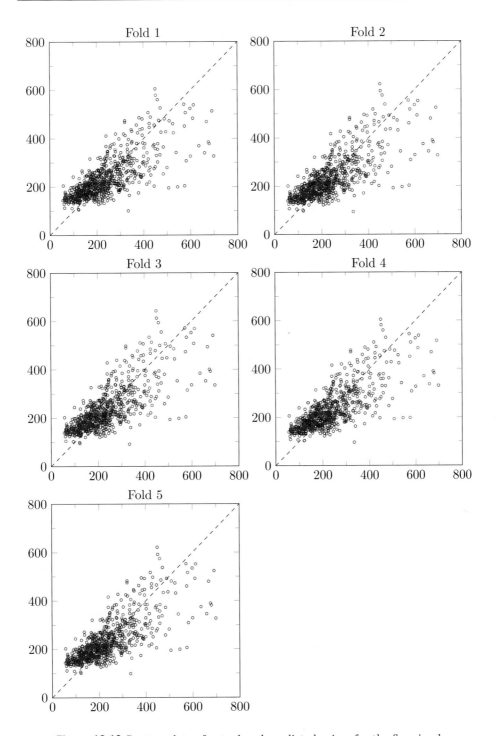

Figure 13.12 Scatter plots of actual and predicted prices for the five simple regression models of table 13.1. The horizontal axis is the actual selling price and the vertical axis is the predicted price, both in thousands of dollars. Blue circles are samples in the training set, red circles samples in the test set.

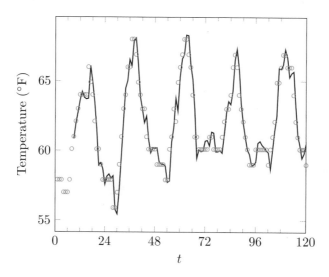

Figure 13.13 Hourly temperature at Los Angeles International Airport between 12:53AM on May 25, 2016, and 11:53PM on May 29, 2016, shown as circles. The solid line is the prediction of an auto-regressive model with eight coefficients, developed using training data from May 1 to May 24.

an AR model are computed using the $(24)(24) - 8 = 568$ samples in the training set. The RMS error on the training set is $1.03°F$. The RMS prediction error on the test set is $0.98°F$, which is similar to the RMS prediction error on the training set, giving us confidence that the AR model is not over-fit. (The fact that the test RMS error is very slightly smaller than the training RMS error has no significance.) Figure 13.13 shows the prediction on the first five days of the test set. The predictions look very similar to those shown in figure 13.9.

Limitations of out-of-sample and cross-validation.   Here we mention a few limitations of out-of-sample and cross-validation. First, the basic assumption that the test data and future data are similar can (and does) fail in some applications. For example, a model that predicts consumer demand, trained and validated on this year's data, can make much poorer predictions next year, simply because consumer tastes shift. In finance, patterns of asset returns periodically shift, so models that predict well on test data from this year need not predict well next year.

Another limitation arises when the data set is small, which makes it harder to interpret out-of-sample and cross-validation results. In this case the out-of-sample test RMS error might be small due to good luck, or large due to bad luck, in the selection of the test set. In cross-validation the test results can vary considerably, due to luck of which data points fall into the different folds. Here too concepts from statistics can make this idea more precise, but we leave it as an informal idea: With small data sets, we can expect to see more variation in test RMS prediction error than with larger data sets.

Despite these limitations, out-of-sample and cross-validation are powerful and useful tools for assessing the generalization ability of a model.

13.3   Feature engineering

In this section we discuss some methods used to find appropriate basis functions or feature mappings $f_1, \ldots, f_p$. We observed above (in §13.1.2) that fitting a linear in the parameters model reduces to regression with new features which are the original features x mapped through the basis (or feature mapping) functions $f_1, \ldots, f_p$. Choosing the feature mapping functions is sometimes called *feature engineering*, since we are generating features to use in regression.

For a given data set we may consider several, or even many, candidate choices of the basis functions. To choose among these candidate choices of basis functions, we use out-of-sample validation or cross-validation.

Adding new features to get a richer model.   In many cases the basis functions include the constant one, *i.e.*, we have $f_1(x) = 1$. (This is equivalent to having the offset in the basic regression model.) It is also very common to include the original features as well, as in $f_i(x) = x_{i-1}$, $i = 2, \ldots, n+1$. If we do this, we are effectively starting with the basic regression model; we can then add new features to get a richer model. In this case we have $p > n$, so there are more mapped features than original features. (Whether or not it is a good idea to add the new features can be determined by out-of-sample validation or cross-validation.)

Dimension reduction.   In some cases, and especially when the number n of the original features is very large, the feature mappings are used to construct a smaller set of $p < n$ features. In this case we can think of the feature mappings or basis functions as a *dimension reduction* or *data aggregation* procedure.

13.3.1   Transforming features

Standardizing features.   Instead of using the original features directly, it is common to apply a scaling and offset to each original feature, say,

$$f_i(x) = (x_i - b_i)/a_i, \quad i = 2, \ldots, n+1,$$

so that across the data set, the average value of $f_i(x)$ is near zero, and the standard deviation is around one. (This is done by choosing b_i to be near the mean of the feature i values over the data set, and choosing a_i to be near the standard deviation of the values.) This is called *standardizing* or *z-scoring* the features. The standardized feature values are easily interpretable since they correspond to z-values; for example, $f_2(x) = +3.3$ means that the value of original feature 2 is quite a bit above the typical value. The standardization of each original feature is typically the first step in feature engineering.

Note that the constant feature $f_1(x) = 1$ is not standardized. (In fact, it cannot be standardized since its standard deviation across the data set is zero.)

Winsorizing features.   When the data include some very large values that are thought to be errors (say, in collecting the data), it is common to *clip* or *winsorize* the data. This means that we set any data values with absolute value larger than some chosen threshold to their sign times the chosen threshold. Assuming, for example, that a feature entry x_5 has already been standardized (so it represents z-scores across the examples), we replace x_5 with its winsorized value (with threshold 3),

$$\tilde{x}_5 = \left\{ \begin{array}{ll} x_5 & |x_5| \leq 3 \\ 3 & x_5 > 3 \\ -3 & x_5 < -3. \end{array} \right.$$

The term winsorize is named after the statistician Charles P. Winsor.

Log transform.   When feature values are positive and vary over a wide range, it is common to replace them with their logarithms. If the feature value also includes the value 0 (so the logarithm is undefined) a common variation on the log transformation is to use $\tilde{x}_k = \log(x_k + 1)$. This compresses the range of values that we encounter. As an example, suppose the original features record the number of visits to websites over some time period. These can easily vary over a range of 10000:1 (or even more) for a very popular website and a less popular one; taking the logarithm of the visit counts gives a feature with less variation, which is possibly more interpretable. (The decision as to whether to use the original feature values or their logarithms can be decided by validation.)

13.3.2   Creating new features

Expanding categoricals.   Some features take on only a few values, such as -1 and 1 or 0 and 1, which might represent some value like presence or absence of some symptom. (Such features are called Boolean.) A Likert scale response (see page 71) naturally only takes on a small number of values, such as -2, -1, 0, 1, 2. Another example is an original feature that takes on the values $1, 2, \ldots, 7$, representing the day of the week. Such features are called *categorical* in statistics, since they specify which category the example is in, and not some real number.

Expanding a categorical feature with l values means replacing it with a set of $l - 1$ new features, each of which is Boolean, and simply records whether or not the original feature has the associated value. (When all these features are zero, it means the original feature had the default value.) As an example, suppose the original feature x_1 takes on only the values -1, 0, and 1. Using the feature value 0 as the default feature value, we replace x_1 with the two mapped features

$$f_1(x) = \left\{ \begin{array}{ll} 1 & x_1 = -1 \\ 0 & \text{otherwise,} \end{array} \right. \qquad f_2(x) = \left\{ \begin{array}{ll} 1 & x_1 = 1 \\ 0 & \text{otherwise.} \end{array} \right.$$

In words, $f_1(x)$ tells us if x_1 has the value -1, and $f_2(x)$ tells us if x_1 has the value 1. (We do not need a new feature for the default value $x_1 = 0$; this corresponds to $f_1(x) = f_2(x) = 0$.) This feature mapping is shown in table 13.3.

Expanding a categorical feature with l values into $l - 1$ features that encode whether the feature has one of the (non-default) values is sometimes called *one-hot encoding*, because for any data example, only one of the new feature values is one, and the others are zero. (When the original feature has the default value, all the new features are zero.)

x_1	$f_1(x)$	$f_2(x)$
-1	1	0
0	0	0
1	0	1

Table 13.3 The original categorical feature x_1 takes on only the three values listed in the first column. This feature is replaced with (expanded into) the two features $f_1(x)$ and $f_2(x)$ shown in the second and third columns.

There is no need to expand an original feature that is Boolean (*i.e.*, takes on two values). If the original Boolean feature is encoded with the values 0 and 1, and 0 is taken as the default value, then the one new feature value will be the same as the original feature value.

As an example of expanding categoricals, consider a model that is used to predict house prices based on various features that include the number of bedrooms, that ranges from 1 to 5 (say). In the basic regression model, we use the number of bedrooms directly as a feature. In the basic model there is one parameter value that corresponds to value per bedroom; we multiply this parameter by the number of bedrooms to get the contribution to our price prediction. In this model, the price prediction increases (or decreases) by the same amount when we change the number of bedrooms from 1 to 2 as it does when we change the number of bedrooms from 4 to 5. If we expand this categorical feature, using 2 bedrooms as the default, we have 4 Boolean features that correspond to a house having 1, 3, 4, and 5 bedrooms. We then have 4 parameters in our model, which assign different amounts to add to our prediction for houses with 1, 3, 4, and 5 bedrooms, respectively. This more flexible model can capture the idea that a change from 1 to 2 bedrooms is different from a change from 4 to 5 bedrooms.

Generalized additive model. We introduce new features that are nonlinear functions of the original features, such as, for each x_i, the functions $\min\{x_i + a, 0\}$ and $\max\{x_i - b, 0\}$, where a and b are parameters. These new features are readily interpreted: $\min\{x_i + a, 0\}$ is the amount by which feature x_i is below $-a$, and $\max\{x_i - b, 0\}$ is the amount by which feature x_i is above b. A common choice, assuming that x_i has already been standardized, is $a = b = 1$. This leads to the predictor

$$\hat{y} = \psi_1(x_1) + \cdots + \psi_n(x_n), \tag{13.5}$$

where ψ_i is the piecewise-linear function

$$\psi_i(x_i) = \theta_{n+i} \min\{x_i + a, 0\} + \theta_i x_i + \theta_{2n+i} \max\{x_i - b, 0\}, \tag{13.6}$$

which has kink or knot points at the values $-a$ and $+b$. The model (13.5) has $3n$ parameters, corresponding to the original features, and the two additional features per original feature. The prediction $\hat{y}$ is a sum of functions of the original features, and is called a *generalized additive model*. (More complex versions add more than two additional functions of each original feature.)

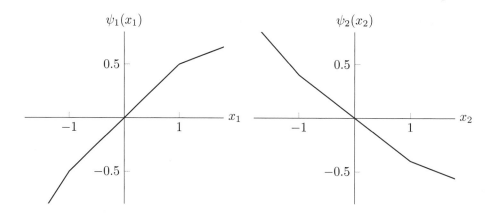

Figure 13.14 The functions ψ_i in (13.6) for $n = 2$, $a = b = 1$, and $\theta_1 = 0.5$, $\theta_2 = -0.4$, $\theta_3 = 0.3$, $\theta_4 = -0.2$, $\theta_5 = -0.3$, $\theta_6 = 0.2$.

Consider an example with $n = 2$ original features. Our prediction $\hat{y}$ is a sum of two piecewise-linear functions, each depending on one of the original features. Figure 13.14 shows an example. In this example model, we can say that increasing x_1 increases our prediction $\hat{y}$; but for high values of x_1 (*i.e.*, above 1) the increase in the prediction is less pronounced, and for low values (*i.e.*, below -1), it is more pronounced.

Products and interactions.   New features can be developed from pairs of original features, for example, their product. From the original features we can add $x_i x_j$, for $i, j = 1, \ldots, n$, $i \le j$. Products are used to model interactions among the features. Product features are easily interpretable when the original features are Boolean, *i.e.*, take the values 0 or 1. Thus $x_i = 1$ means that feature i is present or has occurred, and the new product feature $x_i x_j$ has the value 1 exactly when both feature i and j have occurred.

Stratified models.   In a *stratified model*, we have several different sub-models, and choose the one to use depending on the values of the regressors. For example, instead of treating gender as a regressor in a single model of some medical outcome, we build two different sub-models, one for male patients and one for female patients. In this case we choose the sub-model to use based on one of the original features, gender.

As a more general example, we can carry out clustering of the original feature vectors, and fit a separate model within each cluster. To evaluate $\hat{y}$ for a new x, we first determine which cluster x is in, and then use the associated model. Whether or not a stratified model is a good idea is checked using out-of-sample validation.

13.3.3 Advanced feature generation methods

Custom mappings. In many applications custom mappings of the raw data are used as additional features, in addition to the original features given. For example in a model meant to predict an asset's future price using prior prices, we might also use the highest and lowest prices over the last week. Another well known example in financial models is the price-to-earnings ratio, constructed from the price and (last) earnings features.

In document analysis applications word count features are typically replaced with *term frequency inverse document frequency* (TFIDF) values, which scale the raw count values by a function of the frequency with which the word appears across the given set of documents, usually in such a way that uncommon words are given more weight. (There are many variations on the particular scaling function to use. Which one to use in a given application can be determined by out-of-sample or cross-validation.)

Predictions from other models. In many applications there are existing models for the data. A common trick is to use the predictions of these models as features in your model. In this case you can describe your model as one that combines or blends the raw data available with predictions made from one or more existing models to create a new prediction.

Distance to cluster representatives. We can build new features from a clustering of the data into k groups. One simple method uses the cluster representatives $z_1, \ldots, z_k$, and gives k new features, given by $f(x) = e^{-\|x-z_i\|^2/\sigma^2}$, where σ is a parameter.

Random features. The new features are given by a nonlinear function of a *random* linear combination of the original features. To add K new features of this type, we first generate a random $K \times n$ matrix R. We then generate new features as $(Rx)_+$ or $|Rx|$, where $(\cdot)_+$ and $|\cdot|$ are applied elementwise to the vector Rx. (Other nonlinear functions can be used as well.)

This approach to generating new features is quite counter-intuitive, since you would imagine that feature engineering should be done using detailed knowledge of, and intuition about, the particular application. Nevertheless this method can be very effective in some applications.

Neural network features. A *neural network* computes transformed features using compositions of linear transformations interspersed with nonlinear mappings such as the absolute value. This architecture was originally inspired by biology, as a crude model of how human and animal brains work. The ideas behind neural networks are very old, but their use has accelerated over the last few years due to a combination of new techniques, greatly increased computing power, and access to large amounts of data. Neural networks can find good feature mappings directly from the data, provided there is a very large amount of data available.

13.3.4   Summary

The discussion above makes it clear that there is much art in choosing features to use in a model. But it is important to keep several things in mind when creating new features:

- *Try simple models first.* Start with a constant, then a simple regression model, and so on. You can compare more sophisticated models against these.

- *Compare competing candidate models using validation.* Adding new features will always reduce the RMS error on the training data, but the important question is whether or not it substantially reduces the RMS error on the test or validation data sets. (We add the qualifier 'substantially' here because a small reduction in test set error is not meaningful.)

- *Adding new features can easily lead to over-fit.* (This will show up when validating the model.) The most straightforward way to avoid over-fit is to keep the model simple. We mention here that another approach to avoiding over-fit, called *regularization* (covered in chapter 15), can be very effective when combined with feature engineering.

13.3.5   House price prediction

In this section we use feature engineering to develop a more complicated regression model for the house sales data, illustrating some of the methods described above. As mentioned in §2.3, the data set contains records of 774 house sales in the Sacramento area. For our more complex model we will use four base attributes or original features:

- x_1 is the area of the house (in 1000 square feet),

- x_2 is the number of bedrooms,

- x_3 is equal to one if the property is a condominium, and zero otherwise,

- x_4 is the five-digit ZIP code.

Only the first two attributes were used in the simple regression model

$$\hat{y} = \beta_1 x_1 + \beta_2 x_2 + v$$

given in §2.3. In that model, we do not carry out any feature engineering or modification.

Feature engineering.   Here we examine a more complicated model, with 8 basis functions,

$$\hat{y} = \sum_{i=1}^{8} \theta_i f_i(x).$$

These basis functions are described below.

x_4	$f_6(x)$	$f_7(x)$	$f_8(x)$
95811, 95814, 95816, 95817, 95818, 95819	0	0	0
95608, 95610, 95621, 95626, 95628, 95655, 95660, 95662, 95670, 95673, 95683, 95691, 95742, 95815, 95821, 95825, 95827, 95833, 95834, 95835, 95838, 95841, 95842, 95843, 95864	1	0	0
95624, 95632, 95690, 95693, 95757, 95758, 95820, 95822, 95823, 95824, 95826, 95828, 95829, 95831, 95832	0	1	0
95603, 95614, 95630, 95635, 95648, 95650, 95661, 95663, 95677, 95678, 95682, 95722, 95746, 95747, 95762, 95765	0	0	1

Table 13.4 Definition of basis functions f_6, f_7, f_8 as functions of x_4 (5-digit ZIP code).

The first basis function is the constant $f_1(x) = 1$. The next two are functions of x_1, the area of the house,

$$f_2(x) = x_1, \qquad f_3(x) = \max\{x_1 - 1.5, 0\}.$$

In words, $f_2(x)$ is the area of the house, and $f_3(x)$ is the amount by which the area exceeds 1.5 (*i.e.*, 1500 square feet). The first three basis functions contribute to the price prediction model a piecewise-linear function of the house area,

$$\theta_1 f_1(x) + \theta_2 f_2(x) + \theta_3 f_3(x) = \begin{cases} \theta_1 + \theta_2 x_1 & x_1 \le 1.5 \\ \theta_1 + (\theta_2 + \theta_3)x_1 & x_1 > 1.5, \end{cases}$$

with one knot at 1.5. This is an example of a generalized additive model described on page 271.

The basis function $f_4(x)$ is equal to the number of bedrooms x_2. The basis function $f_5(x)$ is equal to x_3, *i.e.*, one if the property is a condominium, and zero otherwise. In these cases we simply using the original feature value, with no transformation or modification.

The last three basis functions are again Boolean, and indicate or encode the location of the house. We partition the 62 different ZIP codes present in the data set into four groups, corresponding to different areas around the center of Sacramento, as shown in table 13.4. The basis functions f_6, f_7, and f_8 give a one-hot encoding of the four groups of ZIP codes, as described on page 270.

The resulting model.   The coefficients in the least squares fit are

$$\theta_1 = 115.62, \quad \theta_2 = 175.41, \quad \theta_3 = -42.75, \quad \theta_4 = -17.88,$$

$$\theta_5 = -19.05, \quad \theta_6 = -100.91, \quad \theta_7 = -108.79, \quad \theta_8 = -24.77.$$

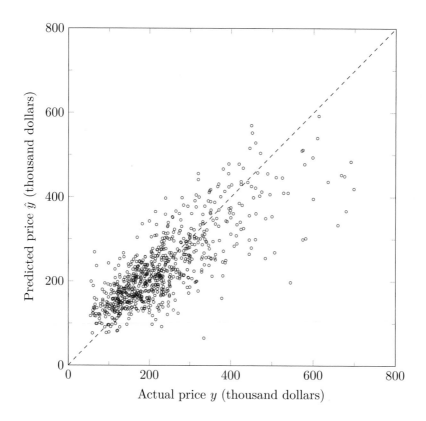

Figure 13.15 Scatter plot of actual and predicted prices for a model with eight parameters.

The RMS fitting error is 68.3, a bit better than the simple regression fit, which achieves RMS fitting error of 74.8. Figure 13.15 shows a scatter plot of predicted and actual prices.

To validate the model, we use 5-fold cross-validation, using the same folds as in table 13.1 and figure 13.12. The results are shown in table 13.5 and figure 13.16. The training and test set errors are similar, so our model is not over-fit. We also see that the test set errors are a bit better than those obtained by our simple regression model; the RMS cross-validation errors are 69.29 and 75.41, respectively. We conclude that our more complex model, that uses feature engineering, gives a modest (around 8%) improvement in prediction ability over the simple regression model based on only house area and number of bedrooms. (With more data, more features, and more feature engineering, a much more accurate model of house price can be developed.)

The table also shows that the model coefficients are reasonably stable across the different folds, giving us more confidence in the model. Another interesting phenomenon we observe is that the test error for fold 5 is a bit *lower* on the test set than on the training set. This occasionally happens, as a consequence of how the original data were split into folds.

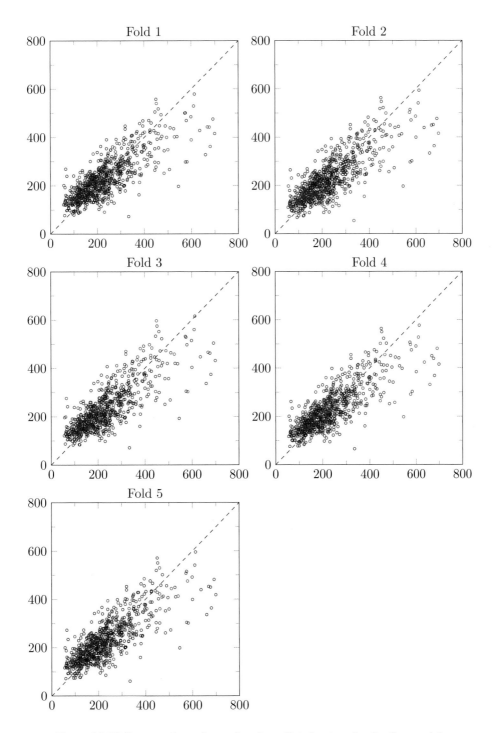

Figure 13.16 Scatter plots of actual and predicted prices for the five models of table 13.5. The horizontal axis is the actual selling price and the vertical axis is the predicted price, both in thousands of dollars. Blue circles are samples in the training set, red circles samples in the test set.

Fold				Model parameters					RMS error	
	θ_1	θ_2	θ_3	θ_4	θ_5	θ_6	θ_7	θ_8	Train	Test
1	122.35	166.87	−39.27	−16.31	−23.97	−100.42	−106.66	−25.98	67.29	72.78
2	100.95	186.65	−55.80	−18.66	−14.81	−99.10	−109.62	−17.94	67.83	70.81
3	133.61	167.15	−23.62	−18.66	−14.71	−109.32	−114.41	−28.46	69.70	63.80
4	108.43	171.21	−41.25	−15.42	−17.68	−94.17	−103.63	−29.83	65.58	78.91
5	114.45	185.69	−52.71	−20.87	−23.26	−102.84	−110.46	−23.43	70.69	58.27

Table 13.5 Five-fold validation on the house sales data set. The RMS cross-validation error is 69.29.

Exercises

13.1 *Error in straight-line fit.* Consider the straight-line fit described on page 249, with data given by the N-vectors x^{d} and y^{d}. Let $r^{\mathrm{d}} = y^{\mathrm{d}} - \hat{y}^{\mathrm{d}}$ denote the residual or prediction error using the straight-line model (13.3). Show that $\mathbf{rms}(r^{\mathrm{d}}) = \mathbf{std}(y^{\mathrm{d}})\sqrt{1 - \rho^2}$, where ρ is the correlation coefficient of x^{d} and y^{d} (assumed non-constant). This shows that the RMS error with the straight-line fit is a factor $\sqrt{1 - \rho^2}$ smaller than the RMS error with a constant fit, which is $\mathbf{std}(y^{\mathrm{d}})$. It follows that when x^{d} and y^{d} are highly correlated ($\rho \approx 1$) or anti-correlated ($\rho \approx -1$), the straight-line fit is much better than the constant fit. *Hint.* From (13.3) we have

$$\hat{y}^{\mathrm{d}} - y^{\mathrm{d}} = \rho\frac{\mathbf{std}(y^{\mathrm{d}})}{\mathbf{std}(x^{\mathrm{d}})}(x^{\mathrm{d}} - \mathbf{avg}(x^{\mathrm{d}})\mathbf{1}) - (y^{\mathrm{d}} - \mathbf{avg}(y^{\mathrm{d}})\mathbf{1}).$$

Expand the norm squared of this expression, and use

$$\rho = \frac{(x^{\mathrm{d}} - \mathbf{avg}(x^{\mathrm{d}})\mathbf{1})^T(y^{\mathrm{d}} - \mathbf{avg}(y^{\mathrm{d}})\mathbf{1})}{\|x^{\mathrm{d}} - \mathbf{avg}(x^{\mathrm{d}})\mathbf{1}\|\|y^{\mathrm{d}} - \mathbf{avg}(y^{\mathrm{d}})\mathbf{1}\|}.$$

13.2 *Regression to the mean.* Consider a data set in which the (scalar) $x^{(i)}$ is the parent's height (average of mother's and father's height), and $y^{(i)}$ is their child's height. Assume that over the data set the parent and child heights have the same mean value μ, and the same standard deviation σ. We will also assume that the correlation coefficient ρ between parent and child heights is (strictly) between zero and one. (These assumptions hold, at least approximately, in real data sets that are large enough.) Consider the simple straight-line fit or regression model given by (13.3), which predicts a child's height from the parent's height. Show that this prediction of the child's height lies (strictly) between the parent's height and the mean height μ (unless the parent's height happens to be exactly the mean μ). For example, if the parents are tall, *i.e.*, have height above the mean, we predict that their child will be shorter, but still tall. This phenomenon, called *regression to the mean*, was first observed by the early statistician Sir Francis Galton (who indeed, studied a data set of parent's and child's heights).

13.3 *Moore's law.* The figure and table below show the number of transistors N in 13 microprocessors, and the year of their introduction.

Year	Transistors
1971	2,250
1972	2,500
1974	5,000
1978	29,000
1982	120,000
1985	275,000
1989	1,180,000
1993	3,100,000
1997	7,500,000
1999	24,000,000
2000	42,000,000
2002	220,000,000
2003	410,000,000

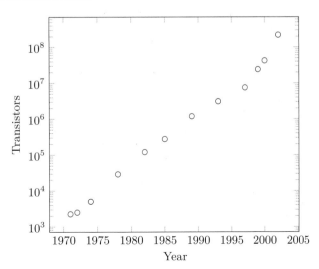

The plot gives the number of transistors on a logarithmic scale. Find the least squares straight-line fit of the data using the model

$$\log_{10} N \approx \theta_1 + \theta_2(t - 1970),$$

where t is the year and N is the number of transistors. Note that θ_1 is the model's prediction of the log of the number of transistors in 1970, and 10^{θ_2} gives the model's prediction of the fractional increase in number of transistors per year.

(a) Find the coefficients θ_1 and θ_2 that minimize the RMS error on the data, and give the RMS error on the data. Plot the model you find along with the data points.

(b) Use your model to predict the number of transistors in a microprocessor introduced in 2015. Compare the prediction to the IBM Z13 microprocessor, released in 2015, which has around 4×10^9 transistors.

(c) Compare your result with Moore's law, which states that the number of transistors per integrated circuit roughly doubles every one and a half to two years.

The computer scientist and Intel corporation co-founder Gordon Moore formulated the law that bears his name in a magazine article published in 1965.

13.4 *Asset α and β and market correlation.* Suppose the T-vectors $r^{\mathrm{ind}} = (r_1^{\mathrm{ind}}, \ldots, r_T^{\mathrm{ind}})$ and $r^{\mathrm{mkt}} = (r_1^{\mathrm{mkt}}, \ldots, r_T^{\mathrm{mkt}})$ are return time series for a specific asset and the whole market, as described on page 251. We let r^{rf} denote the risk-free interest rate, μ^{mkt} and σ^{mkt} the market return and risk (*i.e.*, $\mathbf{avg}(r^{\mathrm{mkt}})$ and $\mathbf{std}(r^{\mathrm{mkt}})$), and μ and σ the return and risk of the asset (*i.e.*, $\mathbf{avg}(r^{\mathrm{ind}})$ and $\mathbf{std}(r^{\mathrm{ind}})$). Let ρ be the correlation coefficient between the market and asset return time series r^{mkt} and r^{ind}. Express the asset α and β in terms of r^{rf}, μ, σ, μ^{mkt}, σ^{mkt}, and ρ.

13.5 *Polynomial model with multiple features.* The idea of polynomial models can be extended from the case discussed on page 255 where there is only one feature. In this exercise we consider a quadratic (degree two) model with 3 features, *i.e.*, x is a 3-vector. This has the form

$$\hat{f}(x) = a + b_1 x_1 + b_2 x_2 + b_3 x_3 + c_1 x_1^2 + c_2 x_2^2 + c_3 x_3^2 + c_4 x_1 x_2 + c_5 x_1 x_3 + c_6 x_2 x_3,$$

where the scalar a, 3-vector b, and 6-vector c are the zeroth, first, and second order coefficients in the model. Put this model into our general linear in the parameters form, by giving p, and the basis functions $f_1, \ldots, f_p$ (which map 2-vectors to scalars).

13.6 *Average prediction error.* Consider a data fitting problem, with first basis function $\phi_1(x) = 1$, and data set $x^{(1)}, \ldots, x^{(N)}$, $y^{(1)}, \ldots, y^{(N)}$. Assume the matrix A in (13.1) has linearly independent columns and let $\hat{\theta}$ denote the parameter values that minimize the mean square prediction error over the data set. Let the N-vector $\hat{r}^{\mathrm{d}}$ denote the prediction errors using the optimal model parameter $\hat{\theta}$. Show that $\mathbf{avg}(\hat{r}^{\mathrm{d}}) = 0$. In other words: With the least squares fit, the mean of the prediction errors over the data set is zero. *Hint.* Use the orthogonality principle (12.9), with $z = e_1$.

13.7 *Data matrix in auto-regressive time series model.* An auto-regressive model with memory M is fit by minimizing the sum of the squares of the predictions errors on a data set with T samples, $z_1, \ldots, z_T$, as described on page 259. Find the matrix A and vector y for which $\|A\beta - y\|^2$ gives the sum of the squares of the prediction errors. Show that A is a Toeplitz matrix (see page 138), *i.e.*, entries A_{ij} with the same value of $i - j$ are the same.

13.8 *Fitting an input-output convolution system.* Let $u_1, \ldots, u_T$ and $y_1, \ldots, y_T$ be observed input and output time series for a system that is thought to be an input-output convolution system, meaning

$$y_t \approx \hat{y}_t = \sum_{j=1}^{n} h_j u_{t-j+1}, \quad t = 1, \ldots, T,$$

where we interpret u_t as zero for $t \leq 0$. Here the n-vector h is the system impulse response; see page 140. This model of the relation between the input and output time series is also called a *moving average* (MA) model. Find a matrix A and vector b for which

$$\|Ah - b\|^2 = (y_1 - \hat{y}_1)^2 + \cdots + (y_T - \hat{y}_T)^2.$$

Show that A is Toeplitz. (See page 138.)

13.9 *Conclusions from 5-fold cross-validation.* You have developed a regression model for predicting a scalar outcome y from a feature vector x of dimension 20, using a collection of $N = 600$ data points. The mean of the outcome variable y across the given data is 1.85, and its standard deviation is 0.32. After running 5-fold cross-validation we get the following RMS test errors (based on forming a model based on the data excluding fold i, and testing it on fold i).

Fold excluded	RMS test error
1	0.13
2	0.11
3	0.09
4	0.11
5	0.14

(a) How would you expect your model to do on new, unseen (but similar) data? Respond briefly and justify your response.

(b) A co-worker observes that the regression model parameters found in the 5 different folds are quite close, but not the same. He says that for the production system, you should use the regression model parameters found when you excluded fold 3 from the original data, since it achieved the best RMS test error. Comment briefly.

13.10 *Augmenting features with the average.* You are fitting a regression model $\hat{y} = x^T \beta + v$ to data, computing the model coefficients β and v using least squares. A friend suggests adding a new feature, which is the average of the original features. (That is, he suggests using the new feature vector $\tilde{x} = (x, \mathbf{avg}(x))$.) He explains that by adding this new feature, you might end up with a better model. (Of course, you would test the new model using validation.) Is this a good idea?

13.11 *Interpreting model fitting results.* Five different models are fit using the same training data set, and tested on the same (separate) test set (which has the same size as the training set). The RMS prediction errors for each model, on the training and test sets, are reported below. Comment briefly on the results for each model. You might mention whether the model's predictions are good or bad, whether it is likely to generalize to unseen data, or whether it is over-fit. You are also welcome to say that you don't believe the results, or think the reported numbers are fishy.

Model	Train RMS	Test RMS
A	1.355	1.423
B	9.760	9.165
C	5.033	0.889
D	0.211	5.072
E	0.633	0.633

13.12 *Standardizing Boolean features.* (See page 269.) Suppose that the N-vector x gives the value of a (scalar) Boolean feature across a set of N examples. (Boolean means that each x_i has the value 0 or 1. This might represent the presence or absence of a symptom, or whether or not a day is a holiday.) How do we standardize such a feature? Express your answer in terms of p, the fraction of x_i that have the value 1. (You can assume that $p > 0$ and $p < 1$; otherwise the feature is constant.)

13.13 *Interaction model with Boolean features.* Consider a data fitting problem in which all n original features are Boolean, *i.e.*, entries of x have the value 0 or 1. These features could be the results of Boolean tests for a patient (or absence or presence of symptoms), or a person's response to a survey with yes/no questions. We wish to use these to predict an outcome, the number y. Our model will include a constant feature 1, the original n Boolean features, and all *interaction terms*, which have the form $x_i x_j$ where $1 \le i < j \le n$.

(a) What is p, the total number of basis functions, in this model? Explicitly list the basis functions for the case $n = 3$. You can decide their order. *Hint.* To count the number of pairs i, j that satisfy $1 \le i < j \le n$, use equation (5.7).

(b) Interpret (together) the following three coefficients of θ: the one associated with the original feature x_3; the one associated with the original feature x_5; and the one associated with the new product feature $x_3 x_5$. *Hint.* Consider the four possible values of the pair x_3, x_5.

13.14 *Least squares timing.* A computer takes around one second to fit a regression model (using least squares) with 20 parameters using 10^6 data points.

(a) About how long do you guess it will take the same computer to fit the same 20-parameter model using 10^7 data points (*i.e.*, 10× more data points)?

(b) About how long do you guess it will take the same computer to fit a 200-parameter model using 10^6 data points (*i.e.*, 10× more model parameters)?

13.15 *Estimating a matrix.* Suppose that the n-vector x and the m-vector y are thought to be approximately related by a linear function, *i.e.*, $y \approx Ax$, where A is an $m \times n$ matrix. We do not know the matrix A, but we do have observed data,

$$x^{(1)}, \dots, x^{(N)}, \qquad y^{(1)}, \dots, y^{(N)}.$$

We can estimate or guess the matrix A by choosing it to minimize

$$\sum_{i=1}^{N} \| A x^{(i)} - y^{(i)} \|^2 = \| AX - Y \|^2,$$

where $X = [x^{(1)} \cdots x^{(N)}]$ and $Y = [y^{(1)} \cdots y^{(N)}]$. We denote this *least squares estimate* as $\hat{A}$. (The notation here can be confusing, since X and Y are known, and A is to be found; it is more conventional to have symbols near the beginning of the alphabet, like A, denote known quantities, and symbols near the end, like X and Y, denote variables or unknowns.)

(a) Show that $\hat{A} = YX^\dagger$, assuming the rows of X are linearly independent. *Hint.* Use $\| AX - Y \|^2 = \| X^T A^T - Y^T \|^2$, which turns the problem into a matrix least squares problem; see page 233.

(b) Suggest a good way to compute $\hat{A}$, and give the complexity in terms of n, m, and N.

13.16 *Relative fitting error and error fitting the logarithm.* (See page 259.) The relative fitting error between a positive outcome y and its positive prediction $\hat{y}$ is given by $\eta = \max\{\hat{y}/y, y/\hat{y}\} - 1$. (This is often given as a percentage.) Let r be the residual between their logarithms, $r = \log y - \log \hat{y}$. Show that $\eta = e^{|r|} - 1$.

13.17 *Fitting a rational function with a polynomial.* Let $x_1, \dots, x_{11}$ be 11 points uniformly spaced in the interval $[-1, 1]$. (This means $x_i = -1.0 + 0.2(i - 1)$ for $i = 1, \dots, 11$.) Take $y_i = (1 + x_i)/(1 + 5x_i^2)$, for $i = 1, \dots, 11$. Find the least squares fit of polynomials of degree $0, 1, \dots, 8$ to these points. Plot the fitting polynomials, and the true function $y = (1 + x)/(1 + 5x^2)$, over the interval $[-1.1, 1.1]$ (say, using 100 points). Note that the interval for the plot, $[-1.1, 1.1]$, extends a bit outside the range of the data used to fit the polynomials, $[-1, 1]$; this gives us an idea of how well the polynomial fits can extrapolate.

Generate a test data set by choosing $u_1, \ldots, u_{10}$ uniformly spaced over $[-1.1, 1.1]$, with $v_i = (1 + u_i)/(1 + 5u_i^2)$. Plot the RMS error of the polynomial fits found above on this test data set. On the same plot, show the RMS error of the polynomial fits on the training data set. Suggest a reasonable value for the degree of the polynomial fit, based on the RMS fits on the training and test data. *Remark.* There is no practical reason to fit a rational function with a polynomial. This exercise is only meant to illustrate the ideas of fitting with different basis functions, over-fit, and validation with a test data set.

13.18 *Vector auto-regressive model.* The auto-regressive time series model (see page 259) can be extended to handle times series $z_1, z_2, \ldots$ where z_t are n-vectors. This arises in applications such as econometrics, where the entries of the vector give different economic quantities. The vector auto-regressive model (already mentioned on page 164) has the same form as a scalar auto-regressive model,

$$\hat{z}_{t+1} = \beta_1 z_t + \cdots + \beta_M z_{t-M+1}, \quad t = M, M+1, \ldots$$

where M is the *memory* of the model; the difference here is that the model parameters $\beta_1, \ldots, \beta_M$ are all $n \times n$ matrices. The total number of (scalar) parameters in the vector auto-regressive model is Mn^2. These parameters are chosen to minimize the sum of the squared norms of the prediction errors over a data set $z_1, \ldots, z_T$,

$$\|z_{M+1} - \hat{z}_{M+1}\|^2 + \cdots + \|z_T - \hat{z}_T\|^2.$$

(a) Give an interpretation of $(\beta_2)_{13}$. What would it mean if this coefficient is very small?

(b) Fitting a vector auto-regressive model can be expressed as a matrix least squares problem (see page 233), *i.e.*, the problem of choosing the matrix X to minimize $\|AX - B\|^2$, where A and B are matrices. Find the matrices A and B. You can consider the case $M = 2$. (It will be clear how to generalize your answer to larger values of M.)

Hints. Use the $(2n) \times n$ matrix variable $X = [\ \beta_1 \quad \beta_2\ ]^T$, and right-hand side $((T - 2) \times n$ matrix) $B = [\ z_3 \quad \cdots \quad z_T\ ]^T$. Your job is to find the matrix A so that $\|AX - B\|^2$ is the sum of the squared norms of the prediction errors.

13.19 *Sum of sinusoids time series model.* Suppose that $z_1, z_2, \ldots$ is a time series. A very common approximation of the time series is as a sum of K *sinusoids*

$$z_t \approx \hat{z}_t = \sum_{k=1}^{K} a_k \cos(\omega_k t - \phi_k), \quad t = 1, 2, \ldots.$$

The kth term in this sum is called a *sinusoid signal.* The coefficient $a_k \geq 0$ is called the *amplitude*, $\omega_k > 0$ is called the *frequency*, and ϕ_k is called the *phase* of the kth sinusoid. (The phase is usually chosen to lie in the range from $-\pi$ to π.) In many applications the frequencies are multiples of ω_1, *i.e.*, $\omega_k = k\omega_1$ for $k = 2, \ldots, K$, in which case the approximation is called a *Fourier approximation*, named for the mathematician Jean-Baptiste Joseph Fourier.

Suppose you have observed the values $z_1, \ldots, z_T$, and wish to choose the sinusoid amplitudes $a_1, \ldots, a_K$ and phases $\phi_1, \ldots, \phi_K$ so as to minimize the RMS value of the approximation error $(\hat{z}_1 - z_1, \ldots, \hat{z}_T - z_T)$. (We assume that the frequencies are given.) Explain how to solve this using least squares model fitting.

Hint. A sinusoid with amplitude a, frequency ω, and phase ϕ can be described by its cosine and sine coefficients α and β, where

$$a \cos(\omega t - \phi) = \alpha \cos(\omega t) + \beta \sin(\omega t),$$

where (using the cosine of sum formula) $\alpha = a \cos \phi$, $\beta = a \sin \phi$. We can recover the amplitude and phase from the cosine and sine coefficients as

$$a = \sqrt{\alpha^2 + \beta^2}, \qquad \phi = \arctan(\beta/\alpha).$$

Express the problem in terms of the cosine and sine coefficients.

13.20 *Fitting with continuous and discontinuous piecewise-linear functions.* Consider a fitting
problem with $n = 1$, so $x^{(1)}, \ldots, x^{(N)}$ and $y^{(1)}, \ldots, y^{(N)}$ are numbers. We consider two
types of closely related models. The first is a piecewise-linear model with knot points
at -1 and 1, as described on page 256, and illustrated in figure 13.8. The second is a
stratified model (see page 272), with three independent affine models, one for $x < -1$,
one for $-1 \leq x \leq 1$, and one for $x > 1$. (In other words, we stratify on x taking low,
middle, or high values.) Are these two models the same? Is one more general than the
other? How many parameters does each model have? *Hint.* See problem title.

What can you say about the training set RMS error and test set RMS error that would
be achieved using least squares with these two models?

13.21 *Efficient cross-validation.* The cost of fitting a model with p basis functions and N
data points (say, using QR factorization) is $2Np^2$ flops. In this exercise we explore the
complexity of carrying out 10-fold cross-validation on the same data set. We divide
the data set into 10 folds, each with $N/10$ data points. The naïve method is to fit 10
different models, each using 9 of the folds, using the QR factorization, which requires
$10 \cdot 2(0.9)Np^2 = 18Np^2$ flops. (To evaluate each of these models on the remaining fold
requires $2(N/10)p$ flops, which can be ignored compared to the cost of fitting the models.)
So the naïve method of carrying out 10-fold cross-validation requires, not surprisingly,
around $10\times$ the number of flops as fitting a single model.

The method below outlines another method to carry out 10-fold cross-validation. Give the
total flop count for each step, keeping only the dominant terms, and compare the total cost
of the method to that of the naïve method. Let $A_1, \ldots, A_{10}$ denote the $(N/10) \times p$ blocks
of the data matrix associated with the folds, and let $b_1, \ldots, b_{10}$ denote the right-hand
sides in the least squares fitting problem.

(a) Form the Gram matrices $G_i = A_i^T A_i$ and the vectors $c_i = A_i^T b_i$.

(b) Form $G = G_1 + \cdots + G_{10}$ and $c = c_1 + \cdots + c_{10}$.

(c) For $k = 1, \ldots, 10$, compute $\theta_k = (G - G_k)^{-1}(c - c_k)$.

13.22 *Prediction contests.* Several companies have run prediction contests open to the public.
Netflix ran the best known contest, offering a \$1M prize for the first prediction of user
movie rating that beat their existing method RMS prediction error by 10% on a test set.
The contests generally work like this (although there are several variations on this format,
and most are more complicated). The company posts a public data set, that includes the
regressors or features and the outcome for a large number of examples. They also post the
features, but not the outcomes, for a (typically smaller) test data set. The contestants,
usually teams with obscure names, submit predictions for the outcomes in the test set.
Usually there is a limit on how many times, or how frequently, each team can submit
a prediction on the test set. The company computes the RMS test set prediction error
(say) for each submission. The teams' prediction performance is shown on a *leaderboard*,
which lists the 100 or so best predictions in order.

Discuss such contests in terms of model validation. How should a team check a set of predictions before submitting it? What would happen if there were no limits on the number
of predictions each team can submit? Suggest an obvious method (typically disallowed
by the contest rules) for a team to get around the limit on prediction submissions. (And
yes, it has been done.)

Chapter 14

Least squares classification

In this chapter we consider the problem of fitting a model to data where the outcome takes on values like TRUE or FALSE (as opposed to being numbers, as in chapter 13). We will see that least squares can be used for this problem as well.

14.1 Classification

In the data fitting problem of chapter 13, the goal is to reproduce or predict the outcome y, which is a (scalar) number, based on an n-vector x. In a *classification problem*, the outcome or dependent variable y takes on only a finite number of values, and for this reason is sometimes called a *label*, or in statistics, a *categorical*. In the simplest case, y has only two values, for example TRUE or FALSE, or SPAM or NOT SPAM. This is called the *two-way classification problem*, the *binary classification problem*, or the *Boolean classification problem*, since the outcome y can take on only two values. We start by considering the Boolean classification problem.

We will encode y as a real number, taking $y = +1$ to mean TRUE and $y = -1$ to mean FALSE. (It is also possible to encode the outcomes using $y = +1$ and $y = 0$, or any other pair of two different numbers.) As in real-valued data fitting, we assume that an approximate relationship of the form $y \approx f(x)$ holds, where $f : \mathbf{R}^n \rightarrow \{-1, +1\}$. (This notation means that the function f takes an n-vector argument, and gives a resulting value that is either $+1$ or -1.) Our model will have the form $\hat{y} = \hat{f}(x)$, where $\hat{f} : \mathbf{R}^n \rightarrow \{-1, +1\}$. The model $\hat{f}$ is also called a *classifier*, since it classifies n-vectors into those for which $\hat{f}(x) = +1$ and those for which $\hat{f}(x) = -1$. As in real-valued data fitting, we choose or construct the classifier $\hat{f}$ using some observed data.

Examples. Boolean classifiers are widely used in many application areas.

- *Email spam detection.* The vector x contains features of an email message. It can include word counts in the body of the email message, other features such as the number of exclamation points or all-capital words, and features related to the origin of the email. The outcome is $+1$ if the message is SPAM, and -1 otherwise. The data used to create the classifier comes from users who have explicitly marked some messages as junk.

- *Fraud detection.* The vector x gives a set of features associated with a credit card holder, such as her average monthly spending levels, median price of purchases over the last week, number of purchases in different categories, average balance, and so on, as well as some features associated with a particular proposed transaction. The outcome y is $+1$ for a fraudulent transaction, and -1 otherwise. The data used to create the classifier is taken from historical data, that includes (some) examples of transactions that were later verified to be fraudulent and (many) that were verified to be bona fide.

- *Boolean document classification.* The vector x is a word count (or histogram) vector for a document, and the outcome y is $+1$ if the document has some specific topic (say, politics) and -1 otherwise. The data used to construct the classifier might come from a corpus of documents with their topics labeled.

- *Disease detection.* The examples correspond to patients, with outcome $y = +1$ meaning the patient has a particular disease, and $y = -1$ meaning they do not. The vector x contains relevant medical features associated with the patient, including for example age, sex, results of tests, and specific symptoms. The data used to build the model come from hospital records or a medical study; the outcome is the associated diagnosis (presence or absence of the disease), confirmed by a doctor.

- *Digital communications receiver.* In a modern electronic communications system, y represents one bit (traditionally represented by the values 0 and 1) that is to be sent from a transmitter to a receiver. The vector x represents n measurements of a received signal. The predictor $\hat{y} = \hat{f}(x)$ is called the *decoded bit.* In communications, the classifier $\hat{f}$ is called a *decoder* or *detector.* The data used to construct the decoder comes from a *training signal*, a sequence of bits known to the receiver, that is transmitted.

Prediction errors.   For a given data point x, y, with predicted outcome $\hat{y} = \hat{f}(x)$, there are only four possibilities:

- *True positive.* $y = +1$ and $\hat{y} = +1$.
- *True negative.* $y = -1$ and $\hat{y} = -1$.
- *False positive.* $y = -1$ and $\hat{y} = +1$.
- *False negative.* $y = +1$ and $\hat{y} = -1$.

In the first two cases the predicted label is correct, and in the last two cases, the predicted label is an error. We refer to the third case as a *false positive* or *type I error*, and we refer to the fourth case as a *false negative* or *type II error*. In some applications we care equally about making the two types of errors; in others we may care more about making one type of error than another.

Outcome	Prediction		Total
	$\hat{y} = +1$	$\hat{y} = -1$	
$y = +1$	N_{tp}	N_{fn}	N_{p}
$y = -1$	N_{fp}	N_{tn}	N_{n}
All	$N_{\text{tp}} + N_{\text{fp}}$	$N_{\text{fn}} + N_{\text{tp}}$	N

Table 14.1 Confusion matrix. The off-diagonal entries N_{fn} and N_{fp} give the numbers of the two types of error.

Error rate and confusion matrix. For a given data set

$$x^{(1)}, \ldots, x^{(N)}, \qquad y^{(1)}, \ldots, y^{(N)},$$

and model $\hat{f}$, we can count the numbers of each of the four possibilities that occur across the data set, and display them in a *contingency table* or *confusion matrix*, which is a 2×2 table with the columns corresponding to the value of $\hat{y}^{(i)}$ and the rows corresponding to the value of $y^{(i)}$. (This is the convention used in machine learning; in statistics, the rows and columns are sometimes reversed.) The entries give the total number of each of the four cases listed above, as shown in table 14.1. The diagonal entries correspond to correct decisions, with the upper left entry the number of true positives, and the lower right entry the number of true negatives. The off-diagonal entries correspond to errors, with the upper right entry the number of false negatives, and the lower left entry the number of false positives. The total of the four numbers is N, the number of examples in the data set. Sometimes the totals of the rows and columns are shown, as in table 14.1.

Various performance metrics are expressed in terms of the numbers in the confusion matrix.

- The *error rate* is the total number of errors (of both kinds) divided by the total number of examples, *i.e.*, $(N_{\text{fp}} + N_{\text{fn}})/N$.

- The *true positive rate* (also known as the *sensitivity* or *recall rate*) is $N_{\text{tp}}/N_{\text{p}}$. This gives the fraction of the data points with $y = +1$ for which we correctly guessed $\hat{y} = +1$.

- The *false positive rate* (also known as the *false alarm rate*) is $N_{\text{fp}}/N_{\text{n}}$. The false positive rate is the fraction of data points with $y = -1$ for which we incorrectly guess $\hat{y} = +1$.

- The *specificity* or *true negative rate* is one minus the false positive rate, *i.e.*, $N_{\text{tn}}/N_{\text{n}}$. The true negative rate is the fraction of the data points with $y = -1$ for which we correctly guess $\hat{y} = -1$.

- The *precision* is $N_{\text{tp}}/(N_{\text{tp}} + N_{\text{fp}})$, the fraction of true predictions that are correct.

	Prediction		
Outcome	$\hat{y} = +1$ (SPAM)	$\hat{y} = -1$ (not SPAM)	Total
$y = +1$ (SPAM)	95	32	127
$y = -1$ (not SPAM)	19	1120	1139
All	114	1152	1266

Table 14.2 Confusion matrix of a SPAM detector on a data set of 1266 examples.

A good classifier will have small (near zero) error rate and false positive rate, and high (near one) true positive rate, true negative rate, and precision. Which of these metrics is more important depends on the particular application.

An example confusion matrix is given in table 14.2 for the performance of a spam detector on a data set of $N = 1266$ examples (emails) of which 127 are SPAM ($y = +1$) and the remaining 1139 are NOT SPAM ($y = -1$). On the data set, this classifier has 95 true positives, 1120 true negatives, 19 false positives, and 32 false negatives. Its error rate is $(19 + 32)/1266 = 4.03\%$. Its true positive rate is $95/127 = 74.8\%$ (meaning it is detecting around 75% of the spam in the data set), and its false positive rate is $19/1139 = 1.67\%$ (meaning it incorrectly labeled around 1.7% of the non-spam messages as spam).

Validation in classification problems.   In classification problems we are concerned with the error, true positive, and false positive rates. So out-of-sample validation and cross-validation are carried out using the performance metric or metrics that we care about, *i.e.*, the error rate or some combination of true positive and false negative rates. We may care more about one of these metrics than the others.

14.2   Least squares classifier

Many sophisticated methods have been developed for constructing a Boolean model or classifier from a data set. *Logistic regression* and *support vector machine* are two methods that are widely used, but beyond the scope of this book. Here we discuss a very simple method, based on least squares, that can work quite well, though not as well as the more sophisticated methods.

We first carry out ordinary real-valued least squares fitting of the outcome, ignoring for the moment that the outcome y takes on only the values -1 and $+1$. We choose basis functions $f_1, \ldots, f_p$, and then choose the parameters $\theta_1, \ldots, \theta_p$ so as to minimize the sum squared error

$$(y^{(1)} - \tilde{f}(x^{(1)}))^2 + \cdots + (y^{(N)} - \tilde{f}(x^{(N)}))^2,$$

where $\tilde{f}(x) = \theta_1 f_1(x) + \cdots + \theta_p f_p(x)$. We use the notation $\tilde{f}$, since this function is not our final model $\hat{f}$. The function $\tilde{f}$ is the least squares fit over our data set, and $\tilde{f}(x)$, for a general vector x, is a number.

Our final classifier is then taken to be

$$\hat{f}(x) = \mathbf{sign}(\tilde{f}(x)), \tag{14.1}$$

where $\mathbf{sign}(a) = +1$ for $a \geq 0$ and -1 for $a < 0$. We call this classifier the *least squares classifier*.

The intuition behind the least squares classifier is simple. The value $\tilde{f}(x)$ is a number, which (ideally) is near $+1$ when $y^{(i)} = +1$, and near -1 when $y^{(i)} = -1$. If we are forced to guess one of the two possible outcomes $+1$ or -1, it is natural to choose $\mathbf{sign}(\tilde{f}(x))$. (Indeed, $\mathbf{sign}(\tilde{f}(x))$ is the nearest neighbor of $\tilde{f}(x)$ among the points -1 and $+1$.) Intuition suggests that the number $\tilde{f}(x)$ can be related to our confidence in our guess $\hat{y} = \mathbf{sign}(\tilde{f}(x))$: When $\tilde{f}(x)$ is near 1 we have confidence in our guess $\hat{y} = +1$; when it is small and negative (say, $\tilde{f}(x) = -0.03$), we guess $\hat{y} = -1$, but our confidence in the guess will be low. We won't pursue this idea further in this book, except in multi-class classifiers, which we discuss in §14.3.

The least squares classifier is often used with a regression model, *i.e.*, $\tilde{f}(x) = x^T\beta + v$, in which case the classifier has the form

$$\hat{f}(x) = \mathbf{sign}(x^T\beta + v). \tag{14.2}$$

We can easily interpret the coefficients in this model. For example, if β_7 is negative, it means that the larger the value of x_7 is, the more likely we are to guess $\hat{y} = -1$. If β_4 is the coefficient with the largest magnitude, then we can say that x_4 is the feature that contributes the most to our classification decision.

14.2.1   Iris flower classification

We illustrate least squares classification with a famous data set, first used in the 1930s by the statistician Ronald Fisher. The data are measurements of four attributes of three types of iris flowers: *Iris Setosa*, *Iris Versicolour*, and *Iris Virginica*. The data set contains 50 examples of each class. The four attributes are:

- x_1 is the sepal length in cm,

- x_2 is the sepal width in cm,

- x_3 is the petal length in cm,

- x_4 is the petal width in cm.

We compute a Boolean classifier of the form (14.2) that distinguishes the class *Iris Virginica* from the other two classes. Using the entire set of 150 examples we find the coefficients

$$v = -2.39, \quad \beta_1 = -0.0918, \quad \beta_2 = 0.406, \quad \beta_3 = 0.00798, \quad \beta_4 = 1.10.$$

The confusion matrix associated with this classifier is shown in table 14.3. The error rate is 7.3%.

	Prediction		
Outcome	$\hat{y}=+1$	$\hat{y}=-1$	Total
$y=+1$	46	4	50
$y=-1$	7	93	100
All	53	97	150

Table 14.3 Confusion matrix for a Boolean classifier of the Iris data set.

	Model parameters					Error rate (%)	
Fold	v	β_1	β_2	β_3	β_4	Train	Test
1	-2.45	0.0240	0.264	-0.00571	0.994	6.7	3.3
2	-2.38	-0.0657	0.398	-0.07593	1.251	5.8	10.0
3	-2.63	0.0340	0.326	-0.08869	1.189	7.5	3.3
4	-1.89	-0.3338	0.577	0.09902	1.151	6.7	16.7
5	-2.42	-0.1464	0.456	0.11200	0.944	8.3	3.3

Table 14.4 Five-fold validation for the Boolean classifier of the Iris data set.

Validation.   To test our least squares classification method, we apply 5-fold cross-validation. We randomly divide the data set into 5 folds of 30 examples (10 for each class). The results are shown in table 14.4. The test data sets contain only 30 examples, so a single prediction error changes the test error rate significantly (*i.e.*, by 3.3%). This explains what would seem to be large variation seen in the test set error rates. We might guess that the classifier will perform on new unseen data with an error rate in the 7–10% range, but our test sets are not large enough to predict future performance more accurately than this. (This is an example of the limitation of cross-validation when the data set is small; see the discussion on page 268.)

14.2.2   Handwritten digit classification

We now consider a much larger example, the MNIST data set described in §4.4.1. The (training) data set contains 60000 images of size 28 by 28. (A few samples are shown in figure 4.6.) The number of examples per digit varies between 5421 (for digit five) and 6742 (for digit one). The pixel intensities are scaled to lie between 0 and 1. We remove the pixels that are nonzero in fewer than 600 training examples. The remaining 493 pixels are shown as the white area in figure 14.1. There is also a separate test set containing 10000 images. Here we will consider classifiers to distinguish the digit zero from the other nine digits.

In this first experiment, we use the 493 pixel intensities, plus an additional feature with value 1, as the $n = 494$ features in the least squares classifier (14.1).

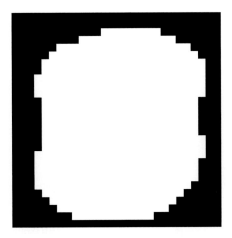

Figure 14.1 Location of the pixels used as features in the handwritten digit classification example.

| | Prediction | | |
| | $\hat{y} = +1$ | $\hat{y} = -1$ | Total |
Outcome			
$y = +1$	5158	765	5923
$y = -1$	167	53910	54077
All	5325	54675	60000

Table 14.5 Confusion matrix for a classifier for recognizing the digit zero, on a training set of 60000 examples.

The performance on the (training) data set is shown in the confusion matrix in table 14.5. The error rate is 1.6%, the true positive rate is 87.1%, and the false positive rate is 0.3%.

Figure 14.2 shows the distribution of the values of $\tilde{f}(x^{(i)})$ for the two classes in the training set. The interval $[-2.1, 2.1]$ is divided in 100 intervals of equal width. For each interval, the height of the blue bar is the fraction of the total number of training examples $x^{(i)}$ from class $+1$ (digit zero) that have a value $\tilde{f}(x^{(i)})$ in the interval. The height of the red bar is the fraction of the total number of training examples from class -1 (digits 1–9) with $\tilde{f}(x^{(i)})$ in the interval. The vertical dashed line shows the decision boundary: For $\tilde{f}(x^{(i)})$ to the left (*i.e.*, negative) we guess that digit i is from class -1, *i.e.*, digits 1–9; for $\tilde{f}(x^{(i)})$ to the right of the dashed line, we guess that digit i is from class $+1$, *i.e.*, digit 0. False positives correspond to red bars to the right of the dashed line, and false negatives correspond to blue bars to the left of the line.

Figure 14.3 shows the values of the coefficients β_k, displayed as an image. We can interpret this image as a map of the sensitivity of our classifier to the pixel

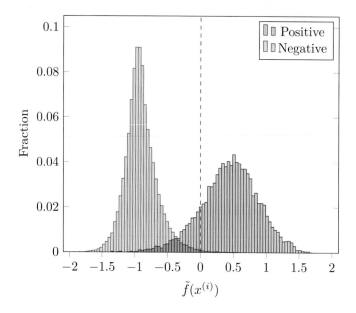

Figure 14.2 The distribution of the values of $\tilde{f}(x^{(i)})$ in the Boolean classifier (14.1) for recognizing the digit zero, over all elements $x^{(i)}$ of the training set. The red bars correspond to the digits from class -1, *i.e.*, the digits 1–9; the blue bars correspond to the digits from class $+1$, *i.e.*, the digit zero.

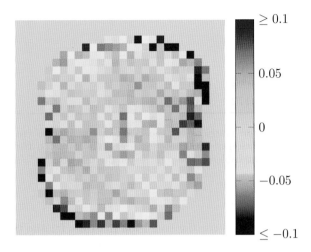

Figure 14.3 The coefficients β_k in the least squares classifier that distinguishes the digit zero from the other nine digits.

	Prediction		
Outcome	$\hat{y} = +1$	$\hat{y} = -1$	Total
$y = +1$	864	116	980
$y = -1$	42	8978	9020
All	906	9094	10000

Table 14.6 Confusion matrix for the classier for recognizing the digit zero, on a test set of 10000 examples.

values. Pixels with $\beta_i = 0$ are not used at all; pixels with larger positive values of β_i are locations where the larger the image pixel value, the more likely we are to guess that the image represents the digit zero.

Validation.   The performance of the least squares classifier on the test set is shown in the confusion matrix in table 14.6. For the test set the error rate is 1.6%, the true positive rate is 88.2%, and the false positive rate is 0.5%. These performance metrics are similar to those for the training data, which suggests that our classifier is not over-fit, and gives us some confidence in our classifier.

Feature engineering.   We now do some simple feature engineering (as in §13.3) to improve our classifier. As described on page 273, we add 5000 new features to the original 494 features, as follows. We first generate a 5000×494 matrix R, with randomly chosen entries ± 1. The 5000 new functions are then given by $\max\{0, (Rx)_j\}$, for $j = 1, \ldots, 5000$. After the addition of the 5000 new features (so

	Prediction				Prediction		
Outcome	$\hat{y} = +1$	$\hat{y} = -1$	Total	Outcome	$\hat{y} = +1$	$\hat{y} = -1$	Total
$y = +1$	5813	110	5923	$y = +1$	963	17	980
$y = -1$	15	54062	54077	$y = -1$	7	9013	9020
All	5828	54172	60000	All	970	9030	10000

Table 14.7 Confusion matrices for the Boolean classifier to recognize the digit zero after addition of 5000 new features. The table on the left is for the training set. The table on the right is for the test set.

the total number is 5494), we get the confusion matrices for the training and test data sets shown in table 14.7. The error rates are consistent, and equal to 0.21% for the training set and 0.24% for the test set, a very substantial improvement compared to the 1.6% in the first experiment. A comparison of the distributions in figures 14.4 and 14.2 also shows how much better the new classifier distinguishes between the two classes of the training set. We conclude that this was a successful exercise in feature engineering.

14.2.3   Receiver operating characteristic

One useful modification of the least squares classifier (14.1) is to skew the decision boundary, by subtracting a constant α from $\tilde{f}(x)$ before taking the sign:

$$\hat{f}(x) = \textbf{sign}(\tilde{f}(x) - \alpha). \tag{14.3}$$

The classifier is then

$$\hat{f}(x) = \begin{cases} +1 & \tilde{f}(x) \geq \alpha \\ -1 & \tilde{f}(x) < \alpha. \end{cases}$$

We call α the *decision threshold* for the modified classifier. The basic least squares classifier (14.1) has decision threshold $\alpha = 0$.

By choosing α positive, we make the guess $\hat{f}(x) = +1$ less frequently, so the numbers in the first column of the confusion matrix go down, and the numbers in the second column go up (since the sum of the numbers in each row is always the same). This means that choosing α positive decreases the true positive rate (which is bad), but it also decreases the false positive rate (which is good). Choosing α negative has the opposite effect, increasing the true positive rate (which is good) and increasing the false positive rate (which is bad). The parameter α is chosen depending on how much we care about the two competing metrics, in the particular application.

By sweeping α over a range, we obtain a family of classifiers that vary in their true positive and false positive rates. We can plot the false positive and negative rates, as well as the error rate, as a function of α. A more common way to plot this data has the strange name *receiver operating characteristic* (ROC). The ROC shows

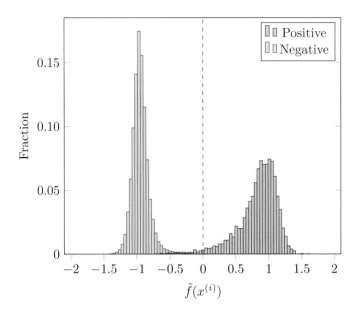

Figure 14.4 The distribution of the values of $\tilde{f}(x^{(i)})$ in the Boolean classifier (14.1) for recognizing the digit zero, after addition of 5000 new features.

the true positive rate on the vertical axis and false positive rate on the horizontal axis. The name comes from radar systems deployed during World War II, where $y = +1$ means that an enemy vehicle (or ship or airplane) is present, and $\hat{y} = +1$ means that an enemy vehicle is detected.

Example.   We examine the skewed threshold least squares classifier (14.3) for the example described above, where we attempt to detect whether or not a handwritten digit is zero. Figure 14.5 shows how the error, true positive, and false positive rates depend on the decision threshold α, for the training set data. We can see that as α increases, the true positive rate decreases, as does the false positive rate. We can see that for this particular case the total error rate is minimized by choosing $\alpha = -0.1$, which gives error rate 1.4%, slightly lower than the basic least squares classifier. The limiting cases when α is negative enough, or positive enough, are readily understood. When α is very negative, the prediction is always $\hat{y} = +1$; our error rate is then the fraction of the data set with $y = -1$. When α is very positive, the prediction is always $\hat{y} = -1$, which gives an error rate equal to the fraction of the data set with $y = +1$.

The same information (without the total error rate) is plotted in the traditional ROC curve shown in figure 14.6. The dots show the basic least squares classifier, with $\alpha = 0$, and the skewed threshold least squares classifiers for $\alpha = -0.25$ and $\alpha = 0.25$. These curves are for the training data; the same curves for the test data look similar, giving us some confidence that our classifiers will have similar performance on new, unseen data.

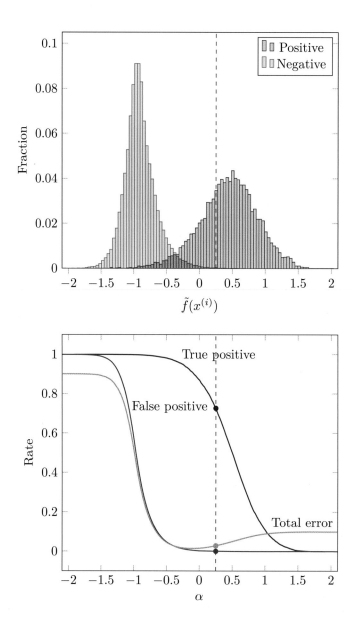

Figure 14.5 True positive, false positive, and total error rate versus decision threshold α. The vertical dashed line is shown for decision threshold $\alpha = 0.25$.

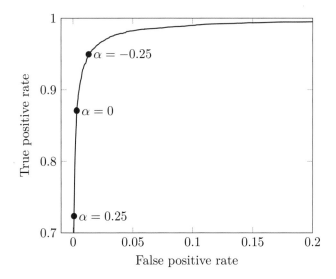

Figure 14.6 ROC curve.

14.3   Multi-class classifiers

In a multi-class classification problem, we have $K > 2$ possible labels. This is sometimes referred to more compactly as K-class classification. (The case $K = 2$ is Boolean classification, discussed above.) For our generic discussion of multi-class classifiers, we will encode the labels as $y = 1, 2, \ldots, K$. In some applications there are more natural encodings; for example, the Likert scale labels *Strongly Disagree*, *Disagree*, *Neutral*, *Agree*, and *Strongly Agree* are typically encoded as $-2, -1, 0, 1, 2$, respectively.

A multi-class classifier is a function $\hat{f} : \mathbf{R}^n \to \{1, \ldots, K\}$. Given a feature vector x, $\hat{f}(x)$ (which is an integer between 1 and K) is our prediction of the associated outcome. A multi-class classifier classifies n-vectors into K groups, corresponding to the values $1, \ldots, K$.

Examples.   Multi-class classifiers are used in many application areas.

- *Handwritten digit classification.* We are given an image of a hand-written digit (and possibly other features generated from the images), and wish to guess which of ten digits it represents. This classifier is used to do automatic (computer-based) reading of handwritten digits.

- *Marketing demographic classification.* Data from purchases made, or web sites visited, is used to train a multi-class classifier for a set of market segments, such as college-educated women aged 25–30, men without college degrees aged 45–55, and so on. This classifier guesses the demographic segment of a new customer, based only on their purchase history. This can be used to select which promotions to offer a customer for whom we only have purchase data. The classifier is trained using data from known customers.

- *Disease diagnosis.* The labels are a set of diseases (including one label that corresponds to disease-free), and the features are medically relevant values, such as patient attributes and the results of tests. Such a classifier carries out diagnosis (of the diseases corresponding to the labels). The classifier is trained on cases in which a definitive diagnosis has been made.

- *Translation word choice.* A machine translation system translates a word in the source language to one of several possible words in the target language. The label corresponds to a particular choice of translation for the word in the source language. The features contain information about the context around the word, for example, words counts or occurrences in the same paragraph. As an example, the English word 'bank' might be translated into another language one way if the word 'river' appears nearby, and another way if 'financial' or 'reserves' appears nearby. The classifier is trained on data taken from translations carried out by (human) experts.

- *Document topic prediction.* Each example corresponds to a document or article, with the feature vector containing word counts or histograms, and the label corresponding to the topic or category, such as POLITICS, SPORTS, ENTERTAINMENT, and so on.

- *Detection in communications.* Many electronic communications systems transmit messages as a sequence of K possible *symbols*. The vector x contains measurements of the received signal. In this context the classifier $\hat{f}$ is called a *detector* or *decoder*; the goal is to correctly determine which of the K symbols was transmitted.

Prediction errors and confusion matrix.   For a multi-class classifier $\hat{f}$ and a given data point (x, y), with predicted outcome $\hat{y} = \hat{f}(x)$, there are K^2 possibilities, corresponding to all the pairs of values of y, the actual outcome, and $\hat{y}$, the predicted outcome. For a given data set (training or validation set) with N elements, the numbers of each of the K^2 occurrences are arranged into a $K \times K$ confusion matrix, where N_{ij} is the number of data points for which $y = i$ and $\hat{y} = j$.

The K diagonal entries $N_{11}, \ldots, N_{KK}$ correspond to the cases when the prediction is correct; the $K^2 - K$ off-diagonal entries N_{ij}, $i \neq j$, correspond to prediction errors. For each i, N_{ii} is the number of data points with label i for which we correctly guessed $\hat{y} = i$. For $i \neq j$, N_{ij} is the number of data points for which we have mistaken label i (its true value) for the label j (our incorrect guess). For $K = 2$ (Boolean classification) there are only two types of prediction errors, false positive and false negative. For $K > 2$ the situation is more complicated, since there are many more types of errors a predictor can make. From the entries of the confusion matrix we can derive various measures of the accuracy of the predictions. We let N_i (with one index) denote the total number of data points for which $y = i$, i.e., $N_i = N_{i1} + \cdots + N_{iK}$. We have $N = N_1 + \cdots + N_K$.

The simplest measure is the overall *error rate*, which is the total number of errors (the sum of all off-diagonal entries in the confusion matrix) divided by the

y	$\hat{y}$		
	Dislike	Neutral	Like
Dislike	183	10	5
Neutral	7	61	8
Like	3	13	210

Table 14.8 Example confusion matrix of a multi-class classifier with three classes.

data set size (the sum of all entries in the confusion matrix):

$$(1/N) \sum_{i \neq j} N_{ij} = 1 - (1/N) \sum_{i} N_{ii}.$$

This measure implicitly assumes that all errors are equally bad. In many applications this is not the case; for example, some medical mis-diagnoses might be worse for a patient than others.

We can also look at the rate with which we predict each label correctly. The quantity N_{ii}/N_i is called the *true label i rate*. It is the fraction of data points with label $y = i$ for which we correctly predicted $\hat{y} = i$. (The true label i rates reduce to the true positive and true negative rates for Boolean classifiers.)

A simple example, with $K = 3$ labels (*Dislike*, *Neutral*, and *Like*), and a total number $N = 500$ data points, is shown in table 14.8. Out of 500 data points, 454 (the sum of the diagonal entries) were classified correctly. The remaining 46 data points (the sum of the off-diagonal entries) correspond to the 6 different types of errors. The overall error rate is $46/500 = 9.2\%$. The true label *Dislike* rate is $183/(183 + 10 + 5) = 92.4\%$, *i.e.*, among the data points with label *Dislike*, we correctly predicted the label on 92.4% of the data. The true label *Neutral* rate is $61/(7 + 61 + 8) = 80.3\%$, and the true label *Like* rate is $210/(3 + 13 + 210) = 92.9\%$.

14.3.1   Least squares multi-class classifier

The idea behind the least squares Boolean classifier can be extended to handle multi-class classification problems. For each possible label value, we construct a new data set with the Boolean label $+1$ if the label has the given value, and -1 otherwise. (This is sometimes called a *one-versus-others* or *one-versus-all* classifier.) From these K Boolean classifiers we must create a classifier that chooses one of the K possible labels. We do this by selecting the label for which the least squares regression fit has the highest value, which roughly speaking is the one with the highest level of confidence. Our classifier is then

$$\hat{f}(x) = \underset{k=1,\ldots,K}{\mathrm{argmax}} \; \tilde{f}_k(x),$$

where $\tilde{f}_k$ is the least squares regression model for label k against the others. The notation argmax means the index of the largest value among the numbers $\tilde{f}_k(x)$, for $k = 1, \ldots, K$. Note that $\tilde{f}_k(x)$ is the *real-valued prediction* for the Boolean classifier for class k versus not class k; it is not the Boolean classifier, which is $\mathbf{sign}(\tilde{f}_k(x))$.

As an example consider a multi-class classification problem with 3 labels. We construct 3 different least squares classifiers, for 1 versus 2 or 3, for 2 versus 1 or 3, and for 3 versus 1 or 2. Suppose for a given feature vector x, we find that

$$\tilde{f}_1(x) = -0.7, \quad \tilde{f}_2(x) = +0.2, \quad \tilde{f}_3(x) = +0.8.$$

The largest of these three numbers is $\tilde{f}_3(x)$, so our prediction is $\hat{f}(x) = 3$. We can interpret these numbers and our final decision. The first classifier is fairly confident that the label is not 1. According to the second classifier, the label could be 2, but it does not have high confidence in this prediction. Finally, the third classifier predicts the label is 3, and moreover has relatively high confidence in this guess. So our final guess is label 3. (This interpretation suggests that if we had to make a second guess, it should be label 2.) Of course here we are anthropomorphizing the individual label classifiers, since they do not have beliefs or levels of confidence in their predictions. But the story is helpful in understanding the motivation behind the classifier above.

Skewed decisions. In a Boolean classifier we can skew the decision threshold (see §14.2.3) to trade off the true positive and false positive rates. In a K-class classifier, an analogous method can be used to trade off the K true label i rates. We apply an offset α_k to $\tilde{f}_l(x)$ before finding the largest value. This gives the predictor

$$\hat{f}(x) = \operatorname*{argmax}_{k=1,\ldots,K} \left(\tilde{f}_k(x) - \alpha_k \right),$$

where α_k are constants chosen to trade off the true label k rates. If we decrease α_k, we predict $\hat{f}(x) = k$ more often, so all entries of the kth column in the confusion matrix increase. This increases our rate of true positives for label k (since N_{kk} increases), which is good. But it can decrease the true positive rates for the other labels.

Complexity. In least squares multi-class classification, we solve K least squares problems, each with N rows and p variables. The naïve method of computing $\theta_1, \ldots, \theta_K$, the coefficients in our one-versus-others classifiers, costs $2KNp^2$ flops. But the K least squares problems we solve all involve the same matrix; only the right-hand side vector changes. This means that we can carry out the QR factorization just once, and use it for computing all K classifier coefficients. Alternatively, we can say that finding the coefficients of all the one-versus-others classifiers can be done by solving a matrix least squares problem (see page 233.) When K (the number of classes or labels) is small compared to p (the number of basis functions or coefficients in the classifier), the cost is about the same as solving just one least squares problem.

Another simplification in K-class least squares classification arises due to the special form of the right-hand sides in the K least squares problems to be solved. The right-hand sides in these K problems are Boolean vectors with entries $+1$ for

Class	Prediction			Total
	Setosa	Versicolour	Virginica	
Setosa	40	0	0	40
Versicolour	0	27	13	40
Virginica	0	4	36	40
All	40	31	49	120

Table 14.9 Confusion matrix for a 3-class classifier of the Iris data set, on a training set of 120 examples.

one of the classes and -1 for all others. It follows that the sum of these K right-hand sides is the vector with all entries equal to $2 - K$, *i.e.*, $(2 - K)\mathbf{1}$. Since the mapping from the right-hand sides to the least squares approximate solutions $\hat{\theta}_k$ is linear (see page 229), we have $\hat{\theta}_1 + \cdots + \hat{\theta}_k = (2 - K)a$, where a is the least squares approximate solution when the right-hand side is $\mathbf{1}$. Assuming that the first basis function is $f_1(x) = 1$, we have $a = e_1$. So we have

$$\hat{\theta}_1 + \cdots + \hat{\theta}_K = (2 - K)e_1,$$

where $\hat{\theta}_k$ is the coefficient vector for distinguishing class k from the others. Once we have computed $\hat{\theta}_1, \ldots, \hat{\theta}_{K-1}$, we can find $\hat{\theta}_K$ by simple vector subtraction.

This explains why for the Boolean classification case we have $K = 2$, but we only have to solve *one* least squares problem. In §14.2 we compute one coefficient vector θ; if the same problem were to be considered a K-class problem with $K = 2$, we would have $\theta_1 = \theta$. (This one distinguishes class 1 versus class 2.) The other coefficient vector is then $\theta_2 = -\theta_1$. (This one distinguishes class 2 versus class 1.)

14.3.2 Iris flower classification

We compute a 3-class classifier for the Iris data set described on page 289. The examples are randomly partitioned into a training set of size 120, containing 40 examples of each class, and a test set of size 30, with 10 examples of each class. The 3×3 confusion matrix for the training set is given in table 14.9. The error rate is 14.2%. The results for the test set are in table 14.10. The error rate is 13.3%, similar enough to the training error rate to give us some confidence in our classifier. The true *Setosa* rate is 100% for both train and test sets, suggesting that our classifier can detect this type well. The true *Versicolour* rate is 67.5% for the training data, and 60% for the test set. The true *Virginica* rate is 90% for the training data, and 100% for the test set. This suggests that our classifier can detect *Virginica* well, but perhaps not as well as *Setosa*. (The 100% true *Virginica* rate on the test set is a matter of luck, due to the very small number of test examples of each type; see the discussion on page 268.)

| Class | Prediction | | | |
	Setosa	Versicolour	Virginica	Total
Setosa	10	0	0	10
Versicolour	0	6	4	10
Virginica	0	0	10	10
All	10	6	14	30

Table 14.10 Confusion matrix for a 3-class classifier of the Iris data set, on a test set of 30 examples.

14.3.3   Handwritten digit classification

We illustrate the least squares multi-class classification method by applying it to the MNIST data set. For each of the ten digits $0, \ldots, 9$ (which we encode as $k = 1, \ldots, 10$) we compute a least squares Boolean classifier

$$\hat{f}_k(x) = \mathbf{sign}(x^T \beta_k + v_k),$$

to distinguish digit k from the other digits. The ten Boolean classifiers are combined into a multi-class classifier

$$\hat{f}(x) = \underset{k=1,\ldots,10}{\mathrm{argmax}} \, (x^T \beta_k + v_k).$$

The 10×10 confusion matrix for the data set and the test set are given in tables 14.11 and 14.12.

The error rate on the training set is 14.5%; on the test set it is 13.9%. The true label rates on the test set range from 73.5% for digit 5 to 97.5% for digit 1. Many of the entries of the confusion matrix make sense. From the first row of the matrix, we see a handwritten 0 was rarely mistakenly classified as a 1, 2, or 9; presumably these digits look different enough that they are easily distinguished. The most common error (80) corresponds to $y = 9$, $\hat{y} = 4$, *i.e.*, mistakenly identifying a handwritten 9 as a 4. This makes sense since these two digits can look very similar.

Feature engineering.   After adding the 5000 randomly generated new features (as described on page 293), the training set error is reduced to about 1.5%, and the test set error to 2.6%. The confusion matrices are given in tables 14.13 and 14.14. Since we have (substantially) reduced the error in the test set, we conclude that adding these 5000 new features was a successful exercise in feature engineering.

You could reasonably wonder how much performance improvement is possible for this example, using feature engineering. For the handwritten digit data set, humans have an error rate around 2% (with the true digits verified by checking actual addresses, ZIP codes, and so on). Further feature engineering (*i.e.*, introducing even more additional random features, or using neural network features) brings the error rate down well below 2%, *i.e.*, *well below human ability*. This should give you some idea of how powerful the ideas in this book are.

Digit	Prediction										Total
	0	1	2	3	4	5	6	7	8	9	
0	5669	8	21	19	25	46	65	4	60	6	5923
1	2	6543	36	17	20	30	14	14	60	6	6742
2	99	278	4757	153	116	17	234	92	190	22	5958
3	38	172	174	5150	31	122	59	122	135	128	6131
4	13	104	41	5	5189	52	45	24	60	309	5842
5	164	94	30	448	103	3974	185	44	237	142	5421
6	104	78	77	2	64	106	5448	0	36	3	5918
7	55	191	36	48	165	9	4	5443	13	301	6265
8	69	492	64	225	102	220	64	21	4417	177	5851
9	67	66	26	115	365	12	4	513	39	4742	5949
All	6280	8026	5262	6182	6180	4588	6122	6277	5247	5836	60000

Table 14.11 Confusion matrix for least squares multi-class classification of handwritten digits (training set).

Digit	Prediction										Total
	0	1	2	3	4	5	6	7	8	9	
0	944	0	1	2	2	8	13	2	7	1	980
1	0	1107	2	2	3	1	5	1	14	0	1135
2	18	54	815	26	16	0	38	22	39	4	1032
3	4	18	22	884	5	16	10	22	20	9	1010
4	0	22	6	0	883	3	9	1	12	46	982
5	24	19	3	74	24	656	24	13	38	17	892
6	17	9	10	0	22	17	876	0	7	0	958
7	5	43	14	6	25	1	1	883	1	49	1028
8	14	48	11	31	26	40	17	13	756	18	974
9	16	10	3	17	80	0	1	75	4	803	1009
All	1042	1330	887	1042	1086	742	994	1032	898	947	10000

Table 14.12 Confusion matrix for least squares multi-class classification of handwritten digits (test set).

Digit	0	1	2	3	4	5	6	7	8	9	Total
					Prediction						
0	5888	1	2	1	3	2	10	0	14	2	5923
1	1	6679	27	6	11	0	0	10	6	2	6742
2	11	7	5866	6	12	0	3	22	26	5	5958
3	1	4	31	5988	0	27	0	24	34	22	6131
4	1	15	3	0	5748	1	13	4	5	52	5842
5	6	2	4	26	7	5335	23	2	9	7	5421
6	8	5	0	0	3	15	5875	0	11	1	5918
7	3	25	23	4	8	0	1	6159	5	37	6265
8	5	16	11	12	9	17	11	7	5749	14	5851
9	10	5	1	29	41	16	2	35	25	5785	5949
All	5934	6759	5968	6072	5842	5413	5938	6263	5884	5927	60000

Table 14.13 Confusion matrix for least squares multi-class classification of handwritten digits, after addition of 5000 features (training set).

Digit	0	1	2	3	4	5	6	7	8	9	Total
					Prediction						
0	972	0	0	2	0	1	1	1	3	0	980
1	0	1126	3	1	1	0	3	0	1	0	1135
2	6	0	998	3	2	0	4	7	11	1	1032
3	0	0	3	977	0	13	0	5	8	4	1010
4	2	1	3	0	953	0	6	3	1	13	982
5	2	0	1	5	0	875	5	0	3	1	892
6	8	3	0	0	4	6	933	0	4	0	958
7	0	8	12	0	2	0	1	992	3	10	1028
8	3	1	3	6	4	3	2	2	946	4	974
9	4	3	1	12	11	7	1	3	3	964	1009
All	997	1142	1024	1006	977	905	956	1013	983	997	10000

Table 14.14 Confusion matrix for least squares multi-class classification of handwritten digits, after addition of 5000 features (test set).

Exercises

14.1 *Chebyshev bound.* Let $\tilde{f}(x)$ denote the continuous prediction of the Boolean outcome y, and $\hat{f}(x) = \mathbf{sign}(\tilde{f}(x))$ the actual classifier. Let σ denote the RMS error in the continuous prediction over some set of data, *i.e.*,

$$\sigma^2 = \frac{(\tilde{f}(x^{(1)}) - y^{(1)})^2 + \cdots + (\tilde{f}(x^{(N)}) - y^{(N)})^2}{N}.$$

Use the Chebyshev bound to argue that the error rate over this data set, *i.e.*, the fraction of data points for which $\hat{f}(x^{(i)}) \neq y^{(i)}$, is no more than σ^2, assuming $\sigma < 1$.

Remark. This bound on the error rate is usually quite bad; that is, the actual error rate in often much lower than this bound. But it does show that if the continuous prediction is good, then the classifier must perform well.

14.2 *Interpreting the parameters in a regression classifier.* Consider a classifier of the form $\hat{y} = \mathbf{sign}(x^T \beta + v)$, where $\hat{y}$ is the prediction, the n-vector x is the feature vector, and the n-vector β and scalar v are the classifier parameters. We will assume here that $v \neq 0$ and $\beta \neq 0$. Evidently $\hat{y} = \mathbf{sign}(v)$ is the prediction when the feature vector x is zero. Show that when $\|x\| < |v|/\|\beta\|$, we have $\hat{y} = \mathbf{sign}(v)$. This shows that when all the features are *small* (*i.e.*, near zero), the classifier makes the prediction $\hat{y} = \mathbf{sign}(v)$. *Hint.* If two numbers a and b satisfy $|a| < |b|$, then $\mathbf{sign}(a + b) = \mathbf{sign}(b)$.

This means that we can interpret $\mathbf{sign}(v)$ as the classifier prediction when the features are small. The ratio $|v|/\|\beta\|$ tells us how big the feature vector must be before the classifier 'changes its mind' and predicts $\hat{y} = -\mathbf{sign}(v)$.

14.3 *Likert classifier.* A response to a question has the options *Strongly Disagree, Disagree, Neutral, Agree,* or *Strongly Agree,* encoded as $-2, -1, 0, 1, 2$, respectively. You wish to build a multi-class classifier that takes a feature vector x and predicts the response. A multi-class least squares classifier builds a separate (continuous) predictor for each response versus the others. Suggest a simpler classifier, based on one continuous regression model $\tilde{f}(x)$ that is fit to the numbers that code the responses, using least squares.

14.4 *Multi-class classifier via matrix least squares.* Consider the least squares multi-class classifier described in §14.3, with a regression model $\tilde{f}_k(x) = x^T \beta_k$ for the one-versus-others classifiers. (We assume that the offset term is included using a constant feature.) Show that the coefficient vectors $\beta_1, \ldots, \beta_K$ can be found by solving the matrix least squares problem of minimizing $\|X^T \beta - Y\|^2$, where β is the $n \times K$ matrix with columns $\beta_1, \ldots, \beta_K$, and Y is an $N \times K$ matrix.

(a) Give Y, *i.e.*, describe its entries. What is the ith row of Y?

(b) Assuming the rows of X (*i.e.*, the data feature vectors) are linearly independent, show that the least squares estimate is given by $\hat{\beta} = (X^T)^\dagger Y$.

14.5 *List classifier.* Consider a multi-class classification problem with K classes. A standard multi-class classifier is a function $\hat{f}$ that returns a class (one of the labels $1, \ldots, K$), when given a feature n-vector x. We interpret $\hat{f}(x)$ as the classifier's guess of the class that corresponds to x. A *list classifier* returns a list of guesses, typically in order from 'most likely' to 'least likely'. For example, for a specific feature vector x, a list classifier might return $3, 6, 2$, meaning (roughly) that its top guess is class 3, its next guess is class 6, and its third guess is class 2. (The lists can have different lengths for different values of the feature vector.) How would you modify the least squares multi-class classifier described in §14.3.1 to create a list classifier? *Remark.* List classifiers are widely used in electronic communication systems, where the feature vector x is the received signal, and the class corresponds to which of K messages was sent. In this context they are called *list decoders*. List decoders produce a list of probable or likely messages, and allow a later processing stage to make the final decision or guess.

14.6 *Polynomial classifier with one feature.* Generate 200 points $x^{(1)}, \ldots, x^{(200)}$, uniformly spaced in the interval $[-1, 1]$, and take

$$y^{(i)} = \begin{cases} +1 & -0.5 \le x^{(i)} < 0.1 \text{ or } 0.5 \le x^{(i)} \\ -1 & \text{otherwise} \end{cases}$$

for $i = 1, \ldots, 200$. Fit polynomial least squares classifiers of degrees $0, \ldots, 8$ to this training data set.

(a) Evaluate the error rate on the training data set. Does the error rate decrease when you increase the degree?

(b) For each degree, plot the polynomial $\tilde{f}(x)$ and the classifier $\hat{f}(x) = \mathbf{sign}(\tilde{f}(x))$.

(c) It is possible to classify this data set perfectly using a classifier $\hat{f}(x) = \mathbf{sign}(\tilde{f}(x))$ and a cubic polynomial

$$\tilde{f}(x) = c(x + 0.5)(x - 0.1)(x - 0.5),$$

for any positive c. Compare this classifier with the least squares classifier of degree 3 that you found and explain why there is a difference.

14.7 *Polynomial classifier with two features.* Generate 200 random 2-vectors $x^{(1)}, \ldots, x^{(200)}$ in a plane, from a standard normal distribution. Define

$$y^{(i)} = \begin{cases} +1 & x_1^{(i)} x_2^{(i)} \ge 0 \\ -1 & \text{otherwise} \end{cases}$$

for $i = 1, \ldots, 200$. In other words, $y^{(i)}$ is $+1$ when $x^{(i)}$ is in the first or third quadrant, and -1 otherwise. Fit a polynomial least squares classifier of degree 2 to the data set, *i.e.*, use a polynomial

$$\tilde{f}(x) = \theta_1 + \theta_2 x_1 + \theta_3 x_2 + \theta_4 x_1^2 + \theta_5 x_1 x_2 + \theta_6 x_2^2.$$

Give the error rate of the classifier. Show the regions in the plane where $\hat{f}(x) = 1$ and $\hat{f}(x) = -1$. Also compare the computed coefficients with the polynomial $\tilde{f}(x) = x_1 x_2$, which classifies the data points with zero error.

14.8 *Author attribution.* Suppose that the N feature n-vectors $x^{(1)}, \ldots, x^{(N)}$ are word count histograms, and the labels $y^{(1)}, \ldots, y^{(N)}$ give the document authors (as one of $1, \ldots, K$). A classifier guesses which of the K authors wrote an unseen document, which is called *author attribution*. A least squares classifier using regression is fit to the data, resulting in the classifier

$$\hat{f}(x) = \underset{k=1,\ldots,K}{\mathrm{argmax}} (x^T \beta_k + v_k).$$

For each author (*i.e.*, $k = 1, \ldots, K$) we find the ten largest (most positive) entries in the n-vector β_k and the ten smallest (most negative) entries. These correspond to two sets of ten words in the dictionary, for each author. Interpret these words, briefly, in English.

14.9 *Nearest neighbor interpretation of multi-class classifier.* We consider the least squares K-class classifier of §14.3.1. We associate with each data point the n-vector x, and the label or class, which is one of $1, \ldots, K$. If the class of the data point is k, we associate it with a K-vector y, whose entries are $y_k = +1$ and $y_j = -1$ for $j \ne k$. (We can write this vector as $y = 2e_k - \mathbf{1}$.) Define $\tilde{y} = (\tilde{f}_1(x), \ldots, \tilde{f}_K(x))$, which is our (real-valued or continuous) prediction of the label y. Our multi-class prediction is given by $\hat{f}(x) = \mathrm{argmax}_{k=1,\ldots,K} \tilde{f}_k(x)$. Show that $\hat{f}(x)$ is also the index of the nearest neighbor of $\tilde{y}$ among the vectors $2e_k - \mathbf{1}$, for $k = 1, \ldots, K$. In other words, our guess $\hat{y}$ for the class is the nearest neighbor of our continuous prediction $\tilde{y}$, among the vectors that encode the class labels.

14.10 *One-versus-one multi-class classifier.* In §14.3.1 we construct a K-class classifier from K Boolean classifiers that attempt to distinguish each class from the others. In this exercise we describe another method for constructing a K-class classifier. We first develop a Boolean classifier for every *pair* of classes i and j, $i < j$. There are $K(K-1)/2$ such pairs of classifiers, called *one-versus-one* classifiers. Given a feature vector x, we let $\hat{y}_{ij}$ be the prediction of the i-versus-j classifier, with $\hat{y}_{ij} = 1$ meaning that the one-versus-one classifier is guessing that $y = i$. We consider $\hat{y}_{ij} = 1$ as one 'vote' for class i, and $\hat{y}_{ij} = -1$ as one 'vote' for class j. We obtain the final estimated class by *majority voting*: We take $\hat{y}$ as the class that has the most votes. (We can break ties in some simple way, like taking the smallest index that achieves the largest number of votes.)

(a) Construct the least squares classifier, and the one-versus-one classifier, for a multi-class (training) data set. Find the confusion matrices, and the error rates, of the two classifiers on both the training data set and a separate test data set.

(b) Compare the complexity of computing the one-versus-one multi-class classifier with the complexity of the least squares multi-class classifier (see page 300). Assume the training set contains N/K examples of each class and that N/K is much greater than the number of features p. Distinguish two methods for the one-versus-one multi-class classifier. The first, naïve, method solves $K(K-1)/2$ least squares problem with N/K rows and p columns. The second, more efficient, method precomputes the Gram matrices $G_i = A_i A_i^T$ for $i = 1, \ldots, K$, where the rows of the $(N/K) \times p$ matrix A_i are the training example for class i, and uses the pre-computed Gram matrices to speed up the solution of the $K(K-1)/2$ least squares problems.

14.11 *Equalizer design from training message.* We consider an electronic communication system, with message to be sent given by an N-vector s, whose entries are -1 or 1, and received signal y, where $y = c*s$, where c is an n-vector, the channel impulse response. The receiver applies equalization to the received signal, which means that it computes $\tilde{y} = h*y = h*c*s$, where h is an n-vector, the equalizer impulse response. The receiver then estimates the original message using $\hat{s} = \mathbf{sign}(\tilde{y}_{1:N})$. This works well if $h * c \approx e_1$. (See exercise 7.15.) If the channel impulse response c is known or can be measured, we can design or choose h using least squares, as in exercise 12.6.

In this exercise we explore a method for choosing h directly, without estimating or measuring c. The sender first sends a message that is *known* to the receiver, called the *training message*, s^{train}. (From the point of view of communications, this is wasted transmission, and is called *overhead*.) The receiver receives the signal $y^{\mathrm{train}} = c * s^{\mathrm{train}}$ from the training message, and then chooses h to minimize $\|(h * y^{\mathrm{train}})_{1:N} - s^{\mathrm{train}}\|^2$. (In practice, this equalizer is used until the bit error rate increases, which means the channel has changed, at which point another training message is sent.) Explain how this method is the same as least squares classification. What are the training data $x^{(i)}$ and $y^{(i)}$? What is the least squares problem that must be solved to determine the equalizer impulse response h?

Chapter 15

Multi-objective least squares

In this chapter we consider the problem of choosing a vector that achieves a compromise in making two or more norm squared objectives small. The idea is widely used in data fitting, image reconstruction, control, and other applications.

15.1 Multi-objective least squares

In the basic least squares problem (12.1), we seek the vector $\hat{x}$ that minimizes the single objective function $\|Ax - b\|^2$. In some applications we have *multiple* objectives, all of which we would like to be small:

$$J_1 = \|A_1 x - b_1\|^2, \quad \ldots, \quad J_k = \|A_k x - b_k\|^2.$$

Here A_i is an $m_i \times n$ matrix, and b_i is an m_i-vector. We can use least squares to find the x that makes any one of these objectives as small as possible (provided the associated matrix has linearly independent columns). This will give us (in general) k different least squares approximate solutions. But we seek a *single* $\hat{x}$ that gives a compromise, and makes them all small, to the extent possible. We call this the *multi-objective* (or *multi-criterion*) least squares problem, and refer to $J_1, \ldots, J_k$ as the k objectives.

Multi-objective least squares via weighted sum. A standard method for finding a value of x that gives a compromise in making all the objectives small is to choose x to minimize a *weighted sum objective:*

$$J = \lambda_1 J_1 + \cdots + \lambda_k J_k = \lambda_1 \|A_1 x - b_1\|^2 + \cdots + \lambda_k \|A_k x - b_k\|^2, \qquad (15.1)$$

where $\lambda_1, \ldots, \lambda_k$ are positive *weights*, that express our relative desire for the terms to be small. If we choose all λ_i to be one, the weighted sum objective is the sum of the objective terms; we give each of them equal weight. If λ_2 is twice as large as λ_1, it means that we attach twice as much weight to the objective J_2 as to J_1. Roughly speaking, we care twice as strongly that J_2 should be small, compared to our desire that J_1 should be small. We will discuss later how to choose these weights.

Scaling all the weights in the weighted sum objective (15.1) by any positive number is the same as scaling the weighted sum objective J by the number, which does not change its minimizers. Since we can scale the weights by any positive number, it is common to choose $\lambda_1 = 1$. This makes the first objective term J_1 our *primary* objective; we can interpret the other weights as being relative to the primary objective.

Weighted sum least squares via stacking. We can minimize the weighted sum objective function (15.1) by expressing it as a standard least squares problem. We start by expressing J as the norm squared of a single vector:

$$J = \left\| \begin{bmatrix} \sqrt{\lambda_1}(A_1 x - b_1) \\ \vdots \\ \sqrt{\lambda_k}(A_k x - b_k) \end{bmatrix} \right\|^2,$$

where we use the property that $\|(a_1, \ldots, a_k)\|^2 = \|a_1\|^2 + \cdots + \|a_k\|^2$ for any vectors $a_1, \ldots, a_k$. So we have

$$J = \left\| \begin{bmatrix} \sqrt{\lambda_1} A_1 \\ \vdots \\ \sqrt{\lambda_k} A_k \end{bmatrix} x - \begin{bmatrix} \sqrt{\lambda_1} b_1 \\ \vdots \\ \sqrt{\lambda_k} b_k \end{bmatrix} \right\|^2 = \|\tilde{A} x - \tilde{b}\|^2,$$

where $\tilde{A}$ and $\tilde{b}$ are the matrix and vector

$$\tilde{A} = \begin{bmatrix} \sqrt{\lambda_1} A_1 \\ \vdots \\ \sqrt{\lambda_k} A_k \end{bmatrix}, \qquad \tilde{b} = \begin{bmatrix} \sqrt{\lambda_1} b_1 \\ \vdots \\ \sqrt{\lambda_k} b_k \end{bmatrix}. \tag{15.2}$$

The matrix $\tilde{A}$ is $m \times n$, and the vector $\tilde{b}$ has length m, where $m = m_1 + \cdots + m_k$.

We have now reduced the problem of minimizing the weighted sum least squares objective to a standard least squares problem. Provided the columns of $\tilde{A}$ are linearly independent, the minimizer is unique, and given by

$$\begin{aligned} \hat{x} &= (\tilde{A}^T \tilde{A})^{-1} \tilde{A}^T \tilde{b} \\ &= (\lambda_1 A_1^T A_1 + \cdots + \lambda_k A_k^T A_k)^{-1} (\lambda_1 A_1^T b_1 + \cdots + \lambda_k A_k^T b_k). \end{aligned} \tag{15.3}$$

This reduces to our standard formula for the solution of a least squares problem when $k = 1$ and $\lambda_1 = 1$. (In fact, when $k = 1$, λ_1 does not matter.) We can compute $\hat{x}$ via the QR factorization of $\tilde{A}$.

Independent columns of stacked matrix. Our assumption (12.2) that the columns of $\tilde{A}$ in (15.2) are linearly independent is not the same as assuming that each of $A_1, \ldots, A_k$ has linearly independent columns. We can state the condition that $\tilde{A}$ has linearly independent columns as: There is no nonzero vector x that satisfies

$A_i x = 0$ for $i = 1, \ldots, k$. This implies that if just *one* of the matrices $A_1, \ldots, A_k$ has linearly independent columns, then $\tilde{A}$ does.

The stacked matrix $\tilde{A}$ can have linearly independent columns even when none of the matrices $A_1, \ldots, A_k$ do. This can happen when $m_i < n$ for all i, *i.e.*, all A_i are wide. However, we must have $m_1 + \cdots + m_k \geq n$, since $\tilde{A}$ must be tall or square for the linearly independent columns assumption to hold.

Optimal trade-off curve.   We start with the special case of two objectives (also called the *bi-criterion problem*), and write the weighted sum objective as

$$J = J_1 + \lambda J_2 = \|A_1 x - b_1\|^2 + \lambda \|A_2 x - b_2\|^2,$$

where $\lambda > 0$ is the relative weight put on the second objective, compared to the first. For small λ, we care much more about J_1 being small than J_2 being small; for large λ, we care much less about J_1 being small than J_2 being small.

Let $\hat{x}(\lambda)$ denote the weighted sum least squares solution $\hat{x}$ as a function of λ, assuming the stacked matrices have linearly independent columns. These points are called *Pareto optimal* (after the economist Vilfredo Pareto) which means there is no point z that satisfies

$$\|A_1 z - b_1\|^2 \leq \|A_1 \hat{x}(\lambda) - b_1\|^2, \qquad \|A_2 z - b_2\|^2 \leq \|A_2 \hat{x}(\lambda) - b_2\|^2,$$

with one of the inequalities holding strictly. In other words, there is no point z that is as good as $\hat{x}(\lambda)$ in one of the objectives, and beats it on the other one. To see why this is the case, we note that any such z would have a value of J that is less than that achieved by $\hat{x}(\lambda)$, which minimizes J, a contradiction.

We can plot the two objectives $\|A_1 \hat{x}(\lambda) - b_1\|^2$ and $\|A_2 \hat{x}(\lambda) - b_2\|^2$ against each other, as λ varies over $(0, \infty)$, to understand the trade-off of the two objectives. This curve is called the *optimal trade-off curve* of the two objectives. There is no point z that achieves values of J_1 and J_2 that lies below and to the left of the optimal trade-off curve.

Simple example.   We consider a simple example with two objectives, with A_1 and A_2 both 10×5 matrices. The entries of the weighted least squares solution $\hat{x}(\lambda)$ are plotted against λ in figure 15.1. On the left, where λ is small, $\hat{x}(\lambda)$ is very close to the least squares approximate solution for A_1, b_1. On the right, where λ is large, $\hat{x}(\lambda)$ is very close to the least squares approximate solution for A_2, b_2. In between the behavior of $\hat{x}(\lambda)$ is very interesting; for instance, we can see that $\hat{x}(\lambda)_3$ first increases with increasing λ before eventually decreasing.

Figure 15.2 shows the values of the two objectives J_1 and J_2 versus λ. As expected, J_1 increases as λ increases, and J_2 decreases as λ increases. (It can be shown that this always holds.) Roughly speaking, as λ increases we put more emphasis on making J_2 small, which comes at the expense of making J_1 bigger. The optimal trade-off curve for this bi-criterion problem is plotted in figure 15.3. The left end-point corresponds to minimizing $\|A_1 x - b_1\|^2$, and the right end-point corresponds to minimizing $\|A_2 x - b_2\|^2$. We can conclude, for example, that there is no vector z that achieves $\|A_1 z - b_1\|^2 \leq 8$ and $\|A_2 z - b_2\|^2 \leq 5$.

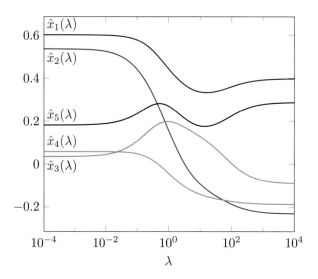

Figure 15.1 Weighted-sum least squares solution $\hat{x}(\lambda)$ as a function of λ for a bi-criterion least squares problem with five variables.

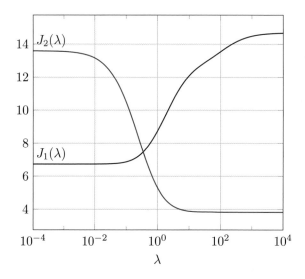

Figure 15.2 Objective functions $J_1 = \|A_1\hat{x}(\lambda) - b_1\|^2$ (blue line) and $J_2 = \|A_2\hat{x}(\lambda) - b_2\|^2$ (red line) as functions of λ for the bi-criterion problem in figure 15.1.

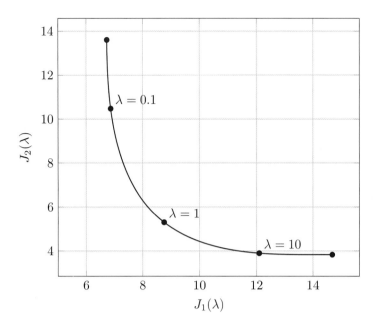

Figure 15.3 Optimal trade-off curve for the bi-criterion least squares problem of figures 15.1 and 15.2.

The steep slope of the optimal trade-off curve near the left end-point means that we can achieve a substantial reduction in J_2 with only a small increase in J_1. The small slope of the optimal trade-off curve near the right end-point means that we can achieve a substantial reduction in J_1 with only a small increase in J_2. This is quite typical, and indeed, is why multi-criterion least squares is useful.

Optimal trade-off surface.   Above we described the case with $k = 2$ objectives. When we have more than 2 objectives, the interpretation is similar, although it is harder to plot the objectives, or the values of $\hat{x}$, versus the weights. For example with $k = 3$ objectives, we have two weights, λ_2 and λ_3, which give the relative weight of J_2 and J_3 compared to J_1. Any solution $\hat{x}(\lambda)$ of the weighted least squares problem is Pareto optimal, which means that there is no point that achieves values of J_1, J_2, J_3 less than or equal to those obtained by $\hat{x}(\lambda)$, with strict inequality holding for at least one of them. As the parameters λ_2 and λ_3 vary over $(0, \infty)$, the values of J_1, J_2, J_3 sweep out the *optimal trade-off surface*.

Using multi-objective least squares.   In the rest of this chapter we will see several specific applications of multi-objective least squares. Here we give some general remarks on how it is used in applications.

First we identify a primary objective J_1 that we would like to be small. The objective J_1 is typically the one that would be used in an ordinary single-objective least squares approach, such as the mean square error of a model on some training data, or the mean square deviation from some target or goal.

We also identify one or more *secondary objectives* $J_2, J_3, \ldots, J_k$, that we would also like to be small. These secondary objectives are typically generic ones, like the desire that some parameters be 'small' or 'smooth', or close to some previous or prior value. In estimation applications these secondary objectives typically correspond to some kind of prior knowledge or assumption about the vector x that we seek. We wish to minimize our primary objective, but are willing to accept an increase in it, if this gives a sufficient decrease in the secondary objectives.

The weights are treated like 'knobs' in our method, that we change ('turn' or 'tune' or 'tweak') to achieve a value of $\hat{x}$ that we like (or can live with). For given candidate values of λ we evaluate the objectives; if we decide that J_2 is larger than we would like, but we can tolerate a somewhat larger J_3, then we increase λ_2 and decrease λ_3, and find $\hat{x}$ and the associated values of J_1, J_2, J_3 using the new weights. This is repeated until a reasonable trade-off among them has been obtained. In some cases we can be principled in how we adjust the weights; for example, in data fitting, we can use validation to help guide us in the choice of the weights. In many other applications, it comes down to (application-specific) judgment or even taste.

The additional terms $\lambda_2 J_2, \ldots, \lambda_k J_k$ that we add to the primary objective J_1, are sometimes called *regularization (terms)*. The secondary objectives are sometimes described by name, as in 'least squares fitting with smoothness regularization'.

In exploring the trade-offs among the objectives, the weights are typically varied over a wide range of values, by choosing a finite number of values (perhaps ten or a few tens) that are *logarithmically spaced*, as in figures 15.1 and 15.2. This means that for N values of λ between $\lambda^{\min}$ and $\lambda^{\max}$, we use the values

$$\lambda^{\min}, \quad \theta\lambda^{\min}, \quad \theta^2\lambda^{\min}, \quad \ldots, \quad \theta^{N-1}\lambda^{\min} = \lambda^{\max},$$

with $\theta = (\lambda^{\max}/\lambda^{\min})^{1/(N-1)}$.

15.2   Control

In control applications, the goal is to decide on a set of actions or inputs, specified by an n-vector x, that achieve some goals. The actions result in some outputs or effects, given by an m-vector y. We consider here the case when the inputs and outputs are related by an affine model

$$y = Ax + b.$$

The $m \times n$ matrix A and m-vector b characterize the *input-output mapping* of the system. The model parameters A and b are found from analytical models, experiments, computer simulations, or fit to past (observed) data. Typically the input or action $x = 0$ has some special meaning. The m-vector b gives the output when the input is zero. In many cases the vectors x and y represent *deviations* of the inputs and outputs from some standard values.

We typically have a desired or target output, denoted by the m-vector y^{des}. The primary objective is

$$J_1 = \|Ax + b - y^{\mathrm{des}}\|^2,$$

the norm squared deviation of the output from the desired output. The main objective is to choose an action x so that the output is as close as possible to the desired value.

There are many possible secondary objectives. The simplest one is the norm squared value of the input, $J_2 = \|x\|^2$, so the problem is to optimally trade off missing the target output (measured by $\|y - y^{\mathrm{des}}\|^2$), and keeping the input small (measured by $\|x\|^2$).

Another common secondary objective has the form $J_2 = \|x - x^{\mathrm{nom}}\|^2$, where x^{nom} is a nominal or standard value for the input. In this case the secondary objective it to keep the input close to the nominal value. This objective is sometimes used when x represents a new choice for the input, and x^{nom} is the current value. In this case the goal is to get the output near its target, while not changing the input much from its current value.

Control of heating and cooling. As an example, x could give the vector of n heating (or cooling) power levels in a commercial building with n air handling units (with $x_i > 0$ meaning heating and $x_i < 0$ meaning cooling) and y could represent the resulting temperature at m locations in the building. The matrix A captures the effect of each of n heating/cooling units on the temperatures in the building at each of m locations; the vector b gives the temperatures at the m locations when no heating or cooling is applied. The desired or target output might be $y^{\mathrm{des}} = T^{\mathrm{des}}\mathbf{1}$, assuming the target temperature is the same at all locations. The primary objective $\|y - y^{\mathrm{des}}\|^2$ is the sum of squares of the deviations of the location temperatures from the target temperature. The secondary objective $J_2 = \|x\|^2$, the norm squared of the vector of heating/cooling powers, would be reasonable, since it is at least roughly related to the energy cost of the heating and cooling.

We find tentative choices of the input by minimizing $J_1 + \lambda_2 J_2$ for various values of λ_2. If for the current value of λ_2 the heating/cooling powers are larger than we'd like, we increase λ_2 and re-compute $\hat{x}$.

Product demand shaping. In *demand shaping*, we adjust or change the prices of a set of n products in order to move the demand for the products towards some given target demand vector, perhaps to better match the available supply of the products. The standard price elasticity of demand model is $\delta^{\mathrm{dem}} = E^{\mathrm{d}}\delta^{\mathrm{price}}$, where δ^{dem} is the vector of fractional demand changes, δ^{price} is the vector of fractional price changes, and E^{d} is the price elasticity of demand matrix. (These are all described on page 150.) In this example the price change vector δ^{price} represents the action that we take; the result is the change in demand, δ^{dem}. The primary objective could be

$$J_1 = \|\delta^{\mathrm{dem}} - \delta^{\mathrm{tar}}\|^2 = \|E^{\mathrm{d}}\delta^{\mathrm{price}} - \delta^{\mathrm{tar}}\|^2,$$

where δ^{tar} is the target change in demand.

At the same time, we want the price changes to be small. This suggests the secondary objective $J_2 = \|\delta^{\mathrm{price}}\|^2$. We then minimize $J_1 + \lambda J_2$ for various values of λ, which trades off how close we come to the target change in demand with how much we change the prices.

Dynamics. The system can also be dynamic, meaning that we take into account time variation of the input and output. In the simplest case x is the time series of a scalar input, so x_i is the action taken in period i, and y_i is the (scalar) output in period i. In this setting, y^{des} is a desired trajectory for the output. A very common model for modeling dynamic systems, with x and y representing scalar input and output time series, is a convolution: $y = h * x$. In this case, A is Toeplitz, and b represents a time series, which is what the output would be with $x = 0$.

As a typical example in this category, the input x_i can represent the torque applied to the drive wheels of a locomotive (say, over one second intervals), and y_i is the locomotive speed.

In addition to the usual secondary objective $J_2 = \|x\|^2$, it is common to have an objective that the input should be smooth, *i.e.*, not vary too rapidly over time. This is achieved with the objective $\|Dx\|^2$, where D is the $(n-1) \times n$ first difference matrix

$$D = \begin{bmatrix} -1 & 1 & 0 & \cdots & 0 & 0 & 0 \\ 0 & -1 & 1 & \cdots & 0 & 0 & 0 \\ \vdots & \vdots & \vdots & & \vdots & \vdots & \vdots \\ 0 & 0 & 0 & \cdots & -1 & 1 & 0 \\ 0 & 0 & 0 & \cdots & 0 & -1 & 1 \end{bmatrix}. \tag{15.4}$$

15.3   Estimation and inversion

In the broad application area of *estimation* (also called *inversion*), the goal is to estimate a set of n values (also called parameters), the entries of the n-vector x. We are given a set of m *measurements*, the entries of an m-vector y. The parameters and measurements are related by

$$y = Ax + v,$$

where A is a known $m \times n$ matrix, and v is an unknown m-vector. The matrix A describes how the measured values (*i.e.*, y_i) depend on the unknown parameters (*i.e.*, x_j). The m-vector v is the *measurement error* or *measurement noise*, and is unknown but presumed to be small. The estimation problem is to make a sensible guess as to what x is, given y (and A), and prior knowledge about x.

If the measurement noise were zero, and A has linearly independent columns, we could recover x exactly, using $x = A^\dagger y$. (This is called *exact inversion*.) Our job here is to *guess* x, even when these strong assumptions do not hold. Of course we cannot expect to find x exactly, when the measurement noise is nonzero, or when A does not have linearly independent columns. This is called *approximate inversion*, or in some contexts, just *inversion*.

The matrix A can be wide, square, or tall; the same methods are used to estimate x in all three cases. When A is wide, we would not have enough measurements to determine x from y, even without the noise (*i.e.*, with $v = 0$). In this case we have to also rely on our prior information about x to make a reasonable guess. When A is square or tall, we would have enough measurements to determine x, if there were no noise present. Even in this case, judicious use of multiple-objective least squares can incorporate our prior knowledge in the estimation, and yield far better results.

15.3.1   Regularized inversion

If we guess that x has the value $\hat{x}$, then we are implicitly making the guess that v has the value $y - A\hat{x}$. If we assume that smaller values of v (measured by $\|v\|$) are more plausible than larger values, then a sensible choice for $\hat{x}$ is the least squares approximate solution, which minimizes $\|A\hat{x} - y\|^2$. We will take this as our primary objective. Our prior information about x enters in one or more secondary objectives. Simple examples are listed below.

- $\|x\|^2$: x should be small. This corresponds to the (prior) assumption that x is more likely to be small than large.

- $\|x - x^{\mathrm{prior}}\|^2$: x should be near x^{prior}. This corresponds to the assumption that x is near some known vector x^{prior}.

- $\|Dx\|^2$, where D is the first difference matrix (15.4). This corresponds to the assumption that x should be smooth, *i.e.*, x_{i+1} should be near x_i. This regularization is often used when x represents a time series.

- The Dirichlet energy $\mathcal{D}(x) = \|A^T x\|^2$, where A is the incidence matrix of a graph (see page 135). This corresponds to the assumption that x varies smoothly across the graph, *i.e.*, x_i is near x_j when i and j are connected by an edge of the graph. When the Dirichlet energy is used as a regularizer, it is sometimes called *Laplacian regularization*. (The previous example, $\|Dx\|^2$, is special case of Dirichlet energy, for the chain graph.)

Finally, we will choose our estimate $\hat{x}$ by minimizing

$$\|Ax - y\|^2 + \lambda_2 J_2(x) + \cdots + \lambda_p J_p(x),$$

where $\lambda_i > 0$ are weights, and $J_2, \ldots, J_p$ are the regularization terms. This is called *regularized inversion* or *regularized estimation*. We may repeat this for several choices of the weights, and choose the best estimate for the particular application.

Tikhonov regularized inversion.   Choosing $\hat{x}$ to minimize

$$\|Ax - y\|^2 + \lambda\|x\|^2$$

for some choice of $\lambda > 0$ is called *Tikhonov regularized inversion*, after the mathematician Andrey Tikhonov. Here we seek a guess $\hat{x}$ that is consistent with the measurements (*i.e.*, $\|A\hat{x} - y\|^2$ is small), but not too big.

The stacked matrix in this case,

$$\tilde{A} = \left[\begin{array}{c} A \\ \sqrt{\lambda} I \end{array} \right],$$

always has linearly independent columns, without any assumption about A, which can have any dimensions, and need not have linearly independent columns. To see this we note that $\tilde{A}x = (Ax, \sqrt{\lambda}x) = 0$ implies that $\sqrt{\lambda}x = 0$, which implies $x = 0$. The Gram matrix associated with $\tilde{A}$,

$$\tilde{A}^T \tilde{A} = A^T A + \lambda I,$$

is therefore always invertible (provided $\lambda > 0$). The Tikhonov regularized approximate solution is then

$$\hat{x} = (A^T A + \lambda I)^{-1} A^T b.$$

Equalization. The vector x represents a transmitted signal or message, consisting of n real values. The matrix A represents the mapping from the transmitted signal to what is received (called the *channel*); $y = Ax + v$ includes noise as well as the action of the channel. Guessing what x is, given y, can be thought of as un-doing the effects of the channel. In this context, estimation is called *equalization*.

15.3.2   Estimating a periodic time series

Suppose that the T-vector y is a (measured) time series, that we believe is a noisy version of a periodic time series, *i.e.*, one that repeats itself every P periods. We might also know or assume that the periodic time series is smooth, *i.e.*, its adjacent values are not too far apart.

Periodicity arises in many time series. For example, we would expect a time series of hourly temperature at some location to approximately repeat itself every 24 hours, or the monthly snowfall in some region to approximately repeat itself every 12 months. (Periodicity with a 24 hour period is called *diurnal*; periodicity with a yearly period is called *seasonal* or *annual*.) As another example, we might expect daily total sales at a restaurant to approximately repeat itself weekly. The goal is to get an estimate of Tuesday's total sales, given some historical daily sales data.

The periodic time series will be represented by a P-vector x, which gives its values over one period. It corresponds to the full time series

$$\hat{y} = (x, x, \dots, x)$$

which just repeats x, where we assume here for simplicity that T is a multiple of P. (If this is not the case, the last x is replaced with a slice of the form $x_{1:k}$.) We can express $\hat{y}$ as $\hat{y} = Ax$, where A is the $T \times P$ selector matrix

$$A = \left[\begin{array}{c} I \\ \vdots \\ I \end{array} \right].$$

Our total square estimation error is $\|Ax - y\|^2$.

We can minimize this objective analytically. The solution $\hat{x}$ is found by averaging the values of y associated with the different entries in x. For example, we estimate Tuesday sales by averaging all the entries in y that correspond to Tuesdays. (See exercise 15.10.) This simple averaging works well if we have many periods worth of data, *i.e.*, if T/P is large.

A more sophisticated estimate can be found by adding regularization for x to be smooth, based on the assumption that

$$x_1 \approx x_2, \quad \ldots, \quad x_{P-1} \approx x_P, \quad x_P \approx x_1.$$

(Note that we include the 'wrap-around' pair x_P and x_1 here.) We measure non-smoothness as $\|D^{\mathrm{circ}}x\|^2$, where D^{circ} is the $P \times P$ *circular difference matrix*

$$D^{\mathrm{circ}} = \begin{bmatrix} -1 & 1 & 0 & \cdots & 0 & 0 & 0 \\ 0 & -1 & 1 & \cdots & 0 & 0 & 0 \\ \vdots & \vdots & \vdots & & \vdots & \vdots & \vdots \\ 0 & 0 & 0 & \cdots & -1 & 1 & 0 \\ 0 & 0 & 0 & \cdots & 0 & -1 & 1 \\ 1 & 0 & 0 & \cdots & 0 & 0 & -1 \end{bmatrix}.$$

We estimate the periodic time series by minimizing

$$\|Ax - y\|^2 + \lambda\|D^{\mathrm{circ}}x\|^2.$$

For $\lambda = 0$ we recover the simple averaging mentioned above; as λ gets bigger, the estimated signal becomes smoother, ultimately converging to a constant (which is the mean of the original time series data).

The time series $A\hat{x}$ is called the *extracted seasonal component* of the given time series data y (assuming we are considering yearly variation). Subtracting this from the original data yields the time series $y - A\hat{x}$, which is called the *seasonally adjusted* time series.

The parameter λ can be chosen using validation. This can be done by selecting a time interval over which to build the estimate, and another one to validate it. For example, with 4 years of data, we might train our model on the first 3 years of data, and test it on the last year of data.

Example. In figure 15.4 we apply this method to a series of hourly ozone measurements. The top figure shows hourly measurements over a period of 14 days (July 1–14, 2014). We represent these values by a 336-vector c, with $c_{24(j-1)+i}$, $i = 1, \ldots, 24$, defined as the hourly values on day j, for $j = 1, \ldots, 14$. As indicated by the gaps in the graph, a number of measurements are missing from the record (only 275 of the $336 = 24 \times 14$ measurements are available). We use the notation $M_j \subseteq \{1, 2, \ldots, 24\}$ to denote the set containing the indices of the available measurements on day j. For example, $M_8 = \{1, 2, 3, 4, 6, 7, 8, 23, 24\}$, because on July 8, the measurements at 4AM, and from 8AM to 9PM are missing. The middle and bottom figures show two periodic time series. The time series are parametrized

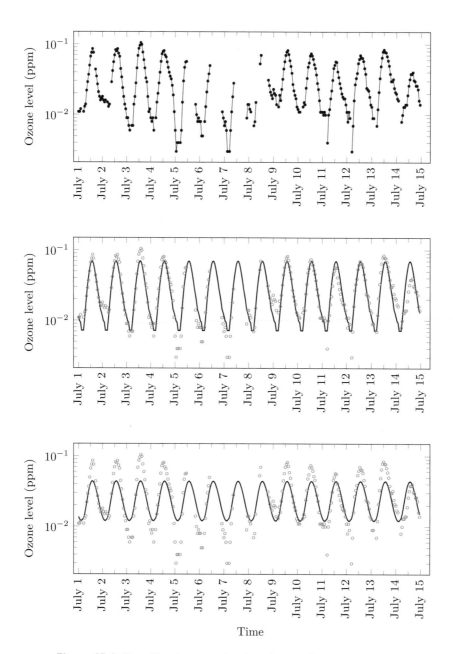

Figure 15.4 *Top.* Hourly ozone level at Azusa, California, during the first 14 days of July 2014 (California Environmental Protection Agency, Air Resources Board, www.arb.ca.gov). Measurements start at 12AM on July 1st, and end at 11PM on July 14. Note the large number of missing measurements. In particular, all 4AM measurements are missing. *Middle.* Smooth periodic least squares fit to logarithmically transformed measurements, using $\lambda = 1$. *Bottom.* Smooth periodic least squares fit using $\lambda = 100$.

by a 24-vector x, repeated 14 times to get the full series $(x, x, \ldots, x)$. The two estimates of x in the figure were computed by minimizing

$$\sum_{j=1}^{14} \sum_{i \in M_j} (x_i - \log(c_{24(j-1)+i}))^2 + \lambda \left(\sum_{i=1}^{23} (x_{i+1} - x_i)^2 + (x_1 - x_{24})^2 \right)$$

for $\lambda = 1$ and $\lambda = 100$.

15.3.3   Image de-blurring

The vector x is an image, and the matrix A gives blurring, so $y = Ax + v$ is a blurred, noisy image. Our prior information about x is that it is smooth; neighboring pixels values are not very different from each other. Estimation is the problem of guessing what x is, and is called *de-blurring*.

In least squares image deblurring we form an estimate $\hat{x}$ by minimizing a cost function of the form

$$\|Ax - y\|^2 + \lambda(\|D_{\mathrm{h}}x\|^2 + \|D_{\mathrm{v}}x\|^2). \tag{15.5}$$

Here D_{v} and D_{h} represent vertical and horizontal differencing operations, and the role of the second term in the weighted sum is to penalize non-smoothness in the reconstructed image. Specifically, suppose the vector x has length MN and contains the pixel intensities of an $M \times N$ image X stored column-wise. Let D_h be the $M(N-1) \times MN$ matrix

$$D_{\mathrm{h}} = \begin{bmatrix} -I & I & 0 & \cdots & 0 & 0 & 0 \\ 0 & -I & I & \cdots & 0 & 0 & 0 \\ \vdots & \vdots & \vdots & & \vdots & \vdots & \vdots \\ 0 & 0 & 0 & \cdots & -I & I & 0 \\ 0 & 0 & 0 & \cdots & 0 & -I & I \end{bmatrix},$$

where all blocks have size $M \times M$, and let D_{v} be the $(M-1)N \times MN$ matrix

$$D_{\mathrm{v}} = \begin{bmatrix} D & 0 & \cdots & 0 \\ 0 & D & \cdots & 0 \\ \vdots & \vdots & \ddots & \vdots \\ 0 & 0 & \cdots & D \end{bmatrix},$$

where each of the N diagonal blocks D is an $(M-1) \times M$ difference matrix

$$D = \begin{bmatrix} -1 & 1 & 0 & \cdots & 0 & 0 & 0 \\ 0 & -1 & 1 & \cdots & 0 & 0 & 0 \\ \vdots & \vdots & \vdots & & \vdots & \vdots & \vdots \\ 0 & 0 & 0 & \cdots & -1 & 1 & 0 \\ 0 & 0 & 0 & \cdots & 0 & -1 & 1 \end{bmatrix}.$$

Figure 15.5 *Left:* Blurred, noisy image. *Right:* Result of regularized least squares deblurring with $\lambda = 0.007$. Image credit: NASA.

With these definitions the penalty term in (15.5) is the sum of squared differences of intensities at adjacent pixels in a row or column:

$$\|D_\mathrm{h}x\|^2 + \|D_\mathrm{v}x\|^2 = \sum_{i=1}^{M}\sum_{j=1}^{N-1}(X_{i,j+1} - X_{ij})^2 + \sum_{i=1}^{M-1}\sum_{j=1}^{N}(X_{i+1,j} - X_{ij})^2.$$

This quantity is the Dirichlet energy (see page 135), for the graph that connects each pixel to its left and right, and up and down, neighbors.

Example.   In figures 15.5 and 15.6 we illustrate this method for an image of size 512×512. The blurred, noisy image is shown in the left part of figure 15.5. Figure 15.6 shows the estimates $\hat{x}$, obtained by minimizing (15.5), for four different values of the parameter λ. The best result (in this case, judged by eye) is obtained for λ around 0.007 and is shown on the right in figure 15.5.

15.3.4   Tomography

In *tomography*, the vector x represents the values of some quantity (such as density) in a region of interest in n voxels (or pixels) over a 3-D (or 2-D) region. The entries of the vector y are measurements obtained by passing a beam of radiation through the region of interest, and measuring the intensity of the beam after it exits the region.

A familiar application is the computer-aided tomography (CAT) scan used in medicine. In this application, beams of X-rays are sent through a patient, and an array of detectors measure the intensity of the beams after passing through the patient. These intensity measurements are related to the integral of the X-ray absorption along the beam. Tomography is also used in applications such as manufacturing, to assess internal damage or certify quality of a welded joint.

$\lambda = 10^{-6}$ $\lambda = 10^{-4}$

$\lambda = 10^{-2}$ $\lambda = 1$

Figure 15.6 Deblurred images for $\lambda = 10^{-6}$, 10^{-4}, 10^{-2}, 1. Image credit: NASA.

Line integral measurements. For simplicity we will assume that each beam is a single line, and that the received value y_i is the integral of the quantity over the region, plus some measurement noise. (The same method can be used when more complex beam shapes are used.) We consider the 2-D case.

Let $d(x, y)$ denote the density (say) at the position (x, y) in the region. (Here x and y are the scalar 2-D coordinates, not the vectors x and y in the estimation problem.) We assume that $d(x, y) = 0$ outside the region of interest. A line through the region is defined by the set of points

$$p(t) = (x_0, y_0) + t(\cos \theta, \sin \theta),$$

where (x_0, y_0) denotes a (base) point on the line, and θ is the angle of the line with respect to the x-axis. The parameter t gives the distance along the line from the point (x_0, y_0). The *line integral* of d is given by

$$\int_{-\infty}^{\infty} d(p(t)) \, dt.$$

We assume that m lines are specified (*e.g.*, by their base points and angles), and the measurement y_i is the line integral of d, plus some noise, which is presumed small.

We divide the region of interest into n pixels (or voxels in the 3-D case), and assume that the density has the constant value x_i over pixel i. Figure 15.7 illustrates this for a simple example with $n = 25$ pixels. (In applications the number of pixels or voxels is in the thousands or millions.) The line integral is then given by the sum of x_i (the density in pixel i) times the length of the intersection of the line with pixel i. In figure 15.7, with the pixels numbered row-wise starting at the top left corner, with width and height one, the line integral for the line shown is

$$1.06x_{16} + 0.80x_{17} + 0.27x_{12} + 1.06x_{13} + 1.06x_{14} + 0.53x_{15} + 0.54x_{10}.$$

The coefficient of x_i is the length of the intersection of the line with pixel i.

Measurement model. We can express the vector of m line integral measurements, without the noise, as Ax, where the $m \times n$ matrix A has entries

$$A_{ij} = \text{length of line } i \text{ in pixel } j, \quad i = 1, \ldots, m, \quad j = 1, \ldots, n,$$

with $A_{ij} = 0$ if line i does not intersect voxel j.

Tomographic reconstruction. In tomography, estimation or inversion is often called *tomographic reconstruction* or *tomographic inversion*.

The objective term $\|Ax - y\|^2$ is the sum of squares of the residual between the predicted (noise-free) line integrals Ax and the actual measured line integrals y. Regularization terms capture prior information or assumptions about the voxel values, for example, that they vary smoothly over the region. A simple regularizer commonly used is the Dirichlet energy (see page 135) associated with the graph that connects each voxel to its 6 neighbors (in the 3-D case) or its 4 neighbors (in the 2-D case). Using the Dirichlet energy as a regularizer is also called Laplacian regularization.

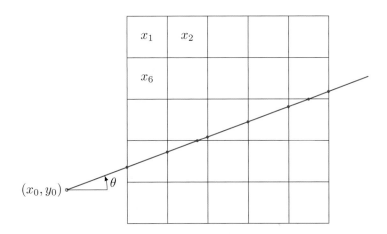

Figure 15.7 A square region of interest divided into 25 pixels, and a line passing through it.

Example.   A simple 2-D example is shown in figures 15.8–15.10. Figure 15.8 shows the geometry of the $m = 4000$ lines and the square region, shown as the square. The square is divided into 100×100 pixels, so $n = 10000$.

The density of the object we are imaging is shown in figure 15.9. In this object the density of each pixel is either 0 or 1 (shown as white or black, respectively). We reconstruct or estimate the object density from the 4000 (noisy) line integral measurements by solving the regularized least squares problem

$$\text{minimize}\quad \|Ax - y\|^2 + \lambda\|Dx\|^2,$$

where $\|Dx\|^2$ is the sum of squares of the differences of the pixel values from their neighbors. Figure 15.10 shows the results for six different values of λ. We can see that for small λ the reconstruction is relatively sharp, but suffers from noise. For large λ the noise in the reconstruction is smaller, but it is too smooth.

15.4   Regularized data fitting

We consider least squares data fitting, as described in chapter 13. In §13.2 we considered the issue of over-fitting, where the model performs poorly on new, unseen data, which occurs when the model is too complicated for the given data set. The remedy is to keep the model simple, *e.g.*, by fitting with a polynomial of not too high a degree.

Regularization is another way to avoid over-fitting, different from simply choosing a model that is simple (*i.e.*, does not have too many basis functions). Regularization is also called *de-tuning*, *shrinkage*, or *ridge regression*, for reasons we will explain below.

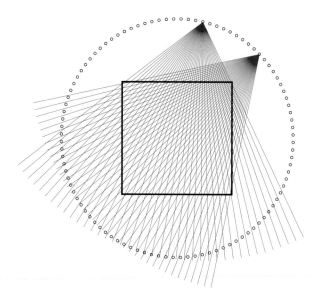

Figure 15.8 The square region at the center of the picture is surrounded by 100 points shown as circles. 40 lines (beams) emanate from each point. (The lines are shown for two points only.) This gives a total of 4000 lines that intersect the region.

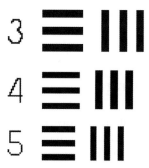

Figure 15.9 Density of object used in the tomography example.

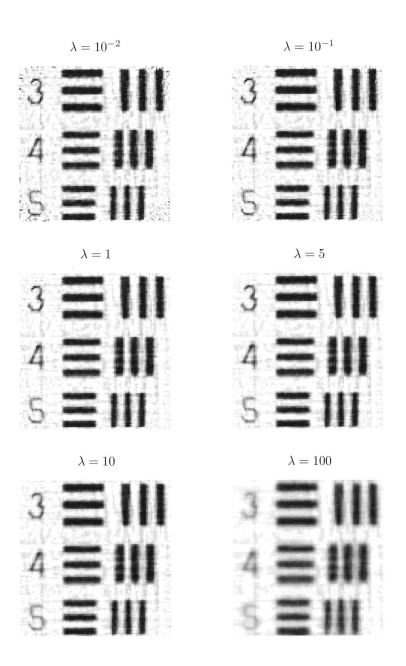

Figure 15.10 Regularized least squares reconstruction for six values of the regularization parameter.

Motivation. To motivate regularization, consider the model

$$\hat{f}(x) = \theta_1 f_1(x) + \cdots + \theta_p f_p(x). \tag{15.6}$$

We can interpret θ_i as the amount by which our prediction depends on $f_i(x)$, so if θ_i is large, our prediction will be very sensitive to changes or variations in the value of $f_i(x)$, such as those we might expect in new, unseen data. This suggests that we should prefer that θ_i be small, so our model is not too sensitive. There is one exception here: if $f_i(x)$ is constant (for example, the number one), then we should not worry about the size of θ_i, since $f_i(x)$ never varies. But we would like all the others to be small, if possible.

This suggests the bi-criterion least squares problem with primary objective $\|y - A\theta\|^2$, the sum of squares of the prediction errors, and secondary objective $\|\theta_{2:p}\|^2$, assuming that f_1 is the constant function one. Thus we should minimize

$$\|y - A\theta\|^2 + \lambda\|\theta_{2:p}\|^2, \tag{15.7}$$

where $\lambda > 0$ is called the *regularization parameter*.

For the regression model, this weighted objective can be expressed as

$$\|y - X^T\beta - v\mathbf{1}\|^2 + \lambda\|\beta\|^2.$$

Here we penalize β being large (because this leads to sensitivity of the model), but not the offset v. Choosing β to minimize this weighted objective is called *ridge regression*.

Effect of regularization. The effect of the regularization is to accept a worse value of sum square fit ($\|y - A\theta\|^2$) in return for a smaller value of $\|\theta_{2:p}\|^2$, which measures the size of the parameters (except θ_1, which is associated with the constant basis function). This explains the name shrinkage: The parameters are smaller than they would be without regularization, *i.e.*, they are shrunk. The term de-tuned suggests that with regularization, the model is not excessively 'tuned' to the training data (which would lead to over-fit).

Regularization path. We get a different model for every choice of λ. The way the parameters change with λ is called the *regularization path*. When p is small enough (say, less than 15 or so) the parameter values can be plotted, with λ on the horizontal axis. Usually only 30 or 50 values of λ are considered, typically spaced logarithmically over a large range (see page 314).

An appropriate value of λ can be chosen via out-of-sample or cross-validation. As λ increases, the RMS fit on the training data worsens (increases). But (as with model order) the test set RMS prediction error typically decreases as λ increases, and then, when λ gets too big, it increases. A good choice of regularization parameter is one which approximately minimizes the test set RMS prediction error. When multiple values of λ approximately minimize the RMS error, common practice is to take the largest value of λ. The idea here is to use the model of minimum sensitivity, as measured by $\|\beta\|^2$, among those that make good predictions on the test set.

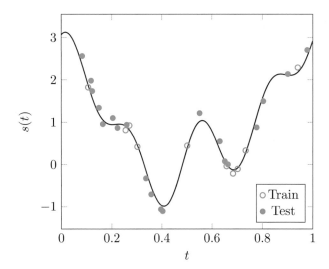

Figure 15.11 A signal $s(t)$ and 30 noisy samples. Ten of the samples are used as training set, the other 20 as test set.

Example. We illustrate these ideas with a small example with synthetic (simulated) data. We start with a signal, shown in figure 15.11, consisting of a constant plus four sinusoids:

$$s(t) = c + \sum_{k=1}^{4} \alpha_k \cos(\omega_k t + \phi_k),$$

with coefficients

$$c = 1.54, \quad \alpha_1 = 0.66, \quad \alpha_2 = -0.90, \quad \alpha_3 = -0.66, \quad \alpha_4 = 0.89. \tag{15.8}$$

(The other parameters are $\omega_1 = 13.69$, $\omega_2 = 3.55$, $\omega_3 = 23.25$, $\omega_4 = 6.03$, and $\phi_1 = 0.21$, $\phi_2 = 0.02$, $\phi_3 = -1.87$, $\phi_4 = 1.72$.) We will fit a model of the form (15.6) with $p = 5$ and

$$f_1(x) = 1, \quad f_{k+1}(x) = \cos(\omega_k x + \phi_k), \quad k = 1, \dots, 4.$$

(Note that the model is exact when the parameters are chosen as $\theta_1 = c$, $\theta_k = \alpha_{k-1}$, $k = 2, \dots, 5$. This rarely occurs in practice.) We fit our model using regularized least squares on 10 noisy samples of the function, shown as the open circles in figure 15.11. We will test the model obtained on the 20 noisy data points shown as the filled circles in figure 15.11.

Figure 15.12 shows the regularization path and the RMS training and test errors as functions of the regularization parameter λ, as it varies over a large range. The regularization path shows that as λ increases, the parameters $\theta_2, \dots, \theta_5$ get smaller (*i.e.*, shrink), converging towards zero as λ gets very large. We can see that the training prediction error increases with increasing λ (since we are trading off model sensitivity for sum square fitting error). The test error follows a typical pattern: It first decreases to a minimum value, then increases again. The minimum test error occurs around $\lambda = 0.079$; any choice between around $\lambda = 0.065$ and 0.100 (say) would be reasonable. The horizontal dashed lines show the 'true' values of the coefficients (*i.e.*, the ones we used to synthesize the data) given in (15.8). We can see that for λ near 0.079, our estimated parameters are close to the 'true' values.

Linear independence of columns. One side benefit of adding regularization to the basic least squares data fitting method, as in (15.7), is that the columns of the associated stacked matrix are *always linearly independent*, even if the columns of the matrix A are not. To see this, suppose that

$$
\begin{bmatrix} A \\ \sqrt{\lambda}B \end{bmatrix} x = 0,
$$

where B is the $(p-1) \times p$ selector matrix

$$
B = (e_2^T, \dots, e_p^T),
$$

so $B\theta = \theta_{2:p}$. From the last $p-1$ entries in the equation above, we get $\sqrt{\lambda}x_i = 0$ for $i = 2, \dots, p$, which implies that $x_2 = \cdots = x_p = 0$. Using these values of $x_2, \dots, x_p$, and the fact that the first column of A is $\mathbf{1}$, the top m equations become $\mathbf{1}x_1 = 0$, and we conclude that $x_1 = 0$ as well. So the columns of the stacked matrix are always linearly independent.

Feature engineering and regularized least squares. The simplest method of avoiding over-fit is to keep the model simple, which usually means that we should not use too many features. A typical and rough rule of thumb is that the number of features should be small compared to the number of data points (say, no more than 10% or 20%). The presence of over-fit can be detected using out-of-sample validation or cross-validation, which is always done when you fit a model to data.

Regularization is a powerful alternative method to avoid over-fitting a model. With regularization, you can fit a model with more features than would be appropriate without regularization. You can even fit a model using more features than you have data points, in which case the matrix A is wide. Regularization is often the key to success in feature engineering, which can greatly increase the number of features.

15.5   Complexity

In the general case we can minimize the weighted sum objective (15.1) by creating the stacked matrix and vector $\tilde{A}$ and $\tilde{b}$ in (15.2), and then using the QR factorization to solve the resulting least squares problem. The complexity of this method is order mn^2 flops, where $m = m_1 + \cdots + m_k$ is the sum of heights of the matrices $A_1, \dots, A_k$.

When using multi-objective least squares, it is common to minimize the weighted sum objective for some, or even many, different choices of weights. Assuming that the weighted sum objective is minimized for L different values of the weights, the total complexity is order Lmn^2 flops.

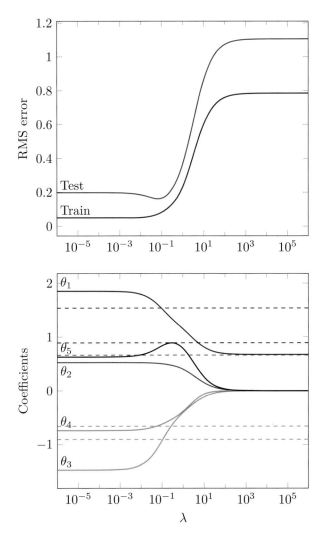

Figure 15.12 *Top.* RMS training and test errors as a function of the regularization parameter λ. *Bottom.* The regularization path. The dashed horizontal lines show the values of the coefficients used to generate the data.

15.5.1   Gram caching

We start from the formula (15.3) for the minimizer of the weighted sum objective,

$$\hat{x} = (\lambda_1 A_1^T A_1 + \cdots + \lambda_k A_k^T A_k)^{-1}(\lambda_1 A_1^T b_1 + \cdots + \lambda_k A_k^T b_k).$$

The matrix appearing in the inverse is a weighted sum of the Gram matrices $G_i = A_i^T A_i$ associated with the matrices A_i. We can compute $\hat{x}$ by forming these Gram matrices G_i, along with the vectors $h_i = A_i^T b_i$, then forming the weighted sums

$$G = \lambda_1 G_1 + \cdots + \lambda_k G_k, \qquad h = \lambda_1 h_1 + \cdots + \lambda_k h_k,$$

and finally, solving the $n \times n$ set of equations $G\hat{x} = h$. Forming G_i and h_i costs $m_i n^2$ and $2m_i n$ flops, respectively. (We save a factor of two in forming the Gram matrix; see page 182.) Ignoring the second term and adding over $i = 1, \ldots, k$ we get a total of mn^2 flops. Forming the weighted sums G and h costs $2kn^2$ flops. Solving $G\hat{x} = h$ costs order $2n^3$ flops.

Gram caching is the simple trick of computing G_i (and h_i) just once, and re-using these matrices and vectors for the L different choices of weights. This leads to a complexity of

$$mn^2 + L(k + 2n)n^2$$

flops. When m is much larger than $k + n$, which is a common occurrence, this cost is smaller than Lmn^2, the cost for the simple method.

As a simple example consider Tikhonov regularization. We will compute

$$\hat{x}^{(i)} = (A^T A + \lambda^{(i)} I)^{-1} A^T b$$

for $i = 1, \ldots, L$, where A is $m \times n$. The cost of the simple method is $2Lmn^2$ flops; using Gram caching the cost is $mn^2 + 2Ln^3 = (m + 2Ln)n^2$ flops. (We drop the term Lkn^2, since $k = 2$ here.) With $m = 100n$ and $L = 100$, Gram caching reduces the computational cost by more than a factor of 50. This means that the entire regularization path (*i.e.*, the solution for 100 values of λ) can be computed in not much more time than it takes to compute the solution for one value of λ.

15.5.2   The kernel trick

In this section we focus on another special case, which arises in many applications:

$$J = \|Ax - b\|^2 + \lambda\|x - x^{\mathrm{des}}\|^2, \tag{15.9}$$

where the $m \times n$ matrix A is wide, *i.e.*, $m < n$, and $\lambda > 0$. (Here we drop the subscripts on A, b, and m since we have only one matrix in this problem.) The associated $(m + n) \times n$ stacked matrix (see (15.2))

$$\tilde{A} = \left[\begin{array}{c} A \\ \sqrt{\lambda} I \end{array} \right]$$

always has linearly independent columns. Using the QR factorization to solve the stacked least squares problem requires $2(m + n)n^2$ flops, which grows like n^3. We will show now how this special problem can be solved far more efficiently when m is much smaller than n, using something called the *kernel trick*. Recall that the minimizer of J is given by (see (15.3))

$$
\begin{aligned}
\hat{x} &= (A^T A + \lambda I)^{-1}(A^T b + \lambda x^{\text{des}}) \\
&= (A^T A + \lambda I)^{-1}(A^T b + (\lambda I + A^T A)x^{\text{des}} - (A^T A)x^{\text{des}}) \\
&= (A^T A + \lambda I)^{-1} A^T (b - A x^{\text{des}}) + x^{\text{des}}.
\end{aligned}
$$

The matrix inverse here has size $n \times n$.

We will use the identity

$$
(A^T A + \lambda I)^{-1} A^T = A^T (A A^T + \lambda I)^{-1} \tag{15.10}
$$

which holds for any matrix A and any $\lambda > 0$. Note that the left-hand side of the identity involves the inverse of an $n \times n$ matrix, whereas the right-hand side involves the inverse of a (smaller) $m \times m$ matrix. (This is a variation on the push-through identity from exercise 11.9.)

To show the identity (15.10), we first observe that the matrices $A^T A + \lambda I$ and $A A^T + \lambda I$ are invertible. We start with the equation

$$
A^T (A A^T + \lambda I) = (A^T A + \lambda I)A^T,
$$

and multiply each side by $(A^T A + \lambda I)^{-1}$ on the left and $(A A^T + \lambda I)^{-1}$ on the right, which yields the identity above.

Using (15.10) we can express the minimizer of J as

$$
\hat{x} = A^T (A A^T + \lambda I)^{-1}(b - A x^{\text{des}}) + x^{\text{des}}.
$$

We can compute the term $(A A^T + \lambda I)^{-1}(b - A x^{\text{des}})$ by computing the QR factorization of the $(m + n) \times m$ matrix

$$
\bar{A} = \begin{bmatrix} A^T \\ \sqrt{\lambda} I \end{bmatrix},
$$

which has a cost of $2(m + n)m^2$ flops. The other operations involve matrix-vector products and have order (at most) mn flops, so we can use this method to compute $\hat{x}$ in around $2(m + n)m^2$ flops. This complexity grows only linearly in n.

To summarize, we can minimize the regularized least squares objective J in (15.9) two different ways. One requires a QR factorization of the $(m + n) \times n$ matrix $\tilde{A}$, which has cost $2(m + n)n^2$ flops. The other (which uses the kernel trick) requires a QR factorization of the $(m + n) \times m$ matrix $\bar{A}$, which has cost $2(m + n)m^2$ flops. We should evidently use the kernel trick when $m < n$. The complexity can then be expressed as

$$
(m + n)\min\{m^2, n^2\} \approx \min\{mn^2, nm^2\} = (\max\{m, n\})(\min\{m, n\})^2.
$$

where $\approx$ means we ignore non-dominant terms.

This is an instance of the *big-times-small-squared* rule or mnemonic, which states that many operations involving a matrix A can be carried out with order

$$
(\text{big}) \times (\text{small})^2 \text{ flops},
$$

where 'big' and 'small' refer to the big and small dimensions of the matrix. Several other examples are listed in appendix B.

Exercises

15.1 *A scalar multi-objective least squares problem.* We consider the special case of the multi-objective least squares problem in which the variable x is a scalar, and the k matrices A_i are all 1×1 matrices with value $A_i = 1$, so $J_i = (x - b_i)^2$. In this case our goal is to choose a number x that is simultaneously close to all the numbers $b_1, \ldots, b_k$. Let $\lambda_1, \ldots, \lambda_k$ be positive weights, and $\hat{x}$ the minimizer of the weighted objective (15.1). Show that $\hat{x}$ is a weighted average (or convex combination; see page 17) of the numbers $b_1, \ldots, b_k$, *i.e.*, it has the form

$$x = w_1 b_1 + \cdots + w_k b_k,$$

where w_i are nonnegative and sum to one. Give an explicit formula for the combination weights w_i in terms of the multi-objective least squares weights λ_i.

15.2 Consider the regularized data fitting problem (15.7). Recall that the elements in the first column of A are one. Let $\hat{\theta}$ be the solution of (15.7), *i.e.*, the minimizer of

$$\|A\theta - y\|^2 + \lambda(\theta_2^2 + \cdots + \theta_p^2),$$

and let $\tilde{\theta}$ be the minimizer of

$$\|A\theta - y\|^2 + \lambda\|\theta\|^2 = \|A\theta - y\|^2 + \lambda(\theta_1^2 + \theta_2^2 + \cdots + \theta_p^2),$$

in which we also penalize θ_1. Suppose columns 2 through p of A have mean zero (for example, because features $2, \ldots, p$ have been standardized on the data set; see page 269). Show that $\hat{\theta}_k = \tilde{\theta}_k$ for $k = 2, \ldots, p$.

15.3 *Weighted Gram matrix.* Consider a multi-objective least squares problems with matrices $A_1, \ldots, A_k$ and positive weights $\lambda_1, \ldots, \lambda_k$. The matrix

$$G = \lambda_1 A_1^T A_1 + \cdots + \lambda_k A_k^T A_k$$

is called the *weighted Gram matrix*; it is the Gram matrix of the stacked matrix $\tilde{A}$ (given in (15.2)) associated with the multi-objective problem. Show that G is invertible provided there is no nonzero vector x that satisfies $A_1 x = 0, \ldots, A_k x = 0$.

15.4 *Robust approximate solution of linear equations.* We wish to solve the square set of n linear equations $Ax = b$ for the n-vector x. If A is invertible the solution is $x = A^{-1}b$. In this exercise we address an issue that comes up frequently: We don't know A exactly. One simple method is to just choose a typical value of A and use it. Another method, which we explore here, takes into account the variation in the matrix A. We find a set of K versions of A, and denote them as $A^{(1)}, \ldots, A^{(K)}$. (These could be found by measuring the matrix A at different times, for example.) Then we choose x so as to minimize

$$\|A^{(1)}x - b\|^2 + \cdots + \|A^{(K)}x - b\|^2,$$

the sum of the squares of residuals obtained with the K versions of A. This choice of x, which we denote x^{rob}, is called a *robust* (approximate) solution. Give a formula for x^{rob}, in terms of $A^{(1)}, \ldots, A^{(K)}$ and b. (You can assume that a matrix you construct has linearly independent columns.) Verify that for $K = 1$ your formula reduces to $x^{\text{rob}} = (A^{(1)})^{-1}b$.

15.5 *Some properties of bi-objective least squares.* Consider the bi-objective least squares problem with objectives

$$J_1(x) = \|A_1 x - b_1\|^2, \qquad J_2(x) = \|A_2 x - b_2\|^2.$$

For $\lambda > 0$, let $\hat{x}(\lambda)$ denote the minimizer of $J_1(x) + \lambda J_2(x)$. (We assume the columns of the stacked matrix are linearly independent.) We define $J_1^\star(\lambda) = J_1(\hat{x}(\lambda))$ and $J_2^\star(\lambda) = J_2(\hat{x}(\lambda))$, the values of the two objectives as functions of the weight parameter. The optimal trade-off curve is the set of points $(J_1^\star(\lambda), J_2^\star(\lambda))$, as λ varies over all positive numbers.

(a) *Bi-objective problems without trade-off.* Suppose μ and γ are different positive weights, and $\hat{x}(\mu) = \hat{x}(\gamma)$. Show that $\hat{x}(\lambda)$ is constant for all $\lambda > 0$. Therefore the point $(J_1^\star(\lambda), J_2^\star(\lambda))$ is the same for all λ and the trade-off curve collapses to a single point.

(b) *Effect of weight on objectives in a bi-objective problem.* Suppose $\hat{x}(\lambda)$ is not constant. Show the following: for $\lambda < \mu$, we have

$$J_1^\star(\lambda) < J_1^\star(\mu), \qquad J_2^\star(\lambda) > J_2^\star(\mu).$$

This means that if you increase the weight (on the second objective), the second objective goes down, and the first objective goes up. In other words the trade-off curve slopes downward.

Hint. Resist the urge to write out any equations or formulas. Use the fact that $\hat{x}(\lambda)$ is the unique minimizer of $J_1(x) + \lambda J_2(x)$, and similarly for $\hat{x}(\mu)$, to deduce the inequalities

$$J_1^\star(\mu) + \lambda J_2^\star(\mu) > J_1^\star(\lambda) + \lambda J_2^\star(\lambda), \qquad J_1^\star(\lambda) + \mu J_2^\star(\lambda) > J_1^\star(\mu) + \mu J_2^\star(\mu).$$

Combine these inequalities to show that $J_1^\star(\lambda) < J_1^\star(\mu)$ and $J_2^\star(\lambda) > J_2^\star(\mu)$.

(c) *Slope of the trade-off curve.* The slope of the trade-off curve at the point $(J_1^\star(\lambda), J_2^\star(\lambda))$ is given by

$$S = \lim_{\mu \to \lambda} \frac{J_2^\star(\mu) - J_2^\star(\lambda)}{J_1^\star(\mu) - J_1^\star(\lambda)}.$$

(This limit is the same if μ approaches λ from below or from above.) Show that $S = -1/\lambda$. This gives another interpretation of the parameter λ: $(J_1^\star(\lambda), J_2^\star(\lambda))$ is the point on the trade-off curve where the curve has slope $-1/\lambda$.

Hint. First assume that μ approaches λ from above (meaning, $\mu > \lambda$) and use the inequalities in the hint for part (b) to show that $S \geq -1/\lambda$. Then assume that μ approaches λ from below and show that $S \leq -1/\lambda$.

15.6 *Least squares with smoothness regularization.* Consider the weighted sum least squares objective

$$\|Ax - b\|^2 + \lambda \|Dx\|^2,$$

where the n-vector x is the variable, A is an $m \times n$ matrix, D is the $(n-1) \times n$ difference matrix, with ith row $(e_{i+1} - e_i)^T$, and $\lambda > 0$. Although it does not matter in this problem, this objective is what we would minimize if we want an x that satisfies $Ax \approx b$, and has entries that are smoothly varying. We can express this objective as a standard least squares objective with a stacked matrix of size $(m + n - 1) \times n$.

Show that the stacked matrix has linearly independent columns if and only if $A\mathbf{1} \neq 0$, *i.e.*, the sum of the columns of A is not zero.

15.7 *Greedy regulation policy.* Consider a linear dynamical system given by $x_{t+1} = Ax_t + Bu_t$, where the n-vector x_t is the state at time t, and the m-vector u_t is the input at time t. The goal in regulation is to choose the input so as to make the state small. (In applications, the state $x_t = 0$ corresponds to the desired operating point, so small x_t means the state is close to the desired operating point.) One way to achieve this goal is to choose u_t so as to minimize

$$\|x_{t+1}\|^2 + \rho \|u_t\|^2,$$

where ρ is a (given) positive parameter that trades off using a small input versus making the (next) state small. Show that choosing u_t this way leads to a state feedback policy $u_t = Kx_t$, where K is an $m \times n$ matrix. Give a formula for K (in terms of A, B, and ρ). If an inverse appears in your formula, state the conditions under which the inverse exists.

Remark. This policy is called *greedy* or *myopic* since it does not take into account the effect of the input u_t on future states, beyond x_{t+1}. It can work very poorly in practice.

15.8 *Estimating the elasticity matrix.* In this problem you create a standard model of how demand varies with the prices of a set of products, based on some observed data. There are n different products, with (positive) prices given by the n-vector p. The prices are held constant over some period, say, a day. The (positive) demands for the products over the day are given by the n-vector d. The demand in any particular day varies, but it is thought to be (approximately) a function of the prices.

The *nominal prices* are given by the n-vector p^{nom}. You can think of these as the prices that have been charged in the past for the products. The *nominal demand* is the n-vector d^{nom}. This is the average value of the demand, when the prices are set to p^{nom}. (The actual daily demand fluctuates around the value d^{nom}.) You know both p^{nom} and d^{nom}. We will describe the prices by their (fractional) variations from the nominal values, and the same for demands. We define δ^p and δ^d as the (vectors of) relative price change and demand change:

$$\delta_i^p = \frac{p_i - p_i^{\mathrm{nom}}}{p_i^{\mathrm{nom}}}, \quad \delta_i^d = \frac{d_i - d_i^{\mathrm{nom}}}{d_i^{\mathrm{nom}}}, \quad i = 1, \ldots, n.$$

So $\delta_3^p = +0.05$ means that the price for product 3 has been increased by 5% over its nominal value, and $\delta_5^d = -0.04$ means that the demand for product 5 in some day is 4% below its nominal value.

Your task is to build a model of the demand as a function of the price, of the form

$$\delta^d \approx E\delta^p,$$

where E is the $n \times n$ elasticity matrix. You don't know E, but you do have the results of some experiments in which the prices were changed a bit from their nominal values for one day, and the day's demands were recorded. This data has the form

$$(p_1, d_1), \ldots, (p_N, d_N),$$

where p_i is the price for day i, and d_i is the observed demand.

Explain how you would estimate E, given this price-demand data. Be sure to explain how you will test for, and (if needed) avoid over-fit. *Hint.* Formulate the problem as a matrix least squares problem; see page 233.

Remark. Note the difference between elasticity estimation and demand shaping, discussed on page 315. In demand shaping, we know the elasticity matrix and are choosing prices; in elasticity estimation, we are guessing the elasticity matrix from some observed price and demand data.

15.9 *Regularizing stratified models.* In a *stratified* model (see page 272), we divide the data into different sets, depending on the value of some (often Boolean) feature, and then fit a separate model for each of these two data sets, using the remaining features. As an example, to develop a model of some health outcome we might build a separate model for women and for men. In some cases better models are obtained when we encourage the different models in a stratified model to be close to each other. For the case of stratifying on one Boolean feature, this is done by choosing the two model parameters $\theta^{(1)}$ and $\theta^{(2)}$ to minimize

$$\|A^{(1)}\theta^{(1)} - y^{(1)}\|^2 + \|A^{(2)}\theta^{(2)} - y^{(2)}\|^2 + \lambda\|\theta^{(1)} - \theta^{(2)}\|^2,$$

where $\lambda \geq 0$ is a parameter. The first term is the least squares residual for the first model on the first data set (say, women); the second term is the least squares residual for the second model on the second data set (say, men); the third term is a regularization term that encourages the two model parameters to be close to each other. Note that when $\lambda = 0$, we simply fit each model separately; when λ is very large, we are basically fitting one model to all the data. Of course the choice of an appropriate value of λ is obtained using out-of-sample validation (or cross-validation).

(a) Give a formula for the optimal $(\hat{\theta}^{(1)}, \hat{\theta}^{(2)})$. (If your formula requires one or more matrices to have linearly independent columns, say so.)

(b) *Stratifying across age groups.* Suppose we fit a model with each data point representing a person, and we stratify over the person's *age group*, which is a range of consecutive ages such as 18–24, 24–32, 33–45, and so on. Our goal is to fit a model for each age of k groups, with the parameters for adjacent age groups similar, or not too far, from each other. Suggest a method for doing this.

15.10 *Estimating a periodic time series.* (See §15.3.2.) Suppose that the T-vector y is a measured time series, and we wish to approximate it with a P-periodic T-vector. For simplicity, we assume that $T = KP$, where K is an integer. Let $\hat{y}$ be the simple least squares fit, with no regularization, *i.e.*, the P-periodic vector that minimizes $\|\hat{y} - y\|^2$. Show that for $i = 1, \ldots, P - 1$, we have

$$\hat{y}_i = \frac{1}{K} \sum_{k=1}^{K} y_{i+(k-1)P}.$$

In other words, each entry of the periodic estimate is the average of the entries of the original vector over the corresponding indices.

15.11 *General pseudo-inverse.* In chapter 11 we encountered the pseudo-inverse of a tall matrix with linearly independent columns, a wide matrix with linearly independent rows, and a square invertible matrix. In this exercise we describe the pseudo-inverse of a general matrix, *i.e.*, one that does not fit these categories. The general pseudo-inverse can be defined in terms of Tikhonov regularized inversion (see page 317). Let A be any matrix, and $\lambda > 0$. The Tikhonov regularized approximate solution of $Ax = b$, *i.e.*, unique minimizer of $\|Ax - b\|^2 + \lambda \|x\|^2$, is given by $(A^T A + \lambda I)^{-1} A^T b$. The pseudo-inverse of A is defined as

$$A^\dagger = \lim_{\lambda \to 0} (A^T A + \lambda I)^{-1} A^T.$$

In other words, $A^\dagger b$ is the limit of the Tikhonov-regularized approximate solution of $Ax = b$, as the regularization parameter converges to zero. (It can be shown that this limit always exists.) Using the kernel trick identity (15.10), we can also express the pseudo-inverse as

$$A^\dagger = \lim_{\lambda \to 0} A^T (AA^T + \lambda I)^{-1}.$$

(a) What is the pseudo-inverse of the $m \times n$ zero matrix?

(b) Suppose A has linearly independent columns. Explain why the limits above reduce to our previous definition, $A^\dagger = (A^T A)^{-1} A^T$.

(c) Suppose A has linearly independent rows. Explain why the limits above reduce to our previous definition, $A^\dagger = A^T (AA^T)^{-1}$.

Hint. For parts (b) and (c), you can use the fact that the matrix inverse is a continuous function, which means that the limit of the inverse of a matrix is the inverse of the limit, provided the limit matrix is invertible.

Chapter 16

Constrained least squares

In this chapter we discuss a useful extension of the least squares problem that includes linear equality constraints. Like least squares, the constrained least squares problem can be reduced to a set of linear equations, which can be solved using the QR factorization.

16.1 Constrained least squares problem

In the basic least squares problem, we seek x that minimizes the objective function $\|Ax - b\|^2$. We now add *constraints* to this problem, by insisting that x satisfy the linear equations $Cx = d$, where the matrix C and the vector d are given. The *linearly constrained least squares problem* (or just constrained least squares problem) is written as

$$
\begin{array}{ll}
\text{minimize} & \|Ax - b\|^2 \\
\text{subject to} & Cx = d.
\end{array}
\tag{16.1}
$$

Here x, the variable to be found, is an n-vector. The problem data (which are given) are the $m \times n$ matrix A, the m-vector b, the $p \times n$ matrix C, and the p-vector d.

We refer to the function $\|Ax - b\|^2$ as the *objective* of the problem, and the set of p linear equality constraints $Cx = d$ as the *constraints* of the problem. They can be written out as p scalar constraints (equations)

$$
c_i^T x = d_i, \quad i = 1, \dots, p,
$$

where c_i^T is the ith row of C.

An n-vector x is called *feasible* (for the problem (16.1)) if it satisfies the constraints, *i.e.*, $Cx = d$. An n-vector $\hat{x}$ is called an *optimal point* or *solution* of the optimization problem (16.1) if it is feasible, and if $\|A\hat{x} - b\|^2 \leq \|Ax - b\|^2$ holds for any feasible x. In other words, $\hat{x}$ solves the problem (16.1) if it is feasible and has the smallest possible value of the objective function among all feasible vectors.

The constrained least squares problem combines the problems of solving a set of linear equations (find x that satisfies $Cx = d$) with the least squares problem

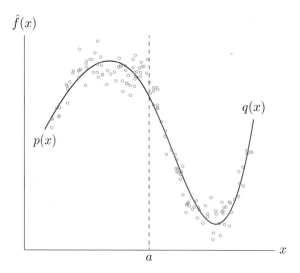

Figure 16.1 Least squares fit of two cubic polynomials to 140 points, with continuity constraints $p(a) = q(a)$ and $p'(a) = q'(a)$.

(find x that minimizes $\|Ax - b\|^2$). Indeed each of these problems can be considered a special case of the constrained least squares problem (16.1).

The constrained least squares problem can also be thought of as a limit of a bi-objective least squares problem, with primary objective $\|Ax - b\|^2$ and secondary objective $\|Cx - d\|^2$. Roughly speaking, we put infinite weight on the second objective, so that any nonzero value is unacceptable (which forces x to satisfy $Cx = d$). So we would expect (and it can be verified) that minimizing the weighted objective

$$\|Ax - b\|^2 + \lambda\|Cx - d\|^2,$$

for a very large value of λ yields a vector close to a solution of the constrained least squares problem (16.1). We will encounter this idea again in chapter 19, when we consider the nonlinear constrained least squares problem.

Example.   In figure 16.1 we fit a *piecewise-polynomial* function $\hat{f}(x)$ to a set of $N = 140$ points (x_i, y_i) in the plane. The function $\hat{f}(x)$ is defined as

$$\hat{f}(x) = \begin{cases} p(x) & x \le a \\ q(x) & x > a, \end{cases}$$

with a given, and $p(x)$ and $q(x)$ polynomials of degree three or less,

$$p(x) = \theta_1 + \theta_2 x + \theta_3 x^2 + \theta_4 x^3, \qquad q(x) = \theta_5 + \theta_6 x + \theta_7 x^2 + \theta_8 x^3.$$

We also impose the condition that $p(a) = q(a)$ and $p'(a) = q'(a)$, so that $\hat{f}(x)$ is continuous and has a continuous first derivative at $x = a$. Suppose the N data points (x_i, y_i) are numbered so that $x_1, \ldots, x_M \leq a$ and $x_{M+1}, \ldots, x_N > a$. The sum of squares of the prediction errors is

$$\sum_{i=1}^{M} (\theta_1 + \theta_2 x_i + \theta_3 x_i^2 + \theta_4 x_i^3 - y_i)^2 + \sum_{i=M+1}^{N} (\theta_5 + \theta_6 x_i + \theta_7 x_i^2 + \theta_8 x_i^3 - y_i)^2.$$

The conditions $p(a) - q(a) = 0$ and $p'(a) - q'(a) = 0$ are two linear equations

$$\begin{aligned}
\theta_1 + \theta_2 a + \theta_3 a^2 + \theta_4 a^3 - \theta_5 - \theta_6 a - \theta_7 a^2 - \theta_8 a^3 &= 0 \\
\theta_2 + 2\theta_3 a + 3\theta_4 a^2 - \theta_6 - 2\theta_7 a - 3\theta_8 a^2 &= 0.
\end{aligned}$$

We can determine the coefficients $\hat{\theta} = (\hat{\theta}_1, \ldots, \hat{\theta}_8)$ that minimize the sum of squares of the prediction errors, subject to the continuity constraints, by solving a constrained least squares problem

$$\begin{aligned}
\text{minimize} \quad & \|A\theta - b\|^2 \\
\text{subject to} \quad & C\theta = d.
\end{aligned}$$

The matrices and vectors A, b, C, d are defined as

$$A = \begin{bmatrix}
1 & x_1 & x_1^2 & x_1^3 & 0 & 0 & 0 & 0 \\
1 & x_2 & x_2^2 & x_2^3 & 0 & 0 & 0 & 0 \\
\vdots & \vdots & \vdots & \vdots & \vdots & \vdots & \vdots & \vdots \\
1 & x_M & x_M^2 & x_M^3 & 0 & 0 & 0 & 0 \\
0 & 0 & 0 & 0 & 1 & x_{M+1} & x_{M+1}^2 & x_{M+1}^3 \\
0 & 0 & 0 & 0 & 1 & x_{M+2} & x_{M+2}^2 & x_{M+2}^3 \\
\vdots & \vdots & \vdots & \vdots & \vdots & \vdots & \vdots & \vdots \\
0 & 0 & 0 & 0 & 1 & x_N & x_N^2 & x_N^3
\end{bmatrix}, \quad b = \begin{bmatrix} y_1 \\ y_2 \\ \vdots \\ y_M \\ y_{M+1} \\ y_{M+2} \\ \vdots \\ y_N \end{bmatrix},$$

and

$$C = \begin{bmatrix}
1 & a & a^2 & a^3 & -1 & -a & -a^2 & -a^3 \\
0 & 1 & 2a & 3a^2 & 0 & -1 & -2a & -3a^2
\end{bmatrix}, \quad d = \begin{bmatrix} 0 \\ 0 \end{bmatrix}.$$

This method is easily extended to piecewise-polynomial functions with more than two intervals. Functions of this kind are called *splines*.

Advertising budget allocation.   We continue the example described on page 234, where the goal is to purchase advertising in n different channels so as to achieve (or approximately achieve) a target set of customer views or impressions in m different demographic groups. We denote the n-vector of channel spending as s; this spending results in a set of views (across the demographic groups) given by the m-vector Rs. We will minimize the sum of squares of the deviation from the target set of views, given by v^{des}. In addition, we fix our total advertising spending, with the constraint $\mathbf{1}^T s = B$, where B is a given total advertising budget. (This can also be described as *allocating* a total budget B across the n different channels.) This leads to the constrained least squares problem

$$\begin{aligned}
\text{minimize} \quad & \|Rs - v^{\text{des}}\|^2 \\
\text{subject to} \quad & \mathbf{1}^T s = B.
\end{aligned}$$

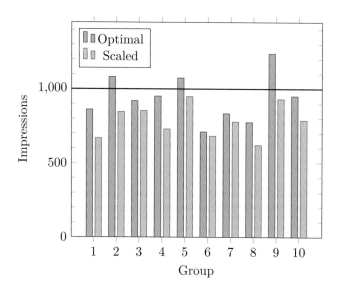

Figure 16.2 Advertising with budget constraint. The 'optimal' views vector is the solution of the constrained least squares problem with budget constraint. The 'scaled' views vector is obtained by scaling the unconstrained least squares solution so that it satisfies the budget constraint. This is a scalar multiple of the views vector of figure 12.4.

(The solution $\hat{s}$ of this problem is not guaranteed to have nonnegative entries, as it must to make sense in this application. But we ignore this aspect of the problem here.)

We consider the same problem instance as on page 234, with $m = 10$ demographic groups and $n = 3$ channels, and reach matrix R given there. The least squares method yields an RMS error of 133 (around 13.3%), with a total budget of $\mathbf{1}^T s^{\mathrm{ls}} = 1605$. We seek a spending plan with a budget that is 20% smaller, $B = 1284$. Solving the associated constrained least squares problem yields the spending vector $s^{\mathrm{cls}} = (315, 110, 859)$, which has RMS error of 161 in the target views. We can compare this spending vector to the one obtained by simply scaling the least squares spending vector by 0.80. The RMS error for this allocation is 239. The resulting impressions for both spending plans are shown in figure 16.2.

16.1.1   Least norm problem

An important special case of the constrained least squares problem (16.1) is when $A = I$ and $b = 0$:

$$
\begin{array}{ll}
\text{minimize} & \|x\|^2 \\
\text{subject to} & Cx = d.
\end{array}
\tag{16.2}
$$

In this problem we seek the vector of smallest or least norm that satisfies the linear equations $Cx = d$. For this reason the problem (16.2) is called the *least norm problem* or *minimum-norm problem*.

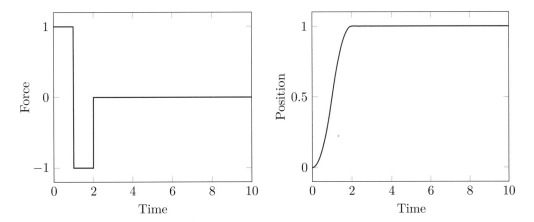

Figure 16.3 *Left:* A force sequence $f^{\mathrm{bb}} = (1, -1, 0, \dots, 0)$ that transfers the mass over a unit distance in 10 seconds. *Right:* The resulting position of the mass $p(t)$.

Example.   The 10-vector f represents a series of forces applied, each for one second, to a unit mass on a surface with no friction. The mass starts with zero velocity and position. By Newton's laws, its final velocity and position are given by

$$
\begin{aligned}
v^{\mathrm{fin}} &= f_1 + f_2 + \cdots + f_{10} \\
p^{\mathrm{fin}} &= (19/2)f_1 + (17/2)f_2 + \cdots + (1/2)f_{10}.
\end{aligned}
$$

(See exercise 2.3.)

Now suppose we want to choose a force sequence that results in $v^{\mathrm{fin}} = 0$, $p^{\mathrm{fin}} = 1$, *i.e.*, a force sequence that moves the mass to a resting position one meter to the right. There are many such force sequences; for example $f^{\mathrm{bb}} = (1, -1, 0, \dots, 0)$. This force sequence accelerates the mass to velocity 0.5 after one second, then decelerates it over the next second, so it arrives after two seconds with velocity 0, at the destination position 1. After that it applies zero force, so the mass stays where it is, at rest at position 1. The superscript 'bb' refers to *bang-bang*, which means that a large force is applied to get the mass moving (the first 'bang') and another large force (the second 'bang') is then applied to slow it to zero velocity. The force and position versus time for this choice of f are shown in figure 16.3.

Now we ask, what is the smallest force sequence that can achieve $v^{\mathrm{fin}} = 0$, $p^{\mathrm{fin}} = 1$, where smallest is measured by the sum of squares of the applied forces, $\|f\|^2 = f_1^2 + \cdots + f_{10}^2$? This problem can be posed as a least norm problem,

$$
\begin{aligned}
\text{minimize} \quad & \|f\|^2 \\
\text{subject to} \quad & \begin{bmatrix} 1 & 1 & \cdots & 1 & 1 \\ 19/2 & 17/2 & \cdots & 3/2 & 1/2 \end{bmatrix} f = \begin{bmatrix} 0 \\ 1 \end{bmatrix},
\end{aligned}
$$

with variable f. The solution f^{ln}, and the resulting position, are shown in figure 16.4. The norm square of the least norm solution f^{ln} is 0.0121; in contrast, the norm square of the bang-bang force sequence is 2, a factor of 165 times larger. (Note the very different vertical axis scales in figures 16.4 and 16.3.)

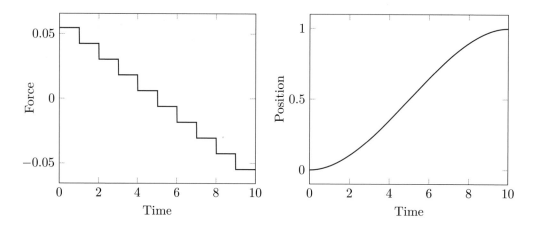

Figure 16.4 *Left:* The smallest force sequence $f^{\ln}$ that transfers the mass over a unit distance in 10 steps. *Right:* The resulting position of the mass $p(t)$.

16.2　Solution

Optimality conditions via Lagrange multipliers.　We will use the *method of Lagrange multipliers* (developed by the mathematician Joseph-Louis Lagrange, and summarized in §C.3) to solve the constrained least squares problem (16.1). Later we give an independent verification, that does not rely on calculus or Lagrange multipliers, that the solution we derive is correct.

We first write the constrained least squares problem with the constraints given as a list of p scalar equality constraints:

$$\begin{array}{ll} \text{minimize} & \|Ax - b\|^2 \\ \text{subject to} & c_i^T x = d_i, \quad i = 1, \ldots, p, \end{array}$$

where c_i^T are the rows of C. We form the *Lagrangian function*

$$L(x, z) = \|Ax - b\|^2 + z_1(c_1^T x - d_1) + \cdots + z_p(c_p^T x - d_p),$$

where z is the p-vector of *Lagrange multipliers*. The method of Lagrange multipliers tells us that if $\hat{x}$ is a solution of the constrained least squares problem, then there is a set of Lagrange multipliers $\hat{z}$ that satisfy

$$\frac{\partial L}{\partial x_i}(\hat{x}, \hat{z}) = 0, \quad i = 1, \ldots, n, \qquad \frac{\partial L}{\partial z_i}(\hat{x}, \hat{z}) = 0, \quad i = 1, \ldots, p. \qquad (16.3)$$

These are the *optimality conditions* for the constrained least squares problem. Any solution of the constrained least squares problem must satisfy them. We will now see that the optimality conditions can be expressed as a set of linear equations. The second set of equations in the optimality conditions can be written as

$$\frac{\partial L}{\partial z_i}(\hat{x}, \hat{z}) = c_i^T \hat{x} - d_i = 0, \quad i = 1, \ldots, p,$$

which states that $\hat{x}$ satisfies the equality constraints $C\hat{x} = d$ (which we already knew). The first set of equations, however, is more informative. Expanding the objective $\|Ax - b\|^2$ as a sum of terms involving the entries of x (as was done on page 229) and taking the partial derivative of L with respect to x_i we obtain

$$\frac{\partial L}{\partial x_i}(\hat{x}, \hat{z}) = 2 \sum_{j=1}^{n} (A^T A)_{ij} \hat{x}_j - 2(A^T b)_i + \sum_{j=1}^{p} \hat{z}_j (c_j)_i = 0.$$

These equations can be written in compact matrix-vector form as

$$2(A^T A)\hat{x} - 2A^T b + C^T \hat{z} = 0.$$

Combining this set of linear equations with the feasibility conditions $C\hat{x} = d$, we can write the optimality conditions (16.3) as one set of $n + p$ linear equations in the variables $(\hat{x}, \hat{z})$:

$$\begin{bmatrix} 2A^T A & C^T \\ C & 0 \end{bmatrix} \begin{bmatrix} \hat{x} \\ \hat{z} \end{bmatrix} = \begin{bmatrix} 2A^T b \\ d \end{bmatrix}. \tag{16.4}$$

These equations are called the *KKT equations* for the constrained least squares problem. (KKT are the initials of the last names of William Karush, Harold Kuhn, and Albert Tucker, the three researchers who derived the optimality conditions for a more general form of constrained optimization problem.) The KKT equations (16.4) are an extension of the normal equations (12.4) for a least squares problem with no constraints. So we have reduced the constrained least squares problem to the problem of solving a (square) set of $n + p$ linear equations in $n + p$ variables $(\hat{x}, \hat{z})$.

Invertibility of KKT matrix.   The $(n + p) \times (n + p)$ coefficient matrix in (16.4) is called the *KKT matrix*. It is invertible if and only if

$$C \text{ has linearly independent rows, and } \begin{bmatrix} A \\ C \end{bmatrix} \text{ has linearly independent columns.}$$
$$\tag{16.5}$$

The first condition requires that C is wide (or square), *i.e.*, that there are fewer constraints than variables. The second condition depends on both A and C, and it can be satisfied even when the columns of A are linearly dependent. The condition (16.5) is the generalization of our assumption (12.2) for unconstrained least squares (*i.e.*, that A has linearly independent columns).

Before proceeding, let us verify that the KKT matrix is invertible if and only if (16.5) holds. First suppose that the KKT matrix is not invertible. This means that there is a nonzero vector $(\bar{x}, \bar{z})$ with

$$\begin{bmatrix} 2A^T A & C^T \\ C & 0 \end{bmatrix} \begin{bmatrix} \bar{x} \\ \bar{z} \end{bmatrix} = 0.$$

Multiply the top block equation $2A^T A\bar{x} + C^T \bar{z} = 0$ on the left by $\bar{x}^T$ to get

$$2\|A\bar{x}\|^2 + \bar{x}^T C^T \bar{z} = 0.$$

The second block equation, $C\bar{x} = 0$, implies (by taking the transpose) $\bar{x}^T C^T = 0$, so the equation above becomes $2\|A\bar{x}\|^2 = 0$, i.e., $A\bar{x} = 0$. We also have $C\bar{x} = 0$, so

$$\begin{bmatrix} A \\ C \end{bmatrix} \bar{x} = 0.$$

Since the matrix on the left has linearly independent columns (by assumption), we conclude that $\bar{x} = 0$. The first block equation above then becomes $C^T \bar{z} = 0$. But by our assumption that the columns of C^T are linearly independent, we have $\bar{z} = 0$. So $(\bar{x}, \bar{z}) = 0$, which is a contradiction.

The converse is also true. First suppose that the rows of C are linearly dependent. Then there is a nonzero vector $\bar{z}$ with $C^T \bar{z} = 0$. Then

$$\begin{bmatrix} 2A^T A & C^T \\ C & 0 \end{bmatrix} \begin{bmatrix} 0 \\ \bar{z} \end{bmatrix} = 0,$$

which shows the KKT matrix is not invertible. Now suppose that the stacked matrix in (16.5) has linearly dependent columns, which means there is a nonzero vector $\bar{x}$ for which

$$\begin{bmatrix} A \\ C \end{bmatrix} \bar{x} = 0.$$

Direct calculation shows that

$$\begin{bmatrix} 2A^T A & C^T \\ C & 0 \end{bmatrix} \begin{bmatrix} \bar{x} \\ 0 \end{bmatrix} = 0,$$

which shows that the KKT matrix is not invertible.

When the conditions (16.5) hold, the constrained least squares problem (16.1) has the (unique) solution $\hat{x}$, given by

$$\begin{bmatrix} \hat{x} \\ \hat{z} \end{bmatrix} = \begin{bmatrix} 2A^T A & C^T \\ C & 0 \end{bmatrix}^{-1} \begin{bmatrix} 2A^T b \\ d \end{bmatrix}. \tag{16.6}$$

(This formula also gives us $\hat{z}$, the set of Lagrange multipliers.) From (16.6), we observe that the solution $\hat{x}$ is a linear function of (b, d).

Direct verification of constrained least squares solution.   We will now show directly, without using calculus, that the solution $\hat{x}$ given in (16.6) is the unique vector that minimizes $\|Ax - b\|^2$ over all x that satisfy the constraints $Cx = d$, when the conditions (16.5) hold. Let $\hat{x}$ and $\hat{z}$ denote the vectors given in (16.6), so they satisfy

$$2A^T A\hat{x} + C^T \hat{z} = 2A^T b, \qquad C\hat{x} = d.$$

Suppose that $x \neq \hat{x}$ is any vector that satisfies $Cx = d$. We will show that $\|Ax - b\|^2 > \|A\hat{x} - b\|^2$.

We proceed in the same way as for the least squares problem:

$$\begin{aligned} \|Ax - b\|^2 &= \|(Ax - A\hat{x}) + (A\hat{x} - b)\|^2 \\ &= \|Ax - A\hat{x}\|^2 + \|A\hat{x} - b\|^2 + 2(Ax - A\hat{x})^T (A\hat{x} - b). \end{aligned}$$

Now we expand the last term:

$$
\begin{aligned}
2(Ax - A\hat{x})^T(A\hat{x} - b) &= 2(x - \hat{x})^T A^T(A\hat{x} - b) \\
&= -(x - \hat{x})^T C^T \hat{z} \\
&= -(C(x - \hat{x}))^T \hat{z} \\
&= 0,
\end{aligned}
$$

where we use $2A^T(A\hat{x} - b) = -C^T\hat{z}$ in the second line and $Cx = C\hat{x} = d$ in the last line. So we have, exactly as in the case of unconstrained least squares,

$$
\|Ax - b\|^2 = \|A(x - \hat{x})\|^2 + \|A\hat{x} - b\|^2,
$$

from which we conclude that $\|Ax - b\|^2 \geq \|A\hat{x} - b\|^2$. So $\hat{x}$ minimizes $\|Ax - b\|^2$ subject to $Cx = d$.

It remains to show that for $x \neq \hat{x}$, we have the strict inequality $\|Ax - b\|^2 > \|A\hat{x} - b\|^2$, which by the equation above is equivalent to $\|A(x - \hat{x})\|^2 > 0$. If this is not the case, then $A(x - \hat{x}) = 0$. We also have $C(x - \hat{x}) = 0$, and so

$$
\begin{bmatrix} A \\ C \end{bmatrix} (x - \hat{x}) = 0.
$$

By our assumption that the matrix on the left has linearly independent columns, we conclude that $x = \hat{x}$.

16.3   Solving constrained least squares problems

We can compute the solution (16.6) of the constrained least squares problem by forming and solving the KKT equations (16.4).

Algorithm 16.1 CONSTRAINED LEAST SQUARES VIA KKT EQUATIONS

given an $m \times n$ matrix A and a $p \times n$ matrix C that satisfy (16.5), an m-vector b, and a p-vector d.

1. *Form Gram matrix.* Compute $A^T A$.

2. *Solve KKT equations.* Solve KKT equations (16.4) by QR factorization and back substitution.

The second step cannot fail, provided the assumption (16.5) holds. Let us analyze the complexity of this algorithm. The first step, forming the Gram matrix, requires mn^2 flops (see page 182). The second step requires the solution of a square system of $n + p$ equations, which costs $2(n + p)^3$ flops, so the total is

$$
mn^2 + 2(n + p)^3
$$

flops. This grows linearly in m and cubicly in n and p. The assumption (16.5) implies $p \leq n$, so in terms of order, $(n + p)^3$ can be replaced with n^3.

Solving constrained least squares problems via QR factorization. We now give a method for solving the constrained least squares problem that generalizes the QR factorization method for least squares problems (algorithm 12.1). We assume that A and C satisfy the conditions (16.5).

We start by rewriting the KKT equations (16.4) as

$$2(A^T A + C^T C)\hat{x} + C^T w = 2A^T b, \qquad C\hat{x} = d \qquad (16.7)$$

with a new variable $w = \hat{z} - 2d$. To obtain (16.7) we multiplied the equation $C\hat{x} = d$ on the left by $2C^T$, then added the result to the first equation of (16.4), and replaced the variable $\hat{z}$ with $w + 2d$.

Next we use the QR factorization

$$\begin{bmatrix} A \\ C \end{bmatrix} = QR = \begin{bmatrix} Q_1 \\ Q_2 \end{bmatrix} R \qquad (16.8)$$

to simplify (16.7). This factorization exists because the stacked matrix has linearly independent columns, by our assumption (16.5). In (16.8) we also partition Q in two blocks Q_1 and Q_2, of size $m \times n$ and $p \times n$, respectively. If we make the substitutions $A = Q_1 R$, $C = Q_2 R$, and $A^T A + C^T C = R^T R$ in (16.7) we obtain

$$2R^T R\hat{x} + R^T Q_2^T w = 2R^T Q_1^T b, \qquad Q_2 R\hat{x} = d.$$

We multiply the first equation on the left by R^{-T} (which we know exists) to get

$$R\hat{x} = Q_1^T b - (1/2)Q_2^T w. \qquad (16.9)$$

Substituting this expression into $Q_2 R\hat{x} = d$ gives an equation in w:

$$Q_2 Q_2^T w = 2Q_2 Q_1^T b - 2d. \qquad (16.10)$$

We now use the second part of the assumption (16.5) to show that the matrix $Q_2^T = R^{-T} C^T$ has linearly independent columns. Suppose $Q_2^T z = R^{-T} C^T z = 0$. Multiplying with R^T gives $C^T z = 0$. Since C has linearly independent rows, this implies $z = 0$, and we conclude that the columns of Q_2^T are linearly independent.

The matrix Q_2^T therefore has a QR factorization $Q_2^T = \tilde{Q}\tilde{R}$. Substituting this into (16.10) gives

$$\tilde{R}^T \tilde{R} w = 2\tilde{R}^T \tilde{Q}^T Q_1^T b - 2d,$$

which we can write as

$$\tilde{R} w = 2\tilde{Q}^T Q_1^T b - 2\tilde{R}^{-T} d.$$

We can use this to compute w, first by computing $\tilde{R}^{-T} d$ (by forward substitution), then forming the right-hand side, and then solving for w using back substitution. Once we know w, we can find $\hat{x}$ from (16.9). The method is summarized in the following algorithm.

Algorithm 16.2 CONSTRAINED LEAST SQUARES VIA QR FACTORIZATION

given an $m \times n$ matrix A and a $p \times n$ matrix C that satisfy (16.5), an m-vector b, and a p-vector d.

1. *QR factorizations.* Compute the QR factorizations

$$\begin{bmatrix} A \\ C \end{bmatrix} = \begin{bmatrix} Q_1 \\ Q_2 \end{bmatrix} R, \qquad Q_2^T = \tilde{Q}\tilde{R}.$$

2. Compute $\tilde{R}^{-T}d$ by forward substitution.

3. Form right-hand side and solve

$$\tilde{R}w = 2\tilde{Q}^T Q_1^T b - 2\tilde{R}^{-T}d$$

via back substitution.

4. *Compute $\hat{x}$.* Form right-hand side and solve

$$R\hat{x} = Q_1^T b - (1/2)Q_2^T w$$

by back substitution.

In the unconstrained case (when $p = 0$), step 1 reduces to computing the QR factorization of A, steps 2 and 3 are not needed, and step 4 reduces to solving $R\hat{x} = Q_1^T b$. This is the same as algorithm 12.1 for solving (unconstrained) least squares problems.

We now give a complexity analysis. Step 1 involves the QR factorizations of an $(m + p) \times n$ and an $n \times p$ matrix, which costs $2(m + p)n^2 + 2np^2$ flops. Step 2 requires p^2 flops. In step 3, we first evaluate $Q_1^T b$ ($2mn$ flops), multiply the result by $\tilde{Q}^T$ ($2pn$ flops), and then solve for w using forward substitution (p^2 flops). Step 4 requires $2mn + 2pn$ flops to form the right-hand side, and n^2 flops to compute $\hat{x}$ via back substitution. The costs of steps 2, 3, and 4 are quadratic in the dimensions, and so are negligible compared to the cost of step 1, so our final complexity is

$$2(m + p)n^2 + 2np^2$$

flops. The assumption (16.5) implies the inequalities

$$p \le n \le m + p,$$

and therefore $(m+p)n^2 \ge np^2$. So the flop count above is no more than $4(m+p)n^2$ flops. In particular, its order is $(m + p)n^2$.

Sparse constrained least squares. Constrained least squares problems with sparse matrices A and C arise in many applications; we will see several examples in the next chapter. Just as for solving linear equations, or (unconstrained) least squares problems, there are methods that exploit the sparsity in A and C to solve constrained least squares problems more efficiently than the generic algorithms 16.1

or 16.2. The simplest such methods follow these basic algorithms, replacing the QR factorizations with sparse QR factorizations (see page 190).

One potential problem with forming the KKT matrix as in algorithm 16.1 is that the Gram matrix $A^T A$ can be far less sparse than the matrix A. This problem can be avoided using a trick analogous to the one used on page 232 to solve sparse (unconstrained) least squares problems. We form the square set of $m+n+p$ linear equations

$$\begin{bmatrix} 0 & A^T & C^T \\ A & -(1/2)I & 0 \\ C & 0 & 0 \end{bmatrix} \begin{bmatrix} \hat{x} \\ \hat{y} \\ \hat{z} \end{bmatrix} = \begin{bmatrix} 0 \\ b \\ d \end{bmatrix}. \tag{16.11}$$

If $(\hat{x}, \hat{y}, \hat{z})$ satisfies these equations, it is easy to see that $(\hat{x}, \hat{z})$ satisfies the KKT equations (16.4); conversely, if $(\hat{x}, \hat{z})$ satisfies the KKT equations (16.4), $(\hat{x}, \hat{y}, \hat{z})$ satisfies the equations above, with $\hat{y} = 2(A\hat{x} - b)$. Provided A and C are sparse, the coefficient matrix above is sparse, and any method for solving a sparse system of linear equations can be used to solve it.

Solution of least norm problem. Here we specialize the solution of the general constrained least squares problem (16.1) given above to the special case of the least norm problem (16.2).

We start with the conditions (16.5). The stacked matrix is in this case

$$\begin{bmatrix} I \\ C \end{bmatrix},$$

which always has linearly independent columns. So the conditions (16.5) reduce to: C has linearly independent rows. We make this assumption now.

For the least norm problem, the KKT equations (16.4) reduce to

$$\begin{bmatrix} 2I & C^T \\ C & 0 \end{bmatrix} \begin{bmatrix} \hat{x} \\ \hat{z} \end{bmatrix} = \begin{bmatrix} 0 \\ d \end{bmatrix}.$$

We can solve this using the methods for general constrained least squares, or derive the solution directly, which we do now. The first block row of this equation is $2\hat{x} + C^T \hat{z} = 0$, so

$$\hat{x} = -(1/2)C^T \hat{z}.$$

We substitute this into the second block equation, $C\hat{x} = d$, to obtain

$$-(1/2)CC^T \hat{z} = d.$$

Since the rows of C are linearly independent, CC^T is invertible, so we have

$$\hat{z} = -2(CC^T)^{-1}d.$$

Substituting this expression for $\hat{z}$ into the formula for $\hat{x}$ above gives

$$\hat{x} = C^T(CC^T)^{-1}d. \tag{16.12}$$

We have seen the matrix in this formula before: It is the pseudo-inverse of a wide matrix with linearly independent rows. So we can express the solution of the least norm problem (16.2) in the very compact form

$$\hat{x} = C^\dagger d.$$

In §11.5, we saw that $C^\dagger$ is a right inverse of C; here we see that not only does $\hat{x} = C^\dagger d$ satisfy $Cx = d$, but it gives the vector of least norm that satisfies $Cx = d$.

In §11.5, we also saw that the pseudo-inverse of C can be expressed as $C^\dagger = QR^{-T}$, where $C^T = QR$ is the QR factorization of C^T. The solution of the least norm problem can therefore be expressed as

$$\hat{x} = QR^{-T}d$$

and this leads to an algorithm for solving the least norm problem via the QR factorization.

Algorithm 16.3 LEAST NORM VIA QR FACTORIZATION

given a $p \times n$ matrix C with linearly independent rows and a p-vector d.

1. *QR factorization.* Compute the QR factorization $C^T = QR$.
2. *Compute $\hat{x}$.* Solve $R^T y = d$ by forward substitution.
3. Compute $\hat{x} = Qy$.

The complexity of this algorithm is dominated by the cost of the QR factorization in step 1, *i.e.*, $2np^2$ flops.

Exercises

16.1 *Smallest right inverse.* Suppose the $m \times n$ matrix A is wide, with linearly independent rows. Its pseudo-inverse $A^\dagger$ is a right inverse of A. In fact, there are many right inverses of A and it turns out that $A^\dagger$ is the smallest one among them, as measured by the matrix norm. In other words, if X satisfies $AX = I$, then $\|X\| \geq \|A^\dagger\|$. You will show this in this problem.

   (a) Suppose $AX = I$, and let $x_1, \ldots, x_m$ denote the columns of X. Let b_j denote the jth column of $A^\dagger$. Explain why $\|x_j\|^2 \geq \|b_j\|^2$. *Hint.* Show that $z = b_j$ is the vector of smallest norm that satisfies $Az = e_j$, for $j = 1, \ldots, m$.

   (b) Use the inequalities from part (a) to establish $\|X\| \geq \|A^\dagger\|$.

16.2 *Matrix least norm problem.* The matrix least norm problem is

$$\begin{array}{ll} \text{minimize} & \|X\|^2 \\ \text{subject to} & CX = D, \end{array}$$

where the variable to be chosen is the $n \times k$ matrix X; the $p \times n$ matrix C and the $p \times k$ matrix D are given. Show that the solution of this problem is $\hat{X} = C^\dagger D$, assuming the rows of C are linearly independent. *Hint.* Show that we can find the columns of X independently, by solving a least norm problem for each one.

16.3 *Closest solution to a given point.* Suppose the wide matrix A has linearly independent rows. Find an expression for the point x that is closest to a given vector y (*i.e.*, minimizes $\|x - y\|^2$) among all vectors that satisfy $Ax = b$.

   Remark. This problem comes up when x is some set of inputs to be found, $Ax = b$ represents some set of requirements, and y is some nominal value of the inputs. For example, when the inputs represent actions that are re-calculated each day (say, because b changes every day), y might be yesterday's action, and the today's action x found as above gives the least change from yesterday's action, subject to meeting today's requirements.

16.4 *Nearest vector with a given average.* Let a be an n-vector and β a scalar. How would you find the n-vector x that is closest to a among all n-vectors that have average value β? Give a formula for x and describe it in English.

16.5 *Checking constrained least squares solution.* Generate a random 20×10 matrix A and a random 5×10 matrix C. Then generate random vectors b and d of appropriate dimensions for the constrained least squares problem

$$\begin{array}{ll} \text{minimize} & \|Ax - b\|^2 \\ \text{subject to} & Cx = d. \end{array}$$

Compute the solution $\hat{x}$ by forming and solving the KKT equations. Verify that the constraints very nearly hold, *i.e.*, $C\hat{x} - d$ is very small. Find the least norm solution x^{ln} of $Cx = d$. The vector x^{ln} also satisfies $Cx = d$ (very nearly). Verify that $\|Ax^{\text{ln}} - b\|^2 > \|A\hat{x} - b\|^2$.

16.6 *Modifying a diet to meet nutrient requirements.* (Continuation of exercise 8.9.) The current daily diet is specified by the n-vector d^{curr}. Explain how to find the closest diet d^{mod} to d^{curr} that satisfies the nutrient requirements given by the m-vector n^{des}, and has the same cost as the current diet d^{curr}.

16.7 *Minimum cost trading to achieve target sector exposures.* A current portfolio is given by the n-vector h^{curr}, with the entries giving the dollar value invested in the n assets. The total value (or net asset value) of the portfolio is $\mathbf{1}^T h^{\text{curr}}$. We seek a new portfolio, given by the n-vector h, with the same total value as h^{curr}. The difference $h - h^{\text{curr}}$ is called the *trade vector*; it gives the amount of each asset (in dollars) that we buy or sell. The n assets are divided into m industry sectors, such as pharmaceuticals or consumer electronics. We let the m-vector s denote the (dollar value) sector exposures to the m sectors. (See exercise 8.13.) These are given by $s = Sh$, where S is the $m \times n$ sector exposure matrix defined by $S_{ij} = 1$ if asset j is in sector i and $S_{ij} = 0$ if asset j is not in sector i. The new portfolio must have a given sector exposure s^{des}. (The given sector exposures are based on forecasts of whether companies in the different sectors will do well or poorly in the future.)

Among all portfolios that have the same value as our current portfolio and achieve the desired exposures, we wish to minimize the trading cost, given by

$$\sum_{i=1}^{n} \kappa_i (h_i - h_i^{\text{curr}})^2,$$

a weighted sum of squares of the asset trades. The weights κ_i are positive. (These depend on the daily trading volumes of assets, as well as other quantities. In general, it is cheaper to trade assets that have high trading volumes.)

Explain how to find h using constrained least squares. Give the KKT equations that you would solve to find h.

16.8 *Minimum energy regulator.* We consider a linear dynamical system with dynamics $x_{t+1} = Ax_t + Bu_t$, where the n-vector x_t is the state at time t and the m-vector u_t is the input at time t. We assume that $x = 0$ represents the desired operating point; the goal is to find an input sequence $u_1, \ldots, u_{T-1}$ that results in $x_T = 0$, given the initial state x_1. Choosing an input sequence that takes the state to the desired operating point at time T is called *regulation*.

Find an explicit formula for the sequence of inputs that yields regulation, and minimizes $\|u_1\|^2 + \cdots + \|u_{T-1}\|^2$, in terms of A, B, T, and x_1. This sequence of inputs is called the *minimum energy regulator*.

Hint. Express x_T in terms of x_1, A, the *controllability matrix*

$$C = \begin{bmatrix} A^{T-2}B & A^{T-3}B & \cdots & AB & B \end{bmatrix},$$

and $(u_1, u_2, \ldots, u_{T-1})$ (which is the input sequence stacked). You may assume that C is wide and has linearly independent rows.

16.9 *Smoothest force sequence to move a mass.* We consider the same setup as the example given on page 343, where the 10-vector f represents a sequence of forces applied to a unit mass over 10 1-second intervals. As in the example, we wish to find a force sequence f that achieves zero final velocity and final position one. In the example on page 343, we choose the smallest f, as measured by its norm (squared). Here, though, we want the *smoothest* force sequence, *i.e.*, the one that minimizes

$$f_1^2 + (f_2 - f_1)^2 + \cdots + (f_{10} - f_9)^2 + f_{10}^2.$$

(This is the sum of the squares of the differences, assuming that $f_0 = 0$ and $f_{11} = 0$.) Explain how to find this force sequence. Plot it, and give a brief comparison with the force sequence found in the example on page 343.

16.10 *Smallest force sequence to move a mass to a given position.* We consider the same setup as the example given on page 343, where the 10-vector f represents a sequence of forces applied to a unit mass over 10 1-second intervals. In that example the goal is to find the smallest force sequence (measured by $\|f\|^2$) that achieves zero final velocity and final position one. Here we ask, what is the smallest force sequence that achieves final position one? (We impose no condition on the final velocity.) Explain how to find this force sequence. Compare it to the force sequence found in the example, and give a brief intuitive explanation of the difference. *Remark.* Problems in which the final position of an object is specified, but the final velocity doesn't matter, generally arise in applications that are not socially positive, for example control of missiles.

16.11 *Least distance problem.* A variation on the least norm problem (16.2) is the least distance problem,

$$\begin{array}{ll} \text{minimize} & \|x - a\|^2 \\ \text{subject to} & Cx = d, \end{array}$$

where the n-vector x is to be determined, the n-vector a is given, the $p \times n$ matrix C is given, and the p-vector d is given. Show that the solution of this problem is

$$\hat{x} = a - C^\dagger(Ca - d),$$

assuming the rows of C are linearly independent. *Hint.* You can argue directly from the KKT equations for the least distance problem, or solve for the variable $y = x - a$ instead of x.

16.12 *Least norm polynomial interpolation.* (Continuation of exercise 8.7.) Find the polynomial of degree 4 that satisfies the interpolation conditions given in exercise 8.7, and minimizes the sum of the squares of its coefficients. Plot it, to verify that if satisfies the interpolation conditions.

16.13 *Steganography via least norm.* In steganography, a secret message is embedded in an image in such a way that the image looks the same, but an accomplice can decode the message. In this exercise we explore a simple approach to steganography that relies on constrained least squares. The secret message is given by a k-vector s with entries that are all either $+1$ or -1 (*i.e.*, it is a Boolean vector). The original image is given by the n-vector x, where n is usually much larger than k. We send (or publish or transmit) the modified message $x + z$, where z is an n-vector of modifications. We would like z to be small, so that the original image x and the modified one $x+z$ look (almost) the same. Our accomplice decodes the message s by multiplying the modified image by a $k \times n$ matrix D, which yields the k-vector $y = D(x + z)$. The message is then decoded as $\hat{s} = \mathbf{sign}(y)$. (We write $\hat{s}$ to show that it is an estimate, and might not be the same as the original.) The matrix D must have linearly independent rows, but otherwise is arbitrary.

(a) *Encoding via least norm.* Let α be a positive constant. We choose z to minimize $\|z\|^2$ subject to $D(x+z) = \alpha s$. (This guarantees that the decoded message is correct, *i.e.*, $\hat{s} = s$.) Give a formula for z in terms of $D^\dagger$, α, and x.

(b) *Complexity.* What is the complexity of encoding a secret message in an image? (You can assume that $D^\dagger$ is already computed and saved.) What is the complexity of decoding the secret message? About how long would each of these take with a computer capable of carrying out 1 Gflop/s, for $k = 128$ and $n = 512^2 = 262144$ (a 512×512 image)?

(c) *Try it out.* Choose an image x, with entries between 0 (black) and 1 (white), and a secret message s with k small compared to n, for example, $k = 128$ for a 512×512 image. (This corresponds to 16 bytes, which can encode 16 characters, *i.e.*, letters, numbers, or punctuation marks.) Choose the entries of D randomly, and compute $D^\dagger$. The modified image $x+z$ may have entries outside the range $[0, 1]$. We replace any negative values in the modified image with zero, and any values greater than one with one. Adjust α until the original and modified images look the same, but the secret message is still decoded correctly. (If α is too small, the clipping of the modified image values, or the round-off errors that occur in the computations, can lead to decoding error, *i.e.*, $\hat{s} \neq s$. If α is too large, the modification will be visually apparent.) Once you've chosen α, send several different secret messages embedded in several different original images.

16.14 *Invertibility of matrix in sparse constrained least squares formulation.* Show that the $(m + n + p) \times (m + n + p)$ coefficient matrix appearing in equation (16.11) is invertible if and only if the KKT matrix is invertible, *i.e.*, the conditions (16.5) hold.

16.15 *Approximating each column of a matrix as a linear combination of the others.* Suppose A is an $m \times n$ matrix with linearly independent columns $a_1, \ldots, a_n$. For each i we consider the problem of finding the linear combination of $a_1, \ldots, a_{i-1}, a_{i+1}, \ldots, a_n$ that is closest to a_i. These are n standard least squares problems, which can be solved using the methods of chapter 12. In this exercise we explore a simple formula that allows us to solve these n least squares problem all at once. Let $G = A^T A$ denote the Gram matrix, and $H = G^{-1}$ its inverse, with columns $h_1, \ldots, h_n$.

(a) Explain why minimizing $\|Ax^{(i)}\|^2$ subject to $x_i^{(i)} = -1$ solves the problem of finding the linear combination of $a_1, \ldots, a_{i-1}, a_{i+1}, \ldots, a_n$ that is closest to a_i. These are n constrained least squares problems.

(b) Solve the KKT equations for these constrained least squares problems,

$$
\begin{bmatrix} 2A^T A & e_i \\ e_i^T & 0 \end{bmatrix} \begin{bmatrix} x^{(i)} \\ z_i \end{bmatrix} = \begin{bmatrix} 0 \\ -1 \end{bmatrix},
$$

to conclude that $x^{(i)} = -(1/H_{ii})h_i$. In words: $x^{(i)}$ is the ith column of $(A^T A)^{-1}$, scaled so its ith entry is -1.

(c) Each of the n original least squares problems has $n-1$ variables, so the complexity is $n(2m(n-1)^2)$ flops, which we can approximate as $2mn^3$ flops. Compare this to the complexity of a method based on the result of part (b): First find the QR factorization of A; then compute H.

(d) Let d_i denote the distance between a_i and the linear combination of the other columns that is closest to it. Show that $d_i = 1/\sqrt{H_{ii}}$.

Remark. When the matrix A is a data matrix, with A_{ij} the value of the jth feature on the ith example, the problem addressed here is the problem of predicting each of the features from the others. The numbers d_i tells us how well each feature can be predicted from the others.

Chapter 17

Constrained least squares applications

In this chapter we discuss several applications of equality constrained least squares.

17.1 Portfolio optimization

In *portfolio optimization* (also known as *portfolio selection*), we invest in different assets, typically stocks, over some investment periods. The goal is to make investments so that the combined return on all our investments is consistently high. (We must accept the idea that for our average return to be high, we must tolerate some variation in the return, *i.e.*, some risk.) The idea of optimizing a portfolio of assets was proposed in 1953 by Harry Markowitz, who won the Nobel prize in economics for this work in 1990. In this section we will show that a version of this problem can be formulated and solved as a linearly constrained least squares problem.

17.1.1 Portfolio risk and return

Portfolio allocation weights. We allocate a total amount of money to be invested in n different assets. The allocation across the n assets is described by an allocation n-vector w, which satisfies $\mathbf{1}^T w = 1$, *i.e.*, its entries sum to one. If a total (dollar) amount V is to be invested in some period, then $V w_j$ is the amount invested in asset j. (This can be negative, meaning a short position of $|V w_j|$ dollars on asset j.) The entries of w are called by various names including *fractional allocations*, *asset weights*, *asset allocations*, or just *weights*.

For example, the asset allocation $w = e_j$ means that we invest everything in asset j. (In this way, we can think of the individual assets as simple portfolios.) The asset allocation $w = (-0.2, 0.0, 1.2)$ means that we take a short position in asset 1 of one fifth of the total amount invested, and put the cash derived from the short position plus our initial amount to be invested into asset 3. We do not invest in asset 2 at all.

The *leverage L* of the portfolio is given by

$$L = |w_1| + \cdots + |w_n|,$$

the sum of the absolute values of the weights. If all entries of w are nonnegative (which is a called a *long-only portfolio*), we have $L = 1$; if some entries are negative, then $L > 1$. If a portfolio has a leverage of 5, it means that for every \$1 of portfolio value, we have \$3 of total long holdings, and \$2 of total short holdings. (Other definitions of leverage are used, for example, $(L - 1)/2$.)

Multi-period investing with allocation weights.   The investments are held for T periods of, say, one day each. (The periods could just as well be hours, weeks, or months). We describe the investment returns by the $T \times n$ matrix R, where R_{tj} is the fractional return of asset j in period t. Thus $R_{61} = 0.02$ means that asset 1 gained 2% in period 6, and $R_{82} = -0.03$ means that asset 2 lost 3%, over period 8. The jth column of R is the return time series for asset j; the tth row of R gives the returns of all assets in period t. It is often assumed that one of the assets is cash, which has a constant (positive) return μ^{rf}, where the superscript stands for *risk-free*. If the risk-free asset is asset n, then the last column of R is $\mu^{\mathrm{rf}}\mathbf{1}$.

Suppose we invest a total (positive) amount V_t at the beginning of period t, so we invest $V_t w_j$ in asset j. At the end of period t, the dollar value of asset j is $V_t w_j (1 + R_{tj})$, and the dollar value of the whole portfolio is

$$V_{t+1} = \sum_{j=1}^{n} V_t w_j (1 + R_{tj}) = V_t (1 + \tilde{r}_t^T w),$$

where $\tilde{r}_t^T$ is the tth row of R. We assume V_{t+1} is positive; if the total portfolio value becomes negative we say that the portfolio has *gone bust* and stop trading.

The total (fractional) return of the portfolio over period t, *i.e.*, its fractional increase in value, is

$$\frac{V_{t+1} - V_t}{V_t} = \frac{V_t(1 + \tilde{r}_t^T w) - V_t}{V_t} = \tilde{r}_t^T w.$$

Note that we invest the total portfolio value in each period according to the weights w. This entails buying and selling assets so that the dollar value fractions are once again given by w. This is called *re-balancing* the portfolio.

The portfolio return in each of the T periods can be expressed compactly using matrix-vector notation as

$$r = Rw,$$

where r is the T-vector of portfolio returns in the T periods, *i.e.*, the time series of portfolio returns. (Note that r is a T-vector, which represents the time series of total portfolio return, whereas $\tilde{r}_t$ is an n-vector, which gives the returns of the n assets in period t.) If asset n is risk-free, and we choose the allocation $w = e_n$, then $r = Re_n = \mu^{\mathrm{rf}}\mathbf{1}$, *i.e.*, we obtain a constant return in each period of μ^{rf}.

We can express the total portfolio value in period t as

$$V_t = V_1(1 + r_1)(1 + r_2) \cdots (1 + r_{t-1}), \qquad (17.1)$$

where V_1 is the total amount initially invested in period $t = 1$. This total value time series is often plotted using $V_1 = \$10000$ as the initial investment by convention. The product in (17.1) arises from re-investing our total portfolio value (including any past gains or losses) in each period. In the simple case when the last asset is risk-free and we choose $w = e_n$, the total value grows as $V_t = V_1(1 + \mu^{\mathrm{rf}})^{t-1}$. This is called *compounded interest* at rate μ^{rf}.

When the returns r_t are small (say, a few percent), and T is not too big (say, a few hundred), we can approximate the product above using the sum or average of the returns. To do this we expand the product in (17.1) into a sum of terms, each of which involves a product of some of the returns. One term involves none of the returns, and is V_1. There are $t - 1$ terms that involve just one return, which have the form $V_1 r_s$, for $s = 1, \ldots, t-1$. All other terms in the expanded product involve the product of at least two returns, and so can be neglected since we assume that the returns are small. This leads to the approximation

$$V_t \approx V_1 + V_1(r_1 + \cdots + r_{t-1}),$$

which for $t = T + 1$ can be written as

$$V_{T+1} \approx V_1 + T \operatorname{\mathbf{avg}}(r) V_1.$$

This approximation suggests that to maximize our total final portfolio value, we should seek high return, *i.e.*, a large value for $\operatorname{\mathbf{avg}}(r)$.

Portfolio return and risk.   The choice of weight vector w is judged by the resulting portfolio return time series $r = Rw$. The portfolio *mean return* (over the T periods), often shortened to just the *return*, is given by $\operatorname{\mathbf{avg}}(r)$. The portfolio *risk* (over the T periods) is the standard deviation of portfolio return, $\operatorname{\mathbf{std}}(r)$.

The quantities $\operatorname{\mathbf{avg}}(r)$ and $\operatorname{\mathbf{std}}(r)$ give the *per-period* return and risk. They are often converted to their equivalent values for one year, which are called the *annualized return and risk*, and reported as percentages. If there are P periods in one year, these are given by

$$P \operatorname{\mathbf{avg}}(r), \qquad \sqrt{P} \operatorname{\mathbf{std}}(r),$$

respectively. For example, suppose each period is one (trading) day. There are about 250 trading days in one year, so the annualized return and risk are given by $250 \operatorname{\mathbf{avg}}(r)$ and $15.81 \operatorname{\mathbf{std}}(r)$. Thus a daily return sequence r with per-period (daily) return 0.05% (0.0005) and risk 0.5% (0.005) has an annualized return and risk of 12.5% and 7.9%, respectively. (The squareroot of P in the risk annualization comes from the assumption that the fluctuations in the returns vary randomly and independently from period to period.)

17.1.2   Portfolio optimization

We want to choose w so that we achieve high return and low risk. This means that we seek portfolio returns r_t that are consistently high. This is an optimization problem with two objectives, return and risk. Since there are two objectives, there is a family of solutions, that trade off return and risk. For example, when the last asset is risk-free, the portfolio weight $w = e_n$ achieves zero risk (which is the smallest possible value), and return μ^{rf}. We will see that other choices of w can lead to higher return, but higher risk as well. Portfolio weights that minimize risk for a given level of return (or maximize return for a given level of risk) are called *Pareto optimal*. The risk and return of this family of weights are typically plotted on a risk-return plot, with risk on the horizontal axis and return on the vertical axis. Individual assets can be considered (very simple) portfolios, corresponding to $w = e_j$. In this case the corresponding portfolio return and risk are simply the return and risk of asset j (over the same T periods).

One approach is to fix the return of the portfolio to be some given value ρ, and minimize the risk over all portfolios that achieve the required return. Doing this for many values of ρ produces (different) portfolio allocation vectors that trade off risk and return. Requiring that the portfolio return be ρ can be expressed as

$$\mathbf{avg}(r) = (1/T)\mathbf{1}^T(Rw) = \mu^T w = \rho,$$

where $\mu = R^T\mathbf{1}/T$ is the n-vector of the average asset returns. This is a single linear equation in w. Assuming that it holds, we can express the square of the risk as

$$\mathbf{std}(r)^2 = (1/T)\|r - \mathbf{avg}(r)\mathbf{1}\|^2 = (1/T)\|r - \rho\mathbf{1}\|^2.$$

Thus to minimize risk (squared), with return value ρ, we must solve the linearly constrained least squares problem

$$
\begin{array}{ll}
\text{minimize} & \|Rw - \rho\mathbf{1}\|^2 \\
\text{subject to} & \begin{bmatrix} \mathbf{1}^T \\ \mu^T \end{bmatrix} w = \begin{bmatrix} 1 \\ \rho \end{bmatrix}.
\end{array}
\tag{17.2}
$$

(We dropped the factor $1/T$ from the objective, which does not affect the solution.) This is a constrained least squares problem with two linear equality constraints. The first constraint sets the sum of the allocation weights to one, and the second requires that the mean portfolio return is ρ.

The portfolio optimization problem has the solution

$$
\begin{bmatrix} w \\ z_1 \\ z_2 \end{bmatrix} =
\begin{bmatrix} 2R^T R & \mathbf{1} & \mu \\ \mathbf{1}^T & 0 & 0 \\ \mu^T & 0 & 0 \end{bmatrix}^{-1}
\begin{bmatrix} 2\rho T\mu \\ 1 \\ \rho \end{bmatrix},
\tag{17.3}
$$

where z_1 and z_2 are Lagrange multipliers for the equality constraints (which we don't care about).

As a historical note, the portfolio optimization problem (17.2) is not exactly the same as the one proposed by Markowitz. His formulation used a statistical model of returns, where instead we are using a set of actual (or *realized*) returns. (See exercise 17.2 for a formulation of the problem that is closer to the original formulation by Markowitz.)

Future returns and the big assumption. The portfolio optimization problem (17.2) suffers from what would appear to be a serious conceptual flaw: It requires us to know the asset returns over the periods $t = 1, \dots, T$, in order to compute the optimal allocation to use over those periods. This is silly: If we knew *any* future returns, we would be able to achieve as large a portfolio return as we like, by simply putting large positive weights on the assets with positive returns and negative weights on those with negative returns. The whole challenge in investing is that we do not know future returns.

Assume the current time is period T, so we know the (so-called *realized*) return matrix R. The portfolio weight w found by solving (17.2), based on the observed returns in periods $t = 1, \dots, T$, can still be useful, when we make one (big) assumption:

$$\textit{Future asset returns are similar to past returns.} \tag{17.4}$$

In other words, if the asset returns for future periods $T + 1, T + 2, \dots$ are similar in nature to the past periods $t = 1, \dots, T$, then the portfolio allocation w found by solving (17.2) could be a wise choice to use in future periods.

Every time you invest, you are warned that the assumption (17.4) need not hold; you are required to acknowledge that past performance is no guarantee of future performance. The assumption (17.4) often holds well enough to be useful, but in times of 'market shift' it need not.

This situation is similar to that encountered when fitting models to observed data, as in chapters 13 and 14. The model is trained on past data that you have observed; but it will be used to make predictions on future data that you have not yet seen. A model is useful only to the extent that future data looks like past data. And this is an assumption which often (but not always) holds reasonably well.

Just as in model fitting, investment allocation vectors can (and should) be validated before being used. For example, we determine the weight vector by solving (17.2) using past returns data over some past training period, and check the performance on some other past testing period. If the portfolio performance over the training and testing periods are reasonably consistent, we gain confidence (but no guarantee) that the weight vector will work in future periods. For example, we might determine the weights using the realized returns from two years ago, and then test these weights by the performance of the portfolio over last year. If the test works out, we use the weights for next year. In portfolio optimization, validation is sometimes called *back-testing*, since you are testing the investment method on previous realized returns, to get an idea of how the method will work on (unknown) future returns.

The basic assumption (17.4) often holds less well than the analogous assumption in data fitting, *i.e.*, that future data looks like past data. For this reason we expect less coherence between the training and test performance of a portfolio, compared to a generic data fitting application. This is especially so when the test period has a small number of periods in it, like 100; see the discussion on page 268.

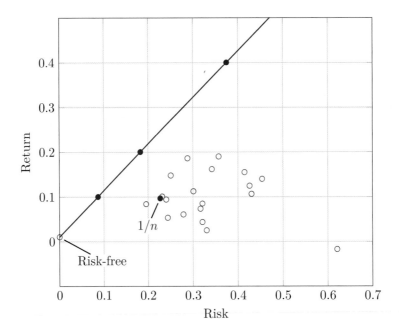

Figure 17.1 The open circles show annualized risk and return for 20 assets (19 stocks and one risk-free asset with a return of 1%). The solid line shows risk and return for the Pareto optimal portfolios. The dots show risk and return for three Pareto optimal portfolios with 10%, 20%, and 40% return, and the portfolio with weights $w_i = 1/n$.

17.1.3   Example

We use daily return data for 19 stocks over a period of 2000 days (8 years). After adding a risk-free asset with a 1% annual return, we obtain a 2000×20 return matrix R. The circles in figure 17.1 show the annualized risk and return for the 20 assets, *i.e.*, the points

$$\begin{bmatrix} \sqrt{250}\,\mathbf{std}(Re_i) \\ 250\,\mathbf{avg}(Re_i) \end{bmatrix}, \quad i = 1, \dots, 20.$$

It also shows the Pareto-optimal risk-return curve, and the risk and return for the uniform portfolio with equal weights $w_i = 1/n$. The annualized risk, return, and leverage for five portfolios (the four Pareto-optimal portfolios indicated in the figure, and the $1/n$ portfolio) are given in table 17.1.

Figure 17.2 shows the total portfolio value (17.1) for the five portfolios. Figure 17.3 shows the portfolio values for a different test period of 500 days (two years).

	Return		Risk		
Portfolio	Train	Test	Train	Test	Leverage
Risk-free	0.01	0.01	0.00	0.00	1.00
10%	0.10	0.08	0.09	0.07	1.96
20%	0.20	0.15	0.18	0.15	3.03
40%	0.40	0.30	0.38	0.31	5.48
$1/n$	0.10	0.21	0.23	0.13	1.00

Table 17.1 Annualized risk, return, and leverage for five portfolios.

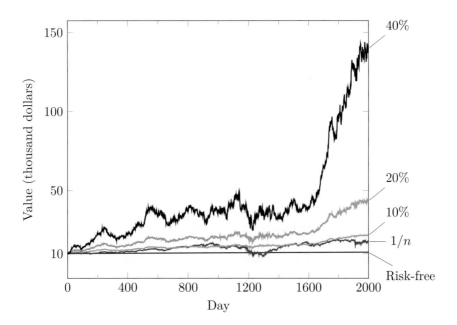

Figure 17.2 Total value over time for five portfolios: the risk-free portfolio with 1% annual return, the Pareto optimal portfolios with 10%, 20%, and 40% return, and the uniform portfolio. The total value is computed using the 2000×20 daily return matrix R.

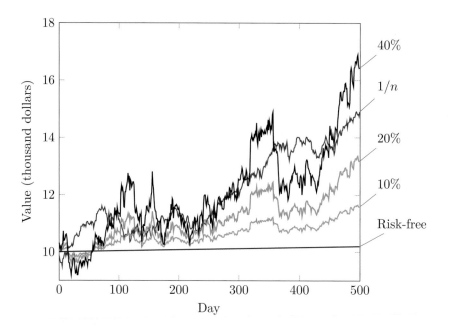

Figure 17.3 Value over time for the five portfolios in figure 17.2 over a test period of 500 days.

17.1.4   Variations

There are many variations on the basic portfolio optimization problem (17.2). We describe a few of them here; a few others are explored in the exercises.

Regularization.   Just as in data fitting, our formulation of portfolio optimization can suffer from over-fit, which means that the chosen weights perform very well on past (realized) returns, but poorly on new (future) returns. Over-fit can be avoided or reduced by adding regularization, which here means to penalize investments in assets other than cash. (This is analogous to regularization in model fitting, where we penalize the size of the model coefficients, except for the coefficient associated with the constant feature.) One natural way to incorporate regularization in the portfolio optimization problem (17.2) is to add a positive multiple λ of the weighted sum of squares term

$$\sigma_1^2 w_1^2 + \cdots + \sigma_{n-1}^2 w_{n-1}^2$$

to the objective in (17.2). Note that we do not penalize w_n, which is the weight associated with the risk-free asset. The constants σ_i are the standard deviations of the (realized) returns, *i.e.*, $\sigma_i = \mathbf{std}(Re_i)$. This regularization penalizes weights associated with risky assets more than those associated with less risky assets. A good choice of λ can be found by back-testing.

Time-varying weights.   Markets do shift, so it is not uncommon to periodically update or change the allocation weights that are used. In one extreme version of this, a new allocation vector is used in every period. The allocation weight for any period is obtained by solving the portfolio optimization problem over the preceding M periods. (This scheme can be modified to include testing periods as well.) The parameter M in this method would be chosen by validation on previous realized returns, *i.e.*, back-testing.

When the allocation weights are changed over time, we can add a (regularization) term of the form $\kappa\|w^{\mathrm{curr}} - w\|^2$ to the objective, where κ is a positive constant. Here w^{curr} is the currently used allocation, and w is the proposed new allocation vector. The additional regularization term encourages the new allocation vector to be near the current one. (When this is not the case, the portfolio will require excessive buying and selling of assets. This is called *turnover*, which leads to trading costs not included in our simple model.) The parameter κ would be chosen by back-testing, taking into account an approximation of trading cost.

17.1.5   Two-fund theorem

We can express the solution (17.3) of the portfolio optimization problem in the form

$$
\begin{bmatrix} w \\ z_1 \\ z_2 \end{bmatrix} = \begin{bmatrix} 2R^T R & \mathbf{1} & \mu \\ \mathbf{1}^T & 0 & 0 \\ \mu^T & 0 & 0 \end{bmatrix}^{-1} \begin{bmatrix} 0 \\ 1 \\ 0 \end{bmatrix} + \rho \begin{bmatrix} 2R^T R & \mathbf{1} & \mu \\ \mathbf{1}^T & 0 & 0 \\ \mu^T & 0 & 0 \end{bmatrix}^{-1} \begin{bmatrix} 2T\mu \\ 0 \\ 1 \end{bmatrix}.
$$

Taking the first n components of this, we obtain

$$
w = w^0 + \rho v, \tag{17.5}
$$

where w^0 and v are the first n components of the $(n+2)$-vectors

$$
\begin{bmatrix} 2R^T R & \mathbf{1} & \mu \\ \mathbf{1}^T & 0 & 0 \\ \mu^T & 0 & 0 \end{bmatrix}^{-1} \begin{bmatrix} 0 \\ 1 \\ 0 \end{bmatrix}, \qquad \begin{bmatrix} 2R^T R & \mathbf{1} & \mu \\ \mathbf{1}^T & 0 & 0 \\ \mu^T & 0 & 0 \end{bmatrix}^{-1} \begin{bmatrix} 2T\mu \\ 0 \\ 1 \end{bmatrix},
$$

respectively. The equation (17.5) shows that the Pareto optimal portfolios form a *line* in weight space, parametrized by the required return ρ. The portfolio w^0 is a point on the line, and the vector v, which satisfies $\mathbf{1}^T v = 0$, gives the direction of the line. This equation tells us that we do not need to solve the equation (17.3) for each value of ρ. We first compute w^0 and v (by factoring the matrix once and using two solve steps), and then form the optimal portfolio with return ρ as $w^0 + \rho v$.

Any point on a line can be expressed as an affine combination of two different points on the line. So if we find two different Pareto optimal portfolios, then we can express a general Pareto optimal portfolio as an affine combination of them. In other words, all Pareto optimal portfolios are affine combinations of just two portfolios (indeed, any two different Pareto optimal portfolios). This is the *two-fund theorem*. (*Fund* is another term for portfolio.)

Now suppose that the last asset is risk-free. The portfolio $w = e_n$ is Pareto optimal, since it achieves return μ^{rf} with zero risk. We then find one other Pareto optimal portfolio, for example, the one w^2 that achieves return $2\mu^{\mathrm{rf}}$, twice the risk-free return. (We could choose here any return other than μ^{rf}.) Then we can express the general Pareto optimal portfolio as

$$w = (1 - \theta)e_n + \theta w^2,$$

where $\theta = \rho/\mu^{\mathrm{rf}} - 1$.

17.2   Linear quadratic control

We consider a time-varying linear dynamical system with state n-vector x_t and input m-vector u_t, with dynamics equations

$$x_{t+1} = A_t x_t + B_t u_t, \quad t = 1, 2, \dots. \tag{17.6}$$

The system has an output, the p-vector y_t, given by

$$y_t = C_t x_t, \quad t = 1, 2, \dots. \tag{17.7}$$

Usually, $m \leq n$ and $p \leq n$, *i.e.*, there are fewer inputs and outputs than states.

In control applications, the input u_t represents quantities that we can choose or manipulate, like control surface deflections or engine thrust on an airplane. The state x_t, input u_t, and output y_t typically represent *deviations* from some standard or desired operating condition, for example, the deviation of aircraft speed and altitude from the desired values. For this reason it is desirable to have x_t, y_t, and u_t small.

Linear quadratic control refers to the problem of choosing the input and state sequences, over a time period $t = 1, \dots, T$, so as to minimize a sum of squares objective, subject to the dynamics equations (17.6), the output equations (17.7), and additional linear equality constraints. (In 'linear quadratic', 'linear' refers to the linear dynamics, and 'quadratic' refers to the objective function, which is a sum of squares.)

Most control problems include an *initial state constraint*, which has the form $x_1 = x^{\mathrm{init}}$, where x^{init} is a given initial state. Some control problems also include a *final state constraint* $x_T = x^{\mathrm{des}}$, where x^{des} is a given ('desired') final (also called terminal or target) state.

The objective function has the form $J = J_{\mathrm{output}} + \rho J_{\mathrm{input}}$, where

$$\begin{aligned}
J_{\mathrm{output}} &= \|y_1\|^2 + \dots + \|y_T\|^2 = \|C_1 x_1\|^2 + \dots + \|C_T x_T\|^2, \\
J_{\mathrm{input}} &= \|u_1\|^2 + \dots + \|u_{T-1}\|^2.
\end{aligned}$$

The positive parameter ρ weights the input objective J_{input} relative to the output objective J_{output}.

The linear quadratic control problem (with initial and final state constraints) is

$$
\begin{array}{ll}
\text{minimize} & J_{\text{output}} + \rho J_{\text{input}} \\
\text{subject to} & x_{t+1} = A_t x_t + B_t u_t, \quad t = 1, \dots, T-1, \\
& x_1 = x^{\text{init}}, \quad x_T = x^{\text{des}},
\end{array}
\tag{17.8}
$$

where the variables to be chosen are $x_1, \dots, x_T$ and $u_1, \dots, u_{T-1}$.

Formulation as constrained least squares problem. We can solve the linear quadratic control problem (17.8) by setting it up as a big linearly constrained least squares problem. We define the vector z of all these variables, stacked:

$$
z = (x_1, \dots, x_T, u_1, \dots, u_{T-1}).
$$

The dimension of z is $Tn + (T-1)m$. The control objective can be expressed as $\|\tilde{A}z - \tilde{b}\|^2$, where $\tilde{b} = 0$ and $\tilde{A}$ is the block matrix

$$
\tilde{A} = \left[
\begin{array}{cccc|cccc}
C_1 & & & & & & & \\
& C_2 & & & & & & \\
& & \ddots & & & & & \\
& & & C_T & & & & \\
\hline
& & & & \sqrt{\rho}I & & & \\
& & & & & \ddots & & \\
& & & & & & \sqrt{\rho}I &
\end{array}
\right].
$$

In this matrix, (block) entries not shown are zero, and the identity matrices in the lower right corner have dimension m. (The lines in the matrix delineate the portions related to the states and the inputs.) The dynamics constraints, and the initial and final state constraints, can be expressed as $\tilde{C}z = \tilde{d}$, with

$$
\tilde{C} = \left[
\begin{array}{ccccc|cccc}
A_1 & -I & & & & B_1 & & & \\
& A_2 & -I & & & & B_2 & & \\
& & \ddots & \ddots & & & & \ddots & \\
& & & A_{T-1} & -I & & & & B_{T-1} \\
\hline
I & & & & & & & & \\
& & & & I & & & &
\end{array}
\right], \quad
\tilde{d} = \left[
\begin{array}{c}
0 \\
0 \\
\vdots \\
0 \\
\hline
x^{\text{init}} \\
x^{\text{des}}
\end{array}
\right],
$$

where (block) entries not shown are zero. (The vertical line separates the portions of the matrix associated with the states and the inputs, and the horizontal lines separate the dynamics equations and the initial and final state constraints.)

The solution $\hat{z}$ of the constrained least squares problem

$$
\begin{array}{ll}
\text{minimize} & \|\tilde{A}z - \tilde{b}\|^2 \\
\text{subject to} & \tilde{C}z = \tilde{d}
\end{array}
\tag{17.9}
$$

gives us the optimal input trajectory and the associated optimal state (and output) trajectory. The solution $\hat{z}$ is a linear function of $\tilde{b}$ and $\tilde{d}$; since here $\tilde{b} = 0$, it is a linear function of x^{init} and x^{des}.

Complexity. The large constrained least squares problem (17.9) has dimensions

$$\tilde{n} = Tn + (T-1)m, \quad \tilde{m} = Tp + (T-1)m, \quad \tilde{p} = (T-1)n + 2n,$$

so using one of the standard methods described in §16.2 would require order

$$(\tilde{p} + \tilde{m})\tilde{n}^2 \approx T^3(m+p+n)(m+n)^2,$$

flops, where the symbol $\approx$ means we have dropped terms with smaller exponents. But the matrices $\tilde{A}$ and $\tilde{C}$ are very sparse, and by exploiting this sparsity (see page 349), the large constrained least squares problem can be solved in order $T(m+p+n)(m+n)^2$ flops, which grows only linearly in T.

17.2.1   Example

We consider the time-invariant linear dynamical system with

$$A = \begin{bmatrix} 0.855 & 1.161 & 0.667 \\ 0.015 & 1.073 & 0.053 \\ -0.084 & 0.059 & 1.022 \end{bmatrix}, \quad B = \begin{bmatrix} -0.076 \\ -0.139 \\ 0.342 \end{bmatrix},$$

$$C = \begin{bmatrix} 0.218 & -3.597 & -1.683 \end{bmatrix},$$

with initial condition $x^{\text{init}} = (0.496, -0.745, 1.394)$, target or desired final state $x^{\text{des}} = 0$, and $T = 100$. In this example, both the input u_t and the output y_t have dimension one, *i.e.*, are scalar. Figure 17.4 shows the output when the input is zero,

$$y_t = CA^{t-1}x^{\text{init}}, \quad t = 1, \ldots, T.$$

which is called the open-loop output. Figure 17.5 shows the optimal trade-off curve of the objectives J_{input} and J_{output}, found by varying the parameter ρ, solving the problem (17.9), and evaluating the objectives J_{input} and J_{output}. The points corresponding to the values $\rho = 0.05$, $\rho = 0.2$, and $\rho = 1$ are shown as circles. As always, increasing ρ has the effect of decreasing J_{input}, at the cost of increasing J_{output}.

The optimal input and output trajectories for these three values of ρ are shown in figure 17.6. Here too we see that for larger ρ, the input is smaller but the output is larger.

17.2.2   Variations

There are many variations on the basic linear quadratic control problem described above. We describe some of them here.

Tracking. We replace y_t in J_{output} with $y_t - y_t^{\text{des}}$, where y_t^{des} is a given desired output trajectory. In this case the objective function J_{output} is called the *tracking error*. Decreasing the parameter ρ leads to better output tracking, at the cost of larger input trajectory. This variation on the linear quadratic control problem can be expressed as a linearly constrained least squares problem with the same big matrices $\tilde{A}$ and $\tilde{C}$, the same vector $\tilde{d}$, and a nonzero vector $\tilde{b}$. The desired trajectory y_t^{des} appears in the vector $\tilde{b}$.

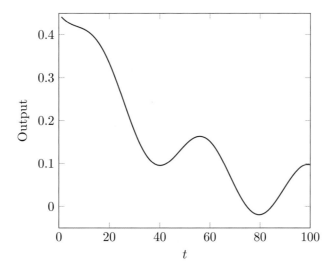

Figure 17.4 Open-loop response $CA^{t-1}x^{\text{init}}$.

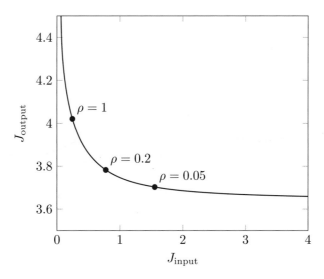

Figure 17.5 Optimal trade-off curve of the objectives J_{input} and J_{output}.

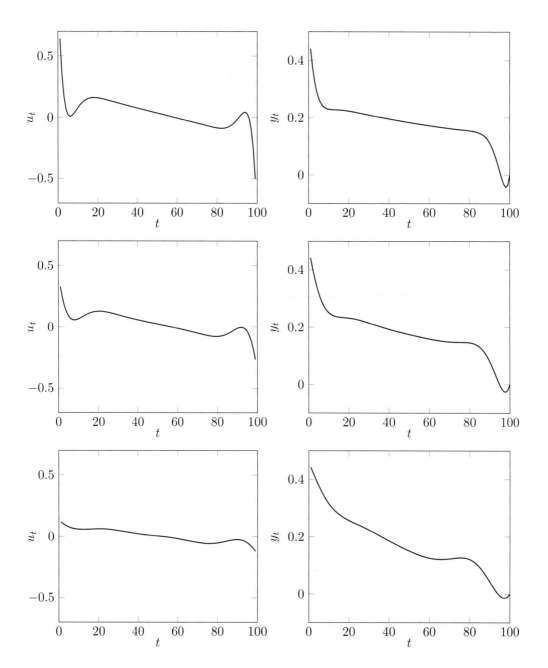

Figure 17.6 Optimal inputs (left) and outputs (right) for $\rho = 0.05$ (top), $\rho = 0.2$ (center), and $\rho = 1$ (bottom).

Time-weighted objective.   We replace J_{output} with

$$J_{\text{output}} = w_1\|y_1\|^2 + \cdots + w_T\|y_T\|^2,$$

where $w_1, \ldots, w_T$ are given positive constants. This allows us to weight earlier or later output values differently. A common choice, called *exponential weighting*, is $w_t = \theta^t$, where $\theta > 0$. For $\theta > 1$ we weight later values of y_t more than earlier values; the opposite is true for $\theta < 1$ (in which case θ is sometimes called the *discount* or *forgetting factor*).

Way-point constraints.   A way-point constraint specifies that $y_\tau = y^{\text{wp}}$, where y^{wp} is a given p-vector, and τ is a given way-point time. This constraint is typically used when y_t represents a position of a vehicle; it requires that the vehicle pass through the position y^{wp} at time $t = \tau$. Way-point constraints can be expressed as linear equality constraints on the big vector z.

17.2.3   Linear state feedback control

In the linear quadratic control problem we work out a sequence of inputs $u_1, \ldots, u_{T-1}$ to apply to the system, by solving the constrained least squares problem (17.8). It is typically used in cases where $t = T$ has some significance, like the time of landing or docking for a vehicle.

We have already mentioned (on page 185) another simpler approach to the control of a linear dynamical system. In *linear state feedback control* we measure the state in each period and use the input

$$u_t = Kx_t$$

for $t = 1, 2, \ldots$. The matrix K is called the *state feedback gain matrix*. State feedback control is very widely used in practical applications, especially ones where there is no fixed future time T when the state must take on some desired value; instead, it is desired that both x_t and u_t should be small and converge to zero. One practical advantage of linear state feedback control is that we can find the state feedback matrix K ahead of time; when the system is operating, we determine the input values using one simple matrix-vector multiply. Here we show how an appropriate state feedback gain matrix K can be found using linear quadratic control.

Let $\hat{z}$ denote the solution of the linear quadratic control problem, *i.e.*, the solution of the linearly constrained least squares problem (17.8), with $x^{\text{des}} = 0$. The solution $\hat{z}$ is a linear function of x^{init} and x^{des}; since here $x^{\text{des}} = 0$, $\hat{z}$ is a linear function of $x^{\text{init}} = x_1$. Since $\hat{u}_1$, the optimal input at $t = 1$, is a slice or subvector of $\hat{z}$, we conclude that $\hat{u}_1$ is a linear function of x_1, and so can be written as $u_1 = Kx_1$ for some $m \times n$ matrix K. The columns of K can be found by solving (17.8) with initial conditions $x^{\text{init}} = e_1, \ldots, e_n$. This can be done efficiently by factoring the coefficient matrix once, and then carrying out n solves.

This matrix generally provides a good choice of state feedback gain matrix. With this choice, the input u_1 with state feedback control and under linear quadratic

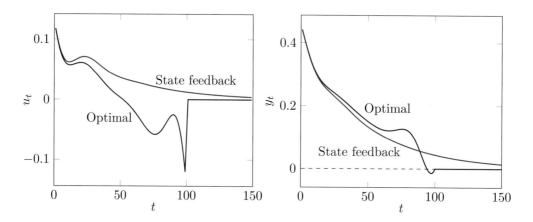

Figure 17.7 The blue curves are the solutions of (17.8) for $\rho = 1$. The red curves are the inputs and outputs that result from the constant state feedback $u_t = K x_t$.

control are the same; for $t > 1$, the two inputs differ. An interesting phenomenon, beyond the scope of this book, is that the state feedback gain matrix K found this way does not depend very much on T, provided it is chosen large enough.

Example.   For the example in §17.2.1 the state feedback gain matrix for $\rho = 1$ is

$$K = \left[\begin{array}{ccc} 0.308 & -2.659 & -1.446 \end{array} \right].$$

In figure 17.7, we plot the trajectories with linear quadratic control (in blue) and using the simpler linear state feedback control $u_t = K x_t$. We can see that the input sequence found using linear quadratic control achieves $y_T = 0$ exactly; the input sequence found by linear state feedback control makes y_T small, but not zero.

17.3   Linear quadratic state estimation

The setting is a linear dynamical system of the form

$$x_{t+1} = A_t x_t + B_t w_t, \qquad y_t = C_t x_t + v_t, \quad t = 1, 2, \dots. \tag{17.10}$$

Here the n-vector x_t is the state of the system, the p-vector y_t is the measurement, the m-vector w_t is the input or process noise, and the p-vector v_t is the measurement noise or residual. The matrices A_t, B_t, and C_t are the dynamics, input, and output matrices, respectively.

In *state estimation*, we know the matrices A_t, B_t, and C_t over the time period $t = 1, \dots, T$, as well as the measurements $y_1, \dots, y_T$, but we do not know the process or measurement noises. The goal is to guess or estimate the state sequence

$x_1, \ldots, x_T$. State estimation is widely used in many application areas, including all guidance and navigation systems, such as the Global Positioning System (GPS).

Since we do not know the process or measurement noises, we cannot exactly deduce the state sequence. Instead we will guess or estimate the state sequence $x_1, \ldots, x_T$ and process noise sequence $w_1, \ldots, w_{T-1}$, subject to the requirement that they satisfy the dynamic system model (17.10). When we guess the state sequence, we implicitly guess that the measurement noise is $v_t = y_t - C_t x_t$. We make one fundamental assumption: The process and measurement noises are both small, or at least, not too large.

Our primary objective is the sum of squares of the norms of the measurement residuals,

$$ J_{\mathrm{meas}} = \|v_1\|^2 + \cdots + \|v_T\|^2 = \|C_1 x_1 - y_1\|^2 + \cdots + \|C_T x_T - y_T\|^2. $$

If this quantity is small, it means that the proposed state sequence guess is consistent with our measurements. Note that the quantities in the squared norms above are the same as $-v_t$.

The secondary objective is the sum of squares of the norms of the process noise,

$$ J_{\mathrm{proc}} = \|w_1\|^2 + \cdots + \|w_{T-1}\|^2. $$

Our prior assumption that the process noise is small corresponds to this objective being small.

Least squares state estimation.   We will make our guesses of $x_1, \ldots, x_T$ and $w_1, \ldots, w_{T-1}$ so as to minimize a weighted sum of our objectives, subject to the dynamics constraints:

$$ \begin{array}{ll} \text{minimize} & J_{\mathrm{meas}} + \lambda J_{\mathrm{proc}} \\ \text{subject to} & x_{t+1} = A_t x_t + B_t w_t, \quad t = 1, \ldots, T-1, \end{array} \tag{17.11} $$

where λ is a positive parameter that allows us to put more emphasis on making our measurement discrepancies small (by choosing λ small), or the process noises small (by choosing λ large). Roughly speaking, small λ means that we trust the measurements more, while large λ means that we trust the measurements less, and put more weight on choosing a trajectory consistent with the dynamics, with small process noise. We will see later how λ can be chosen using validation.

Estimation versus control.   The least squares state estimation problem is very similar to the linear quadratic control problem, but the interpretation is quite different. In the control problem, we can choose the inputs; they are under our control. Once we choose the inputs, we know the state sequence. The inputs are typically actions that we take to affect the state trajectory. In the estimation problem, the inputs (called process noise in the estimation problem) are unknown, and the problem is to guess them. Our job is to guess the state sequence, which we do not know. This is a passive task. We are not choosing inputs to affect the state; rather, we are observing the outputs and hoping to deduce the state sequence. The mathematical formulations of the two problems, however, are very closely related. The close connection between the two problems is sometimes called *control/estimation duality*.

Formulation as constrained least squares problem.   The least squares state estimation problem (17.11) can be formulated as a linearly constrained least squares problem, using stacking. We define the stacked vector

$$z = (x_1, \ldots, x_T, w_1, \ldots, w_{T-1}).$$

The objective in (17.11) can be expressed as $\|\tilde{A}z - \tilde{b}\|^2$, with

$$\tilde{A} = \left[\begin{array}{ccccc|ccc} C_1 & & & & & & & \\ & C_2 & & & & & & \\ & & \ddots & & & & & \\ & & & C_T & & & & \\ \hline & & & & \sqrt{\lambda}I & & & \\ & & & & & \ddots & & \\ & & & & & & \sqrt{\lambda}I & \end{array} \right], \qquad \tilde{b} = \left[\begin{array}{c} y_1 \\ y_2 \\ \vdots \\ y_T \\ 0 \\ \vdots \\ 0 \end{array} \right].$$

The constraints in (17.11) can be expressed as $\tilde{C}z = \tilde{d}$, with $\tilde{d} = 0$ and

$$\tilde{C} = \left[\begin{array}{cccc|cccc} A_1 & -I & & & B_1 & & & \\ & A_2 & -I & & & B_2 & & \\ & & \ddots & \ddots & & & \ddots & \\ & & & A_{T-1} & -I & & & B_{T-1} \end{array} \right].$$

The constrained least squares problem has dimensions

$$\tilde{n} = Tn + (T-1)m, \quad \tilde{m} = Tp + (T-1)m, \quad \tilde{p} = (T-1)n$$

so using one of the standard methods described in §16.2 would require order

$$(\tilde{p} + \tilde{m})\tilde{n}^2 \approx T^3(m + p + n)(m + n)^2$$

flops. As in the case of linear quadratic control, the matrices $\tilde{A}$ and $\tilde{C}$ are very sparse, and by exploiting this sparsity (see page 349), the large constrained least squares problem can be solved in order $T(m + p + n)(m + n)^2$ flops, which grows only linearly in T.

The least squares state estimation problem was formulated in around 1960 by Rudolf Kalman and others (in a statistical framework). He and others developed a particular recursive algorithm for solving the problem, and the whole method has come to be known as *Kalman filtering*. For this work Kalman was awarded the Kyoto Prize in 1985.

17.3.1   Example

We consider a system with $n = 4$, $p = 2$, and $m = 2$, and time-invariant matrices

$$A = \left[\begin{array}{cccc} 1 & 0 & 1 & 0 \\ 0 & 1 & 0 & 1 \\ 0 & 0 & 1 & 0 \\ 0 & 0 & 0 & 1 \end{array} \right], \qquad B = \left[\begin{array}{cc} 0 & 0 \\ 0 & 0 \\ 1 & 0 \\ 0 & 1 \end{array} \right], \qquad C = \left[\begin{array}{cccc} 1 & 0 & 0 & 0 \\ 0 & 1 & 0 & 0 \end{array} \right].$$

This is a very simple model of motion of a mass moving in 2-D. The first two components of x_t represent the position coordinates; components 3 and 4 represent the velocity coordinates. The input w_t acts like a force on the mass, since it adds to the velocity. We think of the 2-vector Cx_t as the exact or true position of the mass at period t. The measurement $y_t = Cx_t + v_t$ is a noisy measurement of the mass position. We will estimate the state trajectory over $t = 1, \ldots, T$, with $T = 100$.

In figure 17.8 the 100 measured positions y_t are shown as circles in 2-D. The solid black lines show Cx_t, i.e., the actual position of the mass. We solve the least squares state estimation problem (17.11) for a range of values of λ. The estimated trajectories $C\hat{x}_t$ for three values of λ are shown as blue lines. We can see that $\lambda = 1$ is too small for this example: The estimated state places too much trust in the measurements, and is following measurement noise. We can also see that $\lambda = 10^5$ is too large: The estimated state is very smooth (since the estimated process noise is small), but the imputed noise measurements are too high. In this example the choice of λ is simple, since we have the true position trajectory. We will see later how λ can be chosen using validation in the general case.

17.3.2   Variations

Known initial state.   There are several interesting variations on the state estimation problem. For example, we might know the initial state x_1. In this case we simply add an equality constraint $x_1 = x_1^{\text{known}}$.

Missing measurements.   Another useful variation on the least squares state estimation problem allows for *missing measurements*, i.e., we only know y_t for $t \in \mathcal{T}$, where $\mathcal{T}$ is the set of times for which we have a measurement. We can handle this variation two (equivalent) ways: We can either replace $\sum_{t=1}^T \|v_t\|^2$ with $\sum_{t \in \mathcal{T}} \|v_t\|^2$, or we can consider y_t for $t \notin \mathcal{T}$ to be optimization variables as well. (Both lead to the same state sequence estimate.) When there are missing measurements, we can estimate what the missing measurements might have been, by taking

$$\hat{y}_t = C_t \hat{x}_t, \quad t \notin \mathcal{T}.$$

(Here we assume that $v_t = 0$.)

17.3.3   Validation

The technique of estimating what a missing measurement might have been directly gives us a method to validate a quadratic state estimation method, and in particular, to choose λ. To do this, we remove some of the measurements (say, 20%), and carry out least squares state estimation pretending that those measurements are missing. Our state estimate produces predicted values for the missing (really, held back) measurements, which we can compare to the actual measurements. We choose a value of λ which approximately minimizes this (test) prediction error.

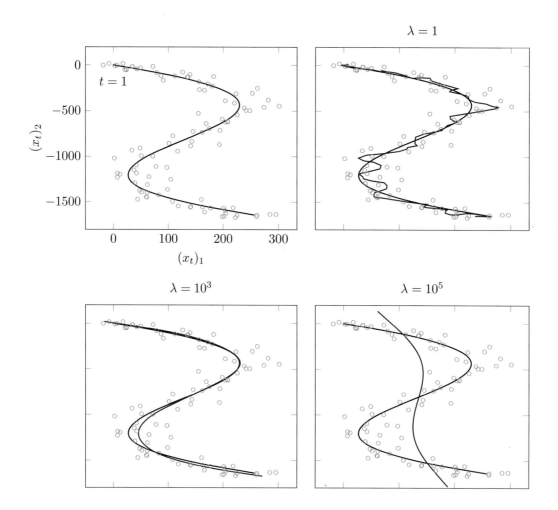

Figure 17.8 The circles show 100 noisy measurements in 2-D. The solid black line in each plot is the exact position Cx_t. The blue lines in plots 2–4 are estimated trajectories $C\hat{x}_t$ for three values of λ.

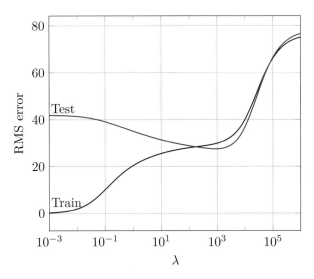

Figure 17.9 Training and test errors for the state estimation example.

Example.   Continuing the previous example, we randomly remove 20 of the 100 measurement points. We solve the same problem (17.11) for a range of values of λ, but with J_{meas} defined as

$$J_{\text{meas}} = \sum_{t \in \mathcal{T}} \|Cx_t - y_t\|^2,$$

i.e., we only sum the measurement errors over the measurements we have. For each value of λ we compute the RMS train and test errors

$$E_{\text{train}} = \frac{1}{\sqrt{80p}} \left(\sum_{t \in \mathcal{T}} \|C\hat{x}_t - y_t\|^2 \right)^{1/2}, \qquad E_{\text{test}} = \frac{1}{\sqrt{20p}} \left(\sum_{t \notin \mathcal{T}} \|C\hat{x}_t - y_t\|^2 \right)^{1/2}.$$

The training error (squared and scaled) appears directly in our minimization problem. The test error, however, is a good test of our estimation method, since it compares predictions of positions (in this example) with measurements of position that were not used to form the estimates. The errors are shown in figure 17.9, as functions of the parameter λ. We can clearly see that for $\lambda < 100$ or so, we are over-fit, since the test RMS error substantially exceeds the train RMS error. We can also see that λ around 10^3 is a good choice.

Exercises

17.1 *A variation on the portfolio optimization formulation.* Consider the following variation on the linearly constrained least squares problem (17.2):

$$\begin{array}{ll}
\text{minimize} & \|Rw\|^2 \\
\text{subject to} & \begin{bmatrix} \mathbf{1}^T \\ \mu^T \end{bmatrix} w = \begin{bmatrix} 1 \\ \rho \end{bmatrix},
\end{array} \tag{17.12}$$

with variable w. (The difference is that here we drop the term $\rho\mathbf{1}$ that appears inside the norm square objective in (17.2).) Show that this problem is equivalent to (17.2). This means w is a solution of (17.12) if and only if it is a solution of (17.2).

Hint. You can argue directly by expanding the objective in (17.2) or via the KKT systems of the two problems.

17.2 *A more conventional formulation of the portfolio optimization problem.* In this problem we derive an equivalent formulation of the portfolio optimization problem (17.2) that appears more frequently in the literature than our version. (Equivalent means that the two problems always have the same solution.) This formulation is based on the *return covariance matrix*, which we define below. (See also exercise 10.16.)

The means of the columns of the asset return matrix R are the entries of the vector μ. The *de-meaned returns matrix* is given by $\tilde{R} = R - \mathbf{1}\mu^T$. (The columns of the matrix $\tilde{R} = R - \mathbf{1}\mu^T$ are the de-meaned return time series for the assets.) The return covariance matrix, traditionally denoted Σ, is its Gram matrix $\Sigma = (1/T)\tilde{R}^T \tilde{R}$.

(a) Show that $\sigma_i = \sqrt{\Sigma_{ii}}$ is the standard deviation (risk) of asset i return. (The symbol σ_i is a traditional one for the standard deviation of asset i.)

(b) Show that the correlation coefficient between asset i and asset j returns is given by $\rho_{ij} = \Sigma_{ij}/(\sigma_i\sigma_j)$. (Assuming neither asset has constant return; if either one does, they are uncorrelated.)

(c) *Portfolio optimization using the return covariance matrix.* Show that the following problem is equivalent to our portfolio optimization problem (17.2):

$$\begin{array}{ll}
\text{minimize} & w^T \Sigma w \\
\text{subject to} & \begin{bmatrix} \mathbf{1}^T \\ \mu^T \end{bmatrix} w = \begin{bmatrix} 1 \\ \rho \end{bmatrix},
\end{array} \tag{17.13}$$

with variable w. This is the form of the portfolio optimization problem that you will find in the literature. *Hint.* Show that the objective is the same as $\|\tilde{R}w\|^2$, and that this is the same as $\|Rw - \rho\mathbf{1}\|^2$ for any feasible w.

17.3 *A simple portfolio optimization problem.*

(a) Find an analytical solution for the portfolio optimization problem with $n = 2$ assets. You can assume that $\mu_1 \neq \mu_2$, i.e., the two assets have different mean returns. *Hint.* The optimal weights depend only on μ and ρ, and not (directly) on the return matrix R.

(b) Find the conditions under which the optimal portfolio takes long positions in both assets, a short position in one and a long position in the other, or a short position in both assets. You can assume that $\mu_1 < \mu_2$, i.e., asset 2 has the higher return. *Hint.* Your answer should depend on whether $\rho < \mu_1$, $\mu_1 < \rho < \mu_2$, or $\mu_2 < \rho$, i.e., how the required return compares to the two asset returns.

17.4 *Index tracking.* Index tracking is a variation on the portfolio optimization problem described in §17.1. As in that problem we choose a portfolio allocation weight vector w that satisfies $\mathbf{1}^T w = 1$. This weight vector gives a portfolio return time series Rw, which is a T-vector. In index tracking, the goal is for this return time series to track (or follow) as closely as possible a given target return time series r^{tar}. We choose w to minimize the RMS deviation between the target return time series r^{tar} and the portfolio return time series r. (Typically the target return is the return of an index, like the Dow Jones Industrial Average or the Russell 3000.) Formulate the index tracking problem as a linearly constrained least squares problem, analogous to (17.2). Give an explicit solution, analogous to (17.3).

17.5 *Portfolio optimization with market neutral constraint.* In the portfolio optimization problem (17.2) the portfolio return time series is the T-vector Rw. Let r^{mkt} denote the T-vector that gives the return of the whole market over the time periods $t = 1, \ldots, T$. (This is the return associated with the total value of the market, *i.e.*, the sum over the assets of asset share price times number of outstanding shares.) A portfolio is said to be *market neutral* if Rw and r^{mkt} are uncorrelated.

Explain how to formulate the portfolio optimization problem, with the additional constraint of market neutrality, as a constrained least squares problem. Give an explicit solution, analogous to (17.3).

17.6 *State feedback control of the longitudinal motions of a Boeing 747 aircraft.* In this exercise we consider the control of the longitudinal motions of a Boeing 747 aircraft in steady level flight, at an altitude of 40000 ft, and speed 774 ft/s, which is 528 MPH or 460 knots, around Mach 0.8 at that altitude. (Longitudinal means that we consider climb rate and speed, but not turning or rolling motions.) For modest deviations from these steady state or *trim* conditions, the dynamics is given by the linear dynamical system $x_{t+1} = Ax_t + Bu_t$, with

$$
A = \begin{bmatrix} 0.99 & 0.03 & -0.02 & -0.32 \\ 0.01 & 0.47 & 4.70 & 0.00 \\ 0.02 & -0.06 & 0.40 & 0.00 \\ 0.01 & -0.04 & 0.72 & 0.99 \end{bmatrix}, \qquad B = \begin{bmatrix} 0.01 & 0.99 \\ -3.44 & 1.66 \\ -0.83 & 0.44 \\ -0.47 & 0.25 \end{bmatrix},
$$

with time unit one second. The state 4-vector x_t consists of deviations from the trim conditions of the following quantities.

- $(x_t)_1$ is the velocity along the airplane body axis, in ft/s, with forward motion positive.
- $(x_t)_2$ is the velocity perpendicular to the body axis, in ft/s, with positive down.
- $(x_t)_3$ is the angle of the body axis above horizontal, in units of 0.01 radian ($0.57°$).
- $(x_t)_4$ is the derivative of the angle of the body axis, called the *pitch rate*, in units of 0.01 radian/s ($0.57°$/s).

The input 2-vector u_t (which we can control) consists of deviations from the trim conditions of the following quantities.

- $(u_t)_1$ is the elevator (control surface) angle, in units of 0.01 radian.
- $(u_t)_2$ is the engine thrust, in units of 10000 lbs.

You do not need to know these details; we mention them only so you know what the entries of x_t and u_t mean.

(a) *Open loop trajectory.* Simulate the motion of the Boeing 747 with initial condition $x_1 = e_4$, in open-loop (*i.e.*, with $u_t = 0$). Plot the state variables over the time interval $t = 1, \ldots, 120$ (two minutes). The oscillation you will see in the open-loop simulation is well known to pilots, and called the *phugoid mode*.

(b) *Linear quadratic control.* Solve the linear quadratic control problem with $C = I$, $\rho = 100$, and $T = 100$, with initial state $x_1 = e_4$, and desired terminal state $x^{\text{des}} = 0$. Plot the state and input variables over $t = 1, \ldots, 120$. (For $t = 100, \ldots, 120$, the state and input variables are zero.)

(c) Find the 2×4 state feedback gain K obtained by solving the linear quadratic control problem with $C = I$, $\rho = 100$, $T = 100$, as described in §17.2.3. Verify that it is almost the same as the one obtained with $T = 50$.

(d) Simulate the motion of the Boeing 747 with initial condition $x_1 = e_4$, under state feedback control (*i.e.*, with $u_t = Kx_t$). Plot the state and input variables over the time interval $t = 1, \ldots, 120$.

17.7 *Bio-mass estimation.* A bio-reactor is used to grow three different bacteria. We let x_t be the 3-vector of the bio-masses of the three bacteria, at time period (say, hour) t, for $t = 1, \ldots, T$. We believe that they each grow, independently, with growth rates given by the 3-vector r (which has positive entries). This means that $(x_{t+1})_i \approx (1 + r_i)(x_t)_i$, for $i = 1, 2, 3$. (These equations are approximate; the real rate is not constant.) At every time sample we measure the total bio-mass in the reactor, *i.e.*, we have measurements $y_t \approx \mathbf{1}^T x_t$, for $t = 1, \ldots, T$. (The measurements are not exactly equal to the total mass; there are small measurement errors.) We do not know the bio-masses $x_1, \ldots, x_T$, but wish to estimate them based on the measurements $y_1, \ldots, y_T$.

Set this up as a linear quadratic state estimation problem as in §17.3. Identify the matrices A_t, B_t, and C_t. Explain what effect the parameter λ has on the estimated bio-mass trajectory $\hat{x}_1, \ldots, \hat{x}_T$.

Chapter 18

Nonlinear least squares

In previous chapters we studied the problems of solving a set of linear equations or finding a least squares approximate solution to them. In this chapter we study extensions of these problems in which linear is replaced with nonlinear. These nonlinear problems are in general hard to solve exactly, but we describe a heuristic algorithm that often works well in practice.

18.1 Nonlinear equations and least squares

18.1.1 Nonlinear equations

Consider a set of m possibly nonlinear equations in n unknowns (or variables) $x = (x_1, \ldots, x_n)$, written as

$$f_i(x) = 0, \qquad i = 1, \ldots, m,$$

where $f_i : \mathbf{R}^n \to \mathbf{R}$ is a scalar-valued function. We refer to $f_i(x) = 0$ as the ith equation. For any x we call $f_i(x)$ the ith *residual*, since it is a quantity we want to be zero. Many interesting practical problems can be expressed as the problem of solving, possibly approximately, a set of nonlinear equations.

We take the right-hand side of the equations to be zero to simplify the problem notation. If we need to solve $f_i(x) = b_i$, $i = 1, \ldots, m$, where b_i are some given nonzero numbers, we define $\tilde{f}_i(x) = f_i(x) - b_i$, and solve $\tilde{f}_i(x) = 0$, $i = 1, \ldots, m$, which gives us a solution of the original equations. Assuming the right-hand sides of the equations are zero will simplify formulas and equations.

We often write the set of equations in the compact vector form

$$f(x) = 0, \tag{18.1}$$

where $f(x) = (f_1(x), \ldots, f_m(x))$ is an m-vector, and the zero vector on the right-hand side has dimension m. We can think of f as a function that maps n-vectors to m-vectors, *i.e.*, $f : \mathbf{R}^n \to \mathbf{R}^m$. We refer to the m-vector $f(x)$ as the residual (vector) associated with the choice of the n-vector x; our goal is to find x with associated residual zero.

When f is an affine function, the set of equations (18.1) is a set of m linear equations in n unknowns, which can be solved (or approximately solved in a least squares sense when $m > n$), using the techniques covered in previous chapters. We are interested here in the case when f is not affine.

We extend the ideas of under-determined, square, and over-determined equations to the nonlinear case. When $m < n$, there are fewer equations than unknowns, and the system of equations (18.1) is called under-determined. When $m = n$, so there are as many equations as unknowns, the system of equations is called square. When $m > n$, there are more equations than unknowns, and the system of equations is called over-determined.

18.1.2   Nonlinear least squares

When we cannot find a solution of the equations (18.1), we can seek an approximate solution, by finding x that minimizes the sum of squares of the residuals,

$$f_1(x)^2 + \cdots + f_m(x)^2 = \|f(x)\|^2.$$

This means finding $\hat{x}$ for which $\|f(x)\|^2 \geq \|f(\hat{x})\|^2$ holds for all x. We refer to such a point as a least squares approximate solution of (18.1), or more directly, as a solution of the *nonlinear least squares problem*

$$\text{minimize} \quad \|f(x)\|^2, \tag{18.2}$$

where the n-vector x is the variable to be found. When the function f is affine, the nonlinear least squares problem (18.2) reduces to the (linear) least squares problem from chapter 12.

The nonlinear least squares problem (18.2) includes the problem of solving nonlinear equations (18.1) as a special case, since any x that satisfies $f(x) = 0$ is also a solution of the nonlinear least squares problem. But as in the case of linear equations, the least squares approximate solution of a set of nonlinear equations is often very useful even when it does not solve the equations. So we will focus on the nonlinear least squares problem (18.2).

18.1.3   Optimality condition

Calculus gives us a necessary condition for $\hat{x}$ to be a solution of (18.2), *i.e.*, to minimize $\|f(x)\|^2$. (This means that the condition must hold for a solution, but it may also hold for other points that are not solutions.) The partial derivative of $\|f(x)\|^2$ with respect to each of $x_1, \ldots, x_n$ must vanish at $\hat{x}$:

$$\frac{\partial}{\partial x_i} \|f(\hat{x})\|^2 = 0, \quad i = 1, \ldots, n,$$

or, in vector form, $\nabla \|f(\hat{x})\|^2 = 0$ (see §C.2). This gradient can be expressed as

$$\nabla \|f(x)\|^2 = \nabla \left(\sum_{i=1}^{m} f_i(x)^2 \right) = 2 \sum_{i=1}^{m} f_i(x) \nabla f_i(x) = 2Df(x)^T f(x),$$

where the $m \times n$ matrix $Df(x)$ is the derivative or Jacobian matrix of the function f at the point x, *i.e.*, the matrix of its partial derivatives (see §8.2.1 and C.1). So if $\hat{x}$ minimizes $\|f(x)\|^2$, it must satisfy

$$2Df(\hat{x})^T f(\hat{x}) = 0. \tag{18.3}$$

This *optimality condition* must hold for any solution of the nonlinear least squares problem (18.2). But the optimality condition can also hold for other points that are not solutions of the nonlinear least squares problem. For this reason the optimality condition (18.3) is called a *necessary condition* for optimality, because it is necessarily satisfied for any solution $\hat{x}$. It is a not a *sufficient condition* for optimality, since the optimality condition (18.3) is not enough (*i.e.*, is not sufficient) to guarantee that the point is a solution of the nonlinear least squares problem.

When the function f is affine, the optimality conditions (18.3) reduce to the normal equations (12.4), the optimality conditions for the (linear) least squares problem.

18.1.4   Difficulty of solving nonlinear equations

Solving a set of nonlinear equations (18.1), or solving the nonlinear least squares problem (18.2), is in general much more difficult than solving a set of linear equations or a linear least squares problem. For nonlinear equations, there can be no solution, or any number of solutions, or an infinite number of solutions. Unlike linear equations, it is a very difficult computational problem to determine which one of these cases holds for a particular set of equations; there is no analog of the QR factorization that we can use for linear equations and least squares problems. Even the simple sounding problem of determining whether or not there are any solutions to a set of nonlinear equations is very difficult computationally. There are advanced non-heuristic algorithms for exactly solving nonlinear equations, or exactly solving nonlinear least squares problems, but they are complicated and very computationally demanding, and rarely used in applications.

Given the difficulty of solving a set of nonlinear equations, or solving a nonlinear least squares problem, we must lower our expectations. We can only hope for an algorithm that often finds a solution (when one exists), or produces a value of x with small residual norm, if not the smallest that is possible. Algorithms like this, that often work, or tend to produce a good if not always the best possible point, are called *heuristics*. The k-means algorithm of chapter 4 is an example of a heuristic algorithm. Solving linear equations or linear least squares problems using the QR factorization are *not* heuristics; these algorithms *always* work.

Many heuristic algorithms for the nonlinear least squares problem, including those we describe later in this chapter, compute a point $\hat{x}$ that satisfies the optimality condition (18.3). Unless $f(\hat{x}) = 0$, however, such a point need not be a solution of the nonlinear least squares problem (18.2).

18.1.5   Examples

In this section we list a few applications that reduce to solving a set of nonlinear equations, or a nonlinear least squares problem.

Computing equilibrium points.   The idea of an equilibrium, where some type of consumption and generation balance each other, arises in many applications. Consumption and generation depend, often nonlinearly, on the values of some parameters, and the goal is to find values of the parameters that lead to equilibrium. These examples typically have $m = n$, $i.e.$, the system of nonlinear equations is square.

- $Equilibrium\ prices.$ We consider n commodities or goods, with associated prices given by the n-vector p. The demand for the n goods (an n-vector) is a nonlinear function of the prices, given by $D(p)$. (In an example on page 150 we described an approximate model for demand that is accurate when the prices change from nominal values by a few percent; here we consider the demand over a large range of prices.) The supply of the goods (an n-vector) also depends on the prices, and is given by $S(p)$. (When the price for a good is high, for example, more producers are willing to produce it, so the supply increases.)

  A set of commodity prices p is an $equilibrium\ price\ vector$ if it results in supply balancing demand, $i.e.$, $S(p) = D(p)$. Finding a set of equilibrium prices is the same as solving the square set of nonlinear equations

  $$f(p) = S(p) - D(p) = 0.$$

  (The vector $f(p)$ is called the excess supply, at the set of prices p.) This is shown in figure 18.1 for a simple case with $n = 1$.

- $Chemical\ equilibrium.$ We consider n chemical species in a solution. The n-vector c denotes the concentrations of the n species. Reactions among the species consume some of them (the reactants) and generate others (the products). The rate of each reaction is a function of the concentrations of its reactants (and other parameters we assume are fixed, like temperature or presence of catalysts). We let $C(c)$ denote the vector of total consumption of the n reactants, over all the reactions, and we let $G(c)$ denote the vector of generation of the n reactants, over all reactions.

  A concentration vector c is in chemical equilibrium if $C(c) = G(c)$, $i.e.$, the rate of consumption of all species balances the rate of generation. Computing a set of equilibrium concentrations is the same as solving the square set of nonlinear equations

  $$f(c) = C(c) - G(c) = 0.$$

- $Mechanical\ equilibrium.$ A mechanical system in 3-D with N nodes is characterized by the positions of the nodes, given by a $3N$-vector q of the stacked node positions, called the $generalized\ position$. The net force on each node is

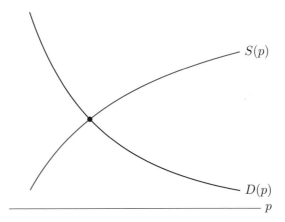

Figure 18.1 Supply and demand as functions of the price, shown on the horizontal axis. They intersect at the point shown as a circle. The corresponding price is the equilibrium price.

a 3-vector, which depends on q, *i.e.*, the node positions. We describe this as a $3N$-vector of forces, $F(q)$.

The system is in mechanical equilibrium if the net force on each node is zero, *i.e.*, $F(q) = 0$, a set of $3N$ nonlinear equations in $3N$ unknowns. (A more complex mechanical equilibrium model takes into account angular displacements and torques at each node.)

- *Nash equilibrium.* We consider a simple setup for a mathematical game. Each of n competing agents or participants chooses a number x_i. Each agent is given a (numerical) reward (say, money) that depends not only on her own choice, but also on the choice of all the other agents. The reward for agent i is given by the function $R_i(x)$, called the payoff function. Each agent wishes to make a choice that maximizes her reward. This is complicated since the reward depends not only on her choice, but the choices of the other agents.

  A *Nash equilibrium* (named after the mathematician John Forbes Nash, Jr.) is a set of choices given by the n-vector x where no agent can improve (increase) her reward by changing her choice. Such a choice is argued to be 'stable' since no agent is incented to change her choice. At a Nash equilibrium x_i maximizes $R_i(x)$, so we must have

$$\frac{\partial R_i}{\partial x_i}(x) = 0, \quad i = 1, \ldots, n.$$

This necessary condition for a Nash equilibrium is a square set of nonlinear equations.

The idea of a Nash equilibrium is widely used in economics, social science, and engineering. Nash was awarded the Nobel Prize in economics for this work in 1994.

Nonlinear least squares examples. Nonlinear least squares problems arise in many of the same settings and applications as linear least squares problems.

- *Location from range measurements.* The 3-vector (or 2-vector) x represents the location of some object or target in 3-D (or 2-D), which we wish to determine or guess. We are given m *range measurements*, *i.e.*, the distance from x to some known locations $a_1, \dots, a_m$,

$$\rho_i = \|x - a_i\| + v_i, \quad i = 1, \dots, m,$$

  where v_i is an unknown measurement error, assumed to be small. Our estimate $\hat{x}$ of the location is found by minimizing the sum of the squares of the range residuals,

$$\sum_{i=1}^{m} (\|x - a_i\| - \rho_i)^2.$$

  A similar method is used in GPS devices, where a_i are the known locations of GPS satellites that are in view.

- *Nonlinear model fitting.* We consider a model $y \approx \hat{f}(x; \theta)$ where x denotes a feature vector, y a scalar outcome, $\hat{f}$ a model of some function relating x and y, and θ is a vector of model parameters that we seek. In chapter 13, $\hat{f}$ is an affine function of the model parameter p-vector θ; but here it need not be. As in chapter 13, we choose the model parameter by minimizing the sum of the squares of the residuals over a data set with N examples,

$$\sum_{i=1}^{N} (\hat{f}(x^{(i)}; \theta) - y^{(i)})^2. \tag{18.4}$$

(As in linear least squares model fitting, we can add a regularization term to this objective function.) This is a nonlinear least squares problem, with variable θ.

18.2　Gauss–Newton algorithm

In this section we describe a powerful heuristic algorithm for the nonlinear least squares problem (18.2) that bears the names of the two famous mathematicians Carl Friedrich Gauss and Isaac Newton. In the next section we will also describe a variation of the Gauss–Newton algorithm known as the *Levenberg–Marquardt algorithm*, which addresses some shortcomings of the basic Gauss–Newton algorithm.

The Gauss–Newton and Levenberg–Marquardt algorithms are iterative algorithms that generate a sequence of points $x^{(1)}, x^{(2)}, \dots$. The vector $x^{(1)}$ is called the *starting point* of the algorithm, and $x^{(k)}$ is called the kth *iterate*. Moving from $x^{(k)}$ to $x^{(k+1)}$ is called an *iteration* of the algorithm. We judge the iterates by the norm of the associated residuals, $\|f(x^{(k)})\|$, or its square. The algorithm is terminated when $\|f(x^{(k)})\|$ is small enough, or $x^{(k+1)}$ is very near $x^{(k)}$, or when a maximum number of iterations is reached.

18.2.1   Basic Gauss–Newton algorithm

The idea behind the Gauss–Newton algorithm is simple: We alternate between finding an affine approximation of the function f at the current iterate, and then solving the associated linear least squares problem to find the next iterate. This combines two of the most powerful ideas in applied mathematics: *Calculus* is used to form an affine approximation of a function near a given point, and *least squares* is used to compute an approximate solution of the resulting affine equations.

We now describe the algorithm in more detail. At each iteration k, we form the affine approximation $\hat{f}$ of f at the current iterate $x^{(k)}$, given by the Taylor approximation

$$\hat{f}(x; x^{(k)}) = f(x^{(k)}) + Df(x^{(k)})(x - x^{(k)}), \qquad (18.5)$$

where the $m \times n$ matrix $Df(x^{(k)})$ is the Jacobian or derivative matrix of f (see §8.2.1 and §C.1). The affine function $\hat{f}(x; x^{(k)})$ is a very good approximation of $f(x)$ provided x is near $x^{(k)}$, *i.e.*, $\|x - x^{(k)}\|$ is small.

The next iterate $x^{(k+1)}$ is then taken to be the minimizer of $\|\hat{f}(x; x^{(k)})\|^2$, the norm squared of the affine approximation of f at $x^{(k)}$. Assuming that the derivative matrix $Df(x^{(k)})$ has linearly independent columns (which requires $m \geq n$), we have

$$x^{(k+1)} = x^{(k)} - \left(Df(x^{(k)})^T Df(x^{(k)}) \right)^{-1} Df(x^{(k)})^T f(x^{(k)}). \qquad (18.6)$$

This iteration gives the basic Gauss–Newton algorithm.

Algorithm 18.1 BASIC GAUSS–NEWTON ALGORITHM FOR NONLINEAR LEAST SQUARES

given a differentiable function $f : \mathbf{R}^n \to \mathbf{R}^m$, an initial point $x^{(1)}$.

For $k = 1, 2, \ldots, k^{\max}$

1. *Form affine approximation at current iterate using calculus.* Evaluate the Jacobian $Df(x^{(k)})$ and define

$$\hat{f}(x; x^{(k)}) = f(x^{(k)}) + Df(x^{(k)})(x - x^{(k)}).$$

2. *Update iterate using linear least squares.* Set $x^{(k+1)}$ as the minimizer of $\|\hat{f}(x; x^{(k)})\|^2$,

$$x^{(k+1)} = x^{(k)} - \left(Df(x^{(k)})^T Df(x^{(k)}) \right)^{-1} Df(x^{(k)})^T f(x^{(k)}).$$

The Gauss–Newton algorithm is terminated early if $f(x)$ is very small, or $x^{(k+1)} \approx x^{(k)}$. It terminates with an error if the columns of $Df(x^{(k)})$ are linearly dependent.

The condition $x^{(k+1)} = x^{(k)}$ (the exact form of our stopping condition) holds when

$$\left(Df(x^{(k)})^T Df(x^{(k)}) \right)^{-1} Df(x^{(k)})^T f(x^{(k)}) = 0,$$

which occurs if and only if $Df(x^{(k)})^T f(x^{(k)}) = 0$ (since we assume that $Df(x^{(k)})$ has linearly independent columns). So the Gauss–Newton algorithm stops only when the optimality condition (18.3) holds.

We can also observe that

$$\|\hat{f}(x^{(k+1)}; x^{(k)})\|^2 \leq \|\hat{f}(x^{(k)}; x^{(k)})\|^2 = \|f(x^{(k)})\|^2 \qquad (18.7)$$

holds, since $x^{(k+1)}$ minimizes $\|\hat{f}(x; x^{(k)})\|^2$, and $\hat{f}(x^{(k)}; x^{(k)}) = f(x^{(k)})$. The norm of the *residual of the approximation* goes down in each iteration. This is *not* the same as

$$\|f(x^{(k+1)})\|^2 \leq \|f(x^{(k)})\|^2, \qquad (18.8)$$

i.e., the norm of the *residual* goes down in each iteration, which is what we would like.

Shortcomings of the basic Gauss–Newton algorithm.   We will see in examples that the Gauss–Newton algorithm can work well, in the sense that the iterates $x^{(k)}$ converge very quickly to a point with small residual. But the Gauss–Newton algorithm has two related serious shortcomings.

The first is that it can fail, by producing a sequence of points with the norm of the residual $\|f(x^{(k)})\|$ increasing to large values, as opposed to decreasing to a small value, which is what we want. (In this case the algorithm is said to *diverge*.) The mechanism behind this failure is related to the difference between (18.7) and (18.8). The approximation

$$\|f(x)\|^2 \approx \|\hat{f}(x; x^{(k)})\|^2$$

is guaranteed to hold only when x is near $x^{(k)}$. So when $x^{(k+1)}$ is not near $x^{(k)}$, $\|f(x^{(k+1)})\|^2$ and $\|\hat{f}(x^{(k+1)}; x^{(k)})\|^2$ can be very different. In particular, the (true) residual at $x^{(k+1)}$ can be *larger* than the residual at $x^{(k)}$.

The second serious shortcoming of the basic Gauss–Newton algorithm is the assumption that the columns of the derivative matrix $Df(x^{(k)})$ are linearly independent. In some applications, this assumption never holds; in others, it can fail to hold at some iterate $x^{(k)}$, in which case the Gauss–Newton algorithm stops, since $x^{(k+1)}$ is not defined.

We will see that a simple modification of the Gauss–Newton algorithm, described below in §18.3, addresses both of these shortcomings.

18.2.2   Newton algorithm

For the special case $m = n$, the Gauss–Newton algorithm reduces to another famous algorithm for solving a set of n nonlinear equations in n variables, called the Newton algorithm. (The algorithm is sometimes called the Newton-Raphson algorithm, since Newton developed the method only for the special case $n = 1$, and Joseph Raphson later extended it to the case $n > 1$.)

When $m = n$, the matrix $Df(x^{(k)})$ is square, so the basic Gauss–Newton update (18.6) can be simplified to

$$\begin{aligned} x^{(k+1)} &= x^{(k)} - (Df(x^{(k)}))^{-1}(Df(x^{(k)})^T)^{-1}Df(x^{(k)})^T f(x^{(k)}) \\ &= x^{(k)} - (Df(x^{(k)}))^{-1}f(x^{(k)}). \end{aligned}$$

This iteration gives the Newton algorithm.

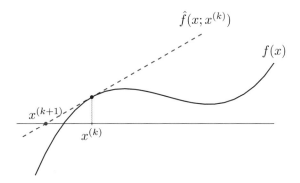

Figure 18.2 One iteration of the Newton algorithm for solving an equation $f(x) = 0$ in one variable.

Algorithm 18.2 NEWTON ALGORITHM FOR SOLVING NONLINEAR EQUATIONS

given a differentiable function $f : \mathbf{R}^n \to \mathbf{R}^n$, an initial point $x^{(1)}$.

For $k = 1, 2, \ldots, k^{\max}$

1. *Form affine approximation at current iterate.* Evaluate the Jacobian $Df(x^{(k)})$ and define
$$\hat{f}(x; x^{(k)}) = f(x^{(k)}) + Df(x^{(k)})(x - x^{(k)}).$$

2. *Update iterate by solving linear equations.* Set $x^{(k+1)}$ as the solution of $\hat{f}(x; x^{(k)}) = 0$,
$$x^{(k+1)} = x^{(k)} - \left(Df(x^{(k)})\right)^{-1} f(x^{(k)}).$$

The basic Newton algorithm shares the same shortcomings as the basic Gauss–Newton algorithm, *i.e.*, it can diverge, and the iterations terminate if the derivative matrix is not invertible.

Newton algorithm for $n = 1$. The Newton algorithm is easily understood for $n = 1$. The iteration is

$$x^{(k+1)} = x^{(k)} - f(x^{(k)})/f'(x^{(k)}) \tag{18.9}$$

and is illustrated in figure 18.2. To update $x^{(k)}$ we form the Taylor approximation

$$\hat{f}(x; x^{(k)}) = f(x^{(k)}) + f'(x^{(k)})(x - x^{(k)})$$

and set it to zero to find the next iterate $x^{(k+1)}$. If $f'(x^{(k)}) \neq 0$, the solution of $\hat{f}(x; x^{(k)}) = 0$ is given by the right-hand side of (18.9). If $f'(x^{(k)}) = 0$, the Newton algorithm terminates with an error.

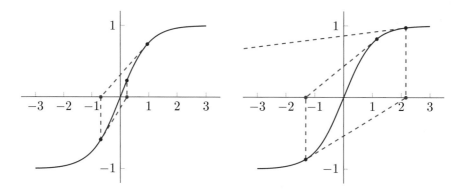

Figure 18.3 The first iterations in the Newton algorithm for solving $f(x) = 0$, for two starting points: $x^{(1)} = 0.95$ and $x^{(1)} = 1.15$.

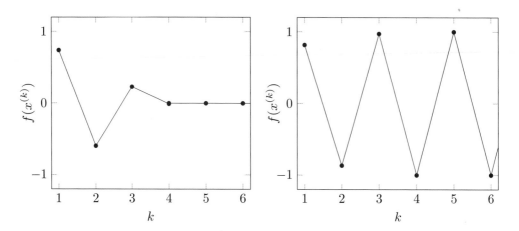

Figure 18.4 Value of $f(x^{(k)})$ versus iteration number k for Newton's method in the example of figure 18.3, started at $x^{(1)} = 0.95$ and $x^{(1)} = 1.15$.

Example. The function

$$f(x) = \frac{e^x - e^{-x}}{e^x + e^{-x}} \tag{18.10}$$

has a unique zero at the origin, *i.e.*, the only solution of $f(x) = 0$ is $x = 0$. (This function is called the *sigmoid function*, and will make another appearance later.) The Newton iteration started at $x^{(1)} = 0.95$ converges quickly to the solution $x = 0$. With $x^{(1)} = 1.15$, however, the iterates diverge. This is shown in figures 18.3 and 18.4.

18.3   Levenberg–Marquardt algorithm

In this section we describe a variation on the basic Gauss–Newton algorithm (as well as the Newton algorithm) that addresses the shortcomings described above. The variation comes directly from ideas we have encountered earlier in this book. It was first proposed by Kenneth Levenberg and Donald Marquardt, and is called the Levenberg–Marquardt algorithm. It is also sometimes called the Gauss–Newton algorithm, since it is a natural extension of the basic Gauss–Newton algorithm described above.

Multi-objective update formulation.   The main problem with the Gauss–Newton algorithm is that the minimizer of the approximation $\|\hat{f}(x; x^{(k)})\|^2$ may be far from the current iterate $x^{(k)}$, in which case the approximation $\hat{f}(x; x^{(k)}) \approx f(x)$ need not hold, which implies that $\|\hat{f}(x; x^{(k)})\|^2 \approx \|f(x)\|^2$ need not hold. In choosing $x^{(k+1)}$, then, we have *two* objectives: We would like $\|\hat{f}(x; x^{(k)})\|^2$ small, and we would also like $\|x - x^{(k)}\|^2$ small. The first objective is an approximation of what we really want to minimize; the second objective expresses the idea that we should not move so far that we cannot trust the affine approximation. This suggests that we should choose $x^{(k+1)}$ as the minimizer of

$$\|\hat{f}(x; x^{(k)})\|^2 + \lambda^{(k)} \|x - x^{(k)}\|^2, \tag{18.11}$$

where $\lambda^{(k)}$ is a positive parameter. We add an iteration superscript to the parameter λ since it can take different values in different iterations. For $\lambda^{(k)}$ small, we primarily minimize the first term, the squared norm of the approximation; for $\lambda^{(k)}$ large, we choose $x^{(k+1)}$ near $x^{(k)}$. (For $\lambda^{(k)} = 0$, this coincides with the next iterate in the basic Gauss–Newton algorithm.) The second term in (18.11) is sometimes called a *trust penalty* term, since it penalizes choices of x that are far from $x^{(k)}$, where we cannot trust the affine approximation. The parameter $\lambda^{(k)}$ is sometimes called the *trust parameter* (although 'distrust parameter' is perhaps more accurate).

Computing the minimizer of (18.11) is a multi-objective least squares or regularized least squares problem, and equivalent to minimizing

$$\left\| \begin{bmatrix} Df(x^{(k)}) \\ \sqrt{\lambda^{(k)}} I \end{bmatrix} x - \begin{bmatrix} Df(x^{(k)})x^{(k)} - f(x^{(k)}) \\ \sqrt{\lambda^{(k)}} x^{(k)} \end{bmatrix} \right\|^2.$$

Since $\lambda^{(k)}$ is positive, the stacked matrix in this least squares problem has linearly independent columns, even when $Df(x^{(k)})$ does not. It follows that the solution of the least squares problem exists and is unique. From the normal equations of the least squares problem we can derive a useful expression for $x^{(k+1)}$:

$$\left(Df(x^{(k)})^T Df(x^{(k)}) + \lambda^{(k)} I \right) x^{(k+1)}$$
$$= Df(x^{(k)})^T \left(Df(x^{(k)})x^{(k)} - f(x^{(k)}) \right) + \lambda^{(k)} x^{(k)}$$
$$= \left(Df(x^{(k)})^T Df(x^{(k)}) + \lambda^{(k)} I \right) x^{(k)} - Df(x^{(k)})^T f(x^{(k)}),$$

and therefore

$$x^{(k+1)} = x^{(k)} - \left(Df(x^{(k)})^T Df(x^{(k)}) + \lambda^{(k)} I\right)^{-1} Df(x^{(k)})^T f(x^{(k)}). \quad (18.12)$$

The matrix inverse here always exists.

From (18.12), we see that $x^{(k+1)} = x^{(k)}$ only if $2Df(x^{(k)})^T f(x^{(k)}) = 0$, *i.e.*, only when the optimality condition (18.3) holds for $x^{(k)}$. So like the Gauss–Newton algorithm, the Levenberg–Marquardt algorithm stops (or more accurately, repeats itself with $x^{(k+1)} = x^{(k)}$) only when the optimality condition (18.3) holds.

Updating the trust parameter. The final issue is how to choose the trust parameter $\lambda^{(k)}$. When $\lambda^{(k)}$ is too small, $x^{(k+1)}$ can be far enough away from $x^{(k)}$ that $\|f(x^{(k+1)})\|^2 > \|f(x^{(k)})\|^2$ can hold, *i.e.*, our true objective function increases, which is not what we want. When $\lambda^{(k)}$ is too large, $x^{(k+1)} - x^{(k)}$ is small, so the affine approximation is good, and the objective decreases (which is good). But in this case we have $x^{(k+1)}$ very near $x^{(k)}$, so the decrease in objective is small, and it will take many iterations to make progress. We want $\lambda^{(k)}$ in between these two cases, big enough that the approximation holds well enough to get a decrease in objective, but not much bigger, which slows convergence.

Several algorithms can be used to adjust λ. One simple method forms $x^{(k+1)}$ using the current value of λ and checks if the objective has decreased. If it has, we accept the new point and decrease λ a bit for the next iteration. If the objective has not decreased, which means λ is too small, we do not update the point $x^{(k+1)}$, and increase the trust parameter λ substantially.

Levenberg–Marquardt algorithm. The ideas above can be formalized as the algorithm given below.

Algorithm 18.3 LEVENBERG–MARQUARDT ALGORITHM FOR NONLINEAR LEAST SQUARES

given a differentiable function $f : \mathbf{R}^n \to \mathbf{R}^m$, an initial point $x^{(1)}$, an initial trust parameter $\lambda^{(1)} > 0$.

For $k = 1, 2, \ldots, k^{\max}$

1. *Form affine approximation at current iterate.* Evaluate the Jacobian $Df(x^{(k)})$ and define
$$\hat{f}(x; x^{(k)}) = f(x^{(k)}) + Df(x^{(k)})(x - x^{(k)}).$$

2. *Compute tentative iterate.* Set $x^{(k+1)}$ as minimizer of
$$\|\hat{f}(x; x^{(k)})\|^2 + \lambda^{(k)} \|x - x^{(k)}\|^2.$$

3. *Check tentative iterate.*
If $\|f(x^{(k+1)})\|^2 < \|f(x^{(k)})\|^2$, accept iterate and reduce λ: $\lambda^{(k+1)} = 0.8\lambda^{(k)}$.
Otherwise, increase λ and do not update x: $\lambda^{(k+1)} = 2\lambda^{(k)}$ and $x^{(k+1)} = x^{(k)}$.

Stopping criteria. The algorithm is stopped before the maximum number of iterations $k^{\max}$ if either of the following two conditions hold.

- *Small residual*: $\|f(x^{(k+1)})\|^2$ is small enough. This means we have (almost) solved the equations $f(x) = 0$, and therefore (almost) minimized $\|f(x)\|^2$.

- *Small optimality condition residual*: $\|2Df(\hat{x})^T f(\hat{x})\|$ is small enough, *i.e.*, the optimality condition (18.3) almost holds.

When the algorithm terminates with small optimality condition residual, we can say very little for sure about the point $x^{(k+1)}$ computed. This point found may be a minimizer of $\|f(x)\|^2$, or perhaps not. Since the algorithm does not always find a minimizer of $\|f(x)\|^2$, it is a heuristic. Like the k-means algorithm, which is also a heuristic, the Levenberg–Marquardt algorithm is widely used in many applications, even when we cannot be sure that it has found a point that gives the smallest possible residual norm.

Warm start. In many applications a sequence of similar or related nonlinear least squares problems are solved. In these cases it is common to start the Levenberg–Marquardt algorithm at the solution of the previously solved problem. If the problem to be solved is not much different from the previous problem, this can greatly reduce the number of iterations required to converge. This technique is called *warm starting*. It is commonly used in nonlinear model fitting, when multiple models are fit as we vary a regularization parameter.

Multiple runs. It is common to run the Levenberg–Marquardt algorithm from several different starting points $x^{(1)}$. If the final points found by running the algorithm from these different starting points are the same, or very close, it increases our confidence that we have found a solution of the nonlinear least squares problem, but we cannot be sure. If the different runs of the algorithm produce different points, we use the best one found, *i.e.*, the one with the smallest value of $\|f(x)\|^2$.

Complexity. Each execution of step 1 requires evaluating the derivative matrix of f. The complexity of this step depends on the particular function f. Each execution of step 2 requires the solution of a regularized least squares problem. Using the QR factorization of the stacked matrix this requires $2(m + n)n^2$ flops (see §15.5). When m is on the order of n, or larger, this is the same order as mn^2. When m is much smaller than n, $x^{(k+1)}$ can be computed using the kernel trick described in §15.5, which requires $2nm^2$ flops.

Levenberg–Marquardt update for $n = 1$. The Newton update for solving $f(x) = 0$ when $n = 1$ is given in (18.9). The Levenberg–Marquardt update for minimizing $f(x)^2$ is

$$x^{(k+1)} = x^{(k)} - \frac{f'(x^{(k)})}{\lambda^{(k)} + (f'(x^{(k)}))^2} f(x^{(k)}). \tag{18.13}$$

For $\lambda^{(k)} = 0$ they agree; but when $f'(x^{(k)}) = 0$, for example, the Levenberg–Marquardt update makes sense (since $\lambda^{(k)} > 0$), whereas the Newton update is undefined.

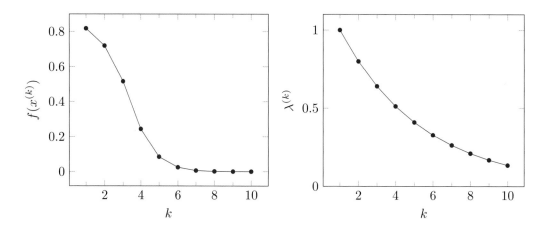

Figure 18.5 Values of $f(x^{(k)})$ and $\lambda^{(k)}$ versus the iteration number k for the Levenberg–Marquardt algorithm applied to $f(x) = (\exp(x) - \exp(-x))/(\exp(x)+\exp(-x))$. The starting point is $x^{(1)} = 1.15$ and $\lambda^{(1)} = 1$.

18.3.1   Examples

Nonlinear equation.   The first example is the sigmoid function (18.10) from the example on page 390. We saw in figures 18.3 and 18.4 that the Gauss–Newton method, which reduces to Newton's method in this case, diverges when the initial value $x^{(1)}$ is 1.15. The Levenberg–Marquardt algorithm, however, solves this problem. Figure 18.5 shows the value of the residual $f(x^{(k)})$, and the value of $\lambda^{(k)}$, for the Levenberg–Marquardt algorithm started from $x^{(1)} = 1.15$ and $\lambda^{(1)} = 1$. It converges to the solution $x = 0$ in around 10 iterations.

Equilibrium prices.   We illustrate algorithm 18.3 with a small instance of the equilibrium price problem, with supply and demand functions

$$
\begin{aligned}
D(p) &= \exp\left(E^{\mathrm{d}}(\log p - \log p^{\mathrm{nom}}) + d^{\mathrm{nom}}\right), \\
S(p) &= \exp\left(E^{\mathrm{s}}(\log p - \log p^{\mathrm{nom}}) + s^{\mathrm{nom}}\right),
\end{aligned}
$$

where E^{d} and E^{s} are the demand and supply elasticity matrices, d^{nom} and s^{nom} are the nominal demand and supply vectors, and the log and exp appearing in the equations apply to vectors elementwise. Figure 18.6 shows the contour lines of $\|f(p)\|^2$, where $f(p) = S(p) - D(p)$ is the excess supply, for

$$
p^{\mathrm{nom}} = (2.8, 10), \qquad d^{\mathrm{nom}} = (3.1, 2.2), \qquad s^{\mathrm{nom}} = (2.2, 0.3)
$$

and

$$
E^{\mathrm{d}} = \begin{bmatrix} -0.5 & 0.2 \\ 0 & -0.5 \end{bmatrix}, \qquad E^{\mathrm{s}} = \begin{bmatrix} 0.5 & -0.3 \\ -0.15 & 0.8 \end{bmatrix}.
$$

Figure 18.7 shows the iterates of the algorithm 18.3, started at $p = (3, 9)$ and $\lambda^{(1)} = 1$. The values of $\|f(p^{(k)})\|^2$ and the trust parameter $\lambda^{(k)}$ versus iteration k are shown in figure 18.8.

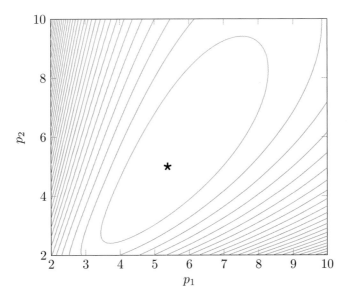

Figure 18.6 Contour lines of the square norm of the excess supply $f(p) = S(p) - D(p)$ for a small example with two commodities. The point marked with a star is the equilibrium prices, for which $f(p) = 0$.

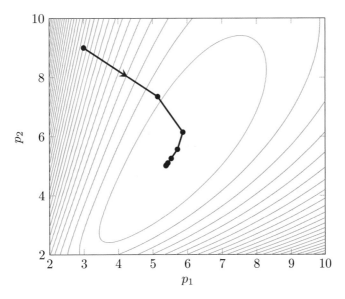

Figure 18.7 Iterates of the Levenberg–Marquardt algorithm started at $p = (3, 9)$.

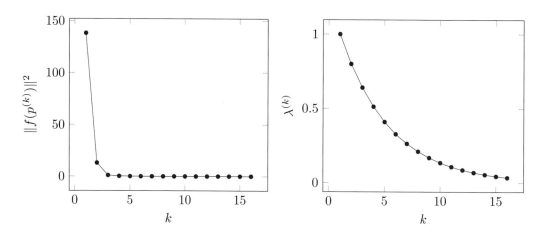

Figure 18.8 Cost function $\|f(p^{(k)}\|^2$ and trust parameter $\lambda^{(k)}$ versus iteration number k in the example of figure 18.7.

Location from range measurements. We illustrate algorithm 18.3 with a small instance of the location from range measurements problem, with five points a_i in a plane, shown in figure 18.9. The range measurements ρ_i are the distances of these points to the 'true' point $(1,1)$, plus some measurement errors. Figure 18.9 also shows the level curves of $\|f(x)\|^2$, and the point $(1.18, 0.82)$ (marked with a star) that minimizes $\|f(x)\|^2$. (This point is close to, but not equal to, the 'true' value $(1, 1)$, due to the noise added to the range measurements.) Figure 18.10 shows the graph of $\|f(x)\|$.

We run algorithm 18.3 from three different starting points,

$$x^{(1)} = (1.8, 3.5), \qquad x^{(1)} = (2.2, 3.5), \qquad x^{(1)} = (3.0, 1.5),$$

with $\lambda^{(1)} = 0.1$. Figure 18.11 shows the iterates $x^{(k)}$ for the three starting points. When started at $(1.8, 3.5)$ (blue circles) or $(3.0, 1.5)$ (brown diamonds) the algorithm converges to $(1.18, 0.82)$, the point that minimizes $\|f(x)\|^2$. When the algorithm is started at $(2.2, 3.5)$ the algorithm converges to a non-optimal point $(2.98, 2.12)$ (which gives a poor estimate of the 'true' location $(1, 1)$).

The values of $\|f(x^{(k)})\|^2$ and the trust parameter $\lambda^{(k)}$ during the iteration are shown in figure 18.12. As can be seen from this figure, in the first run of the algorithm (blue circles), $\lambda^{(k)}$ is increased in the third iteration. Correspondingly, $x^{(3)} = x^{(4)}$ in figure 18.12. For the second starting point (red squares) $\lambda^{(k)}$ decreases monotonically. For the third starting point (brown diamonds) $\lambda^{(k)}$ increases in iterations 2 and 4.

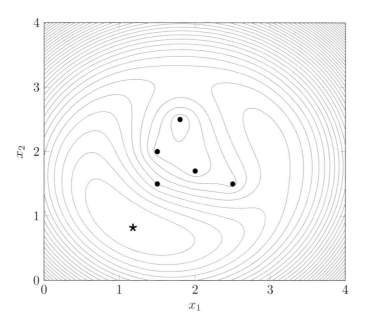

Figure 18.9 Contour lines of $\|f(x)\|^2$ where $f_i(x) = \|x - a_i\| - \rho_i$. The dots show the points a_i, and the point marked with a star is the point that minimizes $\|f(x)\|^2$.

$$\|f(x)\|$$

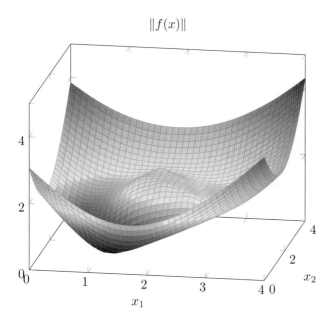

Figure 18.10 Graph of $\|f(x)\|$ in the location from range measurements example.

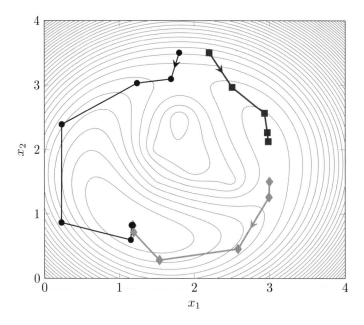

Figure 18.11 Iterates of the Levenberg–Marquardt algorithm started at three different starting points.

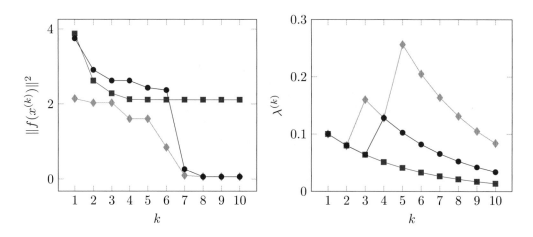

Figure 18.12 Cost function $\|f(x^{(k)})\|^2$ and trust parameter $\lambda^{(k)}$ versus iteration number k for the three starting points.

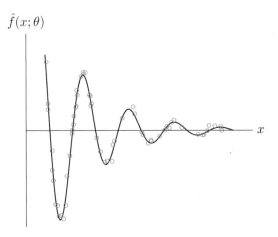

$\hat{f}(x;\theta)$

x

Figure 18.13 Least squares fit of a function $\hat{f}(x;\theta) = \theta_1 e^{\theta_2 x} \cos(\theta_3 x + \theta_4)$ to $N = 60$ points $(x^{(i)}, y^{(i)})$.

18.4   Nonlinear model fitting

The Levenberg–Marquardt algorithm is widely used for *nonlinear model fitting*. As in §13.1 we are given a set of data, $x^{(1)}, \ldots, x^{(N)}$, $y^{(1)}, \ldots, y^{(N)}$, where the n-vectors $x^{(1)}, \ldots, x^{(N)}$ are the feature vectors, and the scalars $y^{(1)}, \ldots, y^{(N)}$ are the associated outcomes. (So here the superscript indexes the given data; previously in this chapter the superscript denoted the iteration number.)

In nonlinear model fitting, we fit a model of the general form $y \approx \hat{f}(x;\theta)$ to the given data, where the p-vector θ contains the model parameters. In linear model fitting, $\hat{f}(x;\theta)$ is a linear function of the parameters, so it has the special form

$$\hat{f}(x;\theta) = \theta_1 f_1(x) + \cdots + \theta_p f_p(x),$$

where $f_1, \ldots, f_p$ are scalar-valued functions, called the basis functions (See §13.1.) In nonlinear model fitting the dependence of $\hat{f}(x;\theta)$ on θ is not linear (or affine), so it does not have the simple form of a linear combination of p basis functions.

As in linear model fitting, we choose the parameter θ by (approximately) minimizing the sum of the squares of the prediction residuals,

$$\sum_{i=1}^{N} (\hat{f}(x^{(i)};\theta) - y^{(i)})^2,$$

which is a nonlinear least squares problem, with variable θ. (We can also add a regularization term to this objective.)

Example.   Figure 18.13 shows a nonlinear model fitting example. The model is an exponentially decaying sinusoid

$$\hat{f}(x;\theta) = \theta_1 e^{\theta_2 x} \cos(\theta_3 x + \theta_4),$$

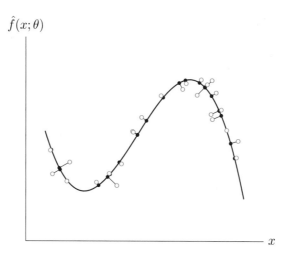

Figure 18.14 The solid line minimizes the sum of the squares of the orthogonal distances of points to the graph of the polynomial.

with four parameters $\theta_1, \theta_2, \theta_3, \theta_4$. (This model is an affine function of θ_1, but it is not an affine function of θ_2, θ_3, or θ_4.) We fit the model to $N = 60$ points $(x^{(i)}, y^{(i)})$ by minimizing the sum of the squared residuals (18.4) over the four parameters.

Orthogonal distance regression.   Consider the linear in the parameters model

$$\hat{f}(x; \theta) = \theta_1 f_1(x) + \cdots + \theta_p f_p(x),$$

with basis functions $f_i : \mathbf{R}^n \to \mathbf{R}$, and a data set of N pairs $(x^{(i)}, y^{(i)})$. The usual objective is the sum of squares of the difference between the model prediction $\hat{f}(x^{(i)})$ and the observed value $y^{(i)}$, which leads to a linear least squares problem. In *orthogonal distance regression* we use another objective, the sum of the squared distances of N points $(x^{(i)}, y^{(i)})$ to the graph of $\hat{f}$, *i.e.*, the set of points of the form $(u, \hat{f}(u))$. This model can be found by solving the nonlinear least squares problem

$$\text{minimize} \quad \sum_{i=1}^{N} (\hat{f}(u^{(i)}; \theta) - y^{(i)})^2 + \sum_{i=1}^{N} \|u^{(i)} - x^{(i)}\|^2$$

with variables $\theta_1, \ldots, \theta_p$, and $u^{(1)}, \ldots, u^{(N)}$. In orthogonal distance regression, we are allowed to choose the parameters in the model, and also, to modify the feature vectors from $x^{(i)}$ from $u^{(i)}$ to obtain a better fit. (Orthogonal distance regression is an example of an *error-in-variables model*, since it takes into account errors in the regressors or independent variables.) Figure 18.14 shows a cubic polynomial fit to 25 points using this method. The open circles are the points $(x^{(i)}, y^{(i)})$. The small circles on the graph of the polynomial are the points $(u^{(i)}, \hat{f}(u^{(i)}; \theta))$. Roughly speaking, we fit a curve that passes near all the data points, as measured by the (minimum) distance from the data points to the curve. In contrast, ordinary least squares regression finds a curve that minimizes the sum of squares of the vertical errors between the curve and the data points.

18.5   Nonlinear least squares classification

In this section we describe a nonlinear extension of the least squares classification method discussed in chapters 14 and 15, that typically out-performs the basic least squares classifier in practice.

The Boolean classifier of chapter 14 fits a linearly parametrized function

$$\tilde{f}(x) = \theta_1 f_1(x) + \cdots + \theta_p f_p(x)$$

to the data points $(x^{(i)}, y^{(i)})$, $i = 1, \ldots, N$, where $y^{(i)} \in \{-1, 1\}$, using linear least squares. The parameters $\theta_1, \ldots, \theta_p$ are chosen to minimize the sum squares objective

$$\sum_{i=1}^{N} (\tilde{f}(x^{(i)}) - y^{(i)})^2, \tag{18.14}$$

plus, optionally, a regularization term. This (hopefully) results in $\tilde{f}(x^{(i)}) \approx y^{(i)}$, which is roughly what we want. We can think of $\tilde{f}(x)$ as the *continuous* prediction of the Boolean outcome y. The classifier itself is given by $\hat{f}(x) = \mathbf{sign}(\tilde{f}(x))$; this is the Boolean prediction of the outcome.

Instead of the sum square prediction error for the continuous prediction, consider the sum square prediction error for the Boolean prediction,

$$\sum_{i=1}^{N} (\hat{f}(x^{(i)}) - y^{(i)})^2 = \sum_{i=1}^{N} (\mathbf{sign}(\tilde{f}(x^{(i)})) - y^{(i)})^2. \tag{18.15}$$

This is 4 times the number of classification errors we make on the training set. To see this, we note that when $\hat{f}(x^{(i)}) = y^{(i)}$, which means that a correct prediction was made on the ith data point, we have $(\hat{f}(x^{(i)}) - y^{(i)})^2 = 0$. When $\hat{f}(x^{(i)}) \neq y^{(i)}$, which means that an incorrect prediction was made on the ith data point, one of the values is $+1$ and the other is -1, so we have $(\hat{f}(x^{(i)}) - y^{(i)})^2 = 4$.

The objective (18.15) is what we really want; the least squares objective (18.14) is a *surrogate* for what we want. But we cannot use the Levenberg–Marquardt algorithm to minimize the objective (18.15), since the sign function is not differentiable. To get around this, we replace the sign function with a differentiable approximation, for example the *sigmoid function*

$$\phi(u) = \frac{e^u - e^{-u}}{e^u + e^{-u}}, \tag{18.16}$$

shown in figure 18.15. We choose θ by solving the nonlinear least squares problem of minimizing

$$\sum_{i=1}^{N} (\phi(\tilde{f}(x^{(i)})) - y^{(i)})^2, \tag{18.17}$$

using the Levenberg–Marquardt algorithm. (We can also add regularization to this objective.) Minimizing the nonlinear least squares objective (18.17) is a good approximation for choosing the parameter vector θ so as to minimize the number of classification errors made on the training set.

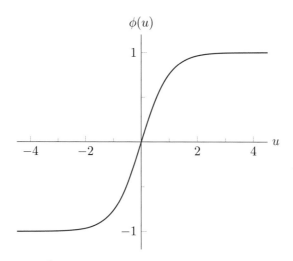

Figure 18.15 The sigmoid function ϕ.

Loss function interpretation. We can interpret the objective functions (18.14), (18.15), and (18.17) in terms of *loss functions* that depend on the continuous prediction $\tilde{f}(x^{(i)})$ and the outcome $y^{(i)}$. Each of the three objectives has the form

$$\sum_{i=1}^{N} \ell(\tilde{f}(x^{(i)}), y^{(i)}),$$

where ℓ is a loss function. The first argument of the loss function is a real number, and the second argument is Boolean, with values -1 or $+1$. For the linear least squares objective (18.14), the loss function is $\ell(u, y) = (u - y)^2$. For the nonlinear least squares objective with the sign function (18.15), the loss function is $\ell(u, y) = (\mathbf{sign}(u) - y)^2$. For the differentiable nonlinear least squares objective (18.15), the loss function is $\ell(u, y) = (\phi(u) - y)^2$. Roughly speaking, the loss function $\ell(u, y)$ tells us how bad it is to have $\tilde{f}(x^{(i)}) = u$ when $y = y^{(i)}$.

Since the outcome y takes on only two values, -1 and $+1$, we can plot the loss functions as functions of u for these two values of y. Figure 18.16 shows these three functions, with the value for $y = -1$ in the left column and the value for $y = +1$ in the right column. We can see that all three loss functions discourage prediction errors, since their values are higher for $\mathbf{sign}(u) \neq y$ than when $\mathbf{sign}(u) = y$. The loss function for nonlinear least squares classification with the sign function (shown in the middle row) assesses a cost of 0 for a correct prediction and 4 for an incorrect prediction. The loss function for nonlinear least squares classification with the sigmoid function (shown in the bottom row) is a smooth approximation of this.

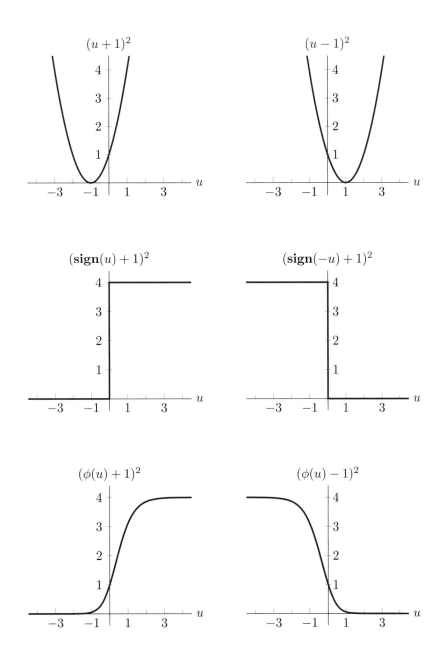

Figure 18.16 The loss functions $\ell(u, y)$ for linear least squares classification (top), nonlinear least squares classification with the sign function (middle), and nonlinear least squares classification with the sigmoid function (bottom). The left column shows $\ell(u, -1)$ and the right columns shows $\ell(u, +1)$.

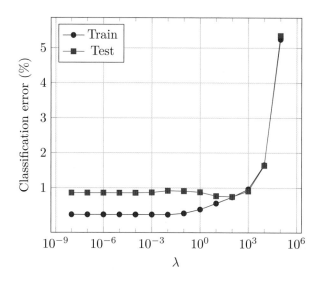

Figure 18.17 Boolean classification error in percent versus λ.

18.5.1   Handwritten digit classification

We apply nonlinear least squares classification on the MNIST set of handwritten digits used in chapter 14. We first consider the Boolean problem of recognizing the digit zero. We use linear features, *i.e.*,

$$\tilde{f}(x) = x^T\beta + v,$$

where x is the 493-vector of pixel intensities. To determine the parameters v and β we solve the nonlinear least squares problem

$$\text{minimize} \sum_{i=1}^{N}(\phi((x^{(i)})^T\beta + v) - y^{(i)})^2 + \lambda\|\beta\|^2, \tag{18.18}$$

where ϕ is the sigmoid function (18.16) and λ is a positive regularization parameter. (This λ is the regularization parameter in the classification problem; it has no relation to the trust parameter $\lambda^{(k)}$ in the iterates of the Levenberg–Marquardt algorithm.)

Figure 18.17 shows the classification error on the training and test sets as a function of the regularization parameter λ. For $\lambda = 100$, the classification errors on the training and test sets are about 0.7%. This is less than half the 1.6% error of the Boolean least squares classifier that used the same features, discussed in chapter 14. This improvement in performance, by more than a factor of two, comes from minimizing an objective that is closer to what we want (*i.e.*, the number of prediction errors on the training set) than the surrogate linear least squares objective. The confusion matrices for the training set and test set are given in table 18.1. Figure 18.18 shows the distribution of the values of $\tilde{f}(x^{(i)})$ for the two classes of the data set.

| | Prediction | | |
Outcome	$\hat{y} = +1$	$\hat{y} = -1$	Total
$y = +1$	5627	296	5923
$y = -1$	148	53929	54077
All	5775	54225	60000

| | Prediction | | |
Outcome	$\hat{y} = +1$	$\hat{y} = -1$	Total
$y = +1$	945	35	980
$y = -1$	40	8980	9020
All	985	9015	10000

Table 18.1 Confusion matrices for a Boolean classifier to recognize the digit zero. The table on the left is for the training set. The table on the right is for the test set.

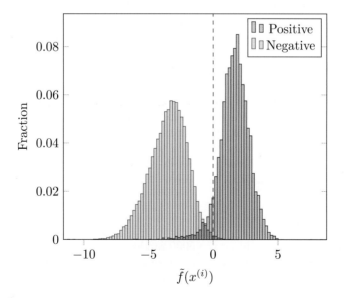

Figure 18.18 The distribution of the values of $\tilde{f}(x^{(i)})$ used in the Boolean classifier (14.1) for recognizing the digit zero. The function $\tilde{f}$ was computed by solving the nonlinear least squares problem (18.17).

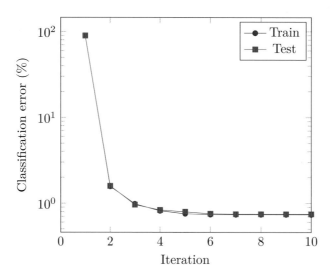

Figure 18.19 Training and test error versus Levenberg–Marquardt iteration for $\lambda = 100$.

Convergence of Levenberg–Marquardt algorithm.   The Levenberg–Marquardt algorithm is used to compute the parameters in the nonlinear least squares classifier. In this example the algorithm takes several tens of iterations to converge, *i.e.*, until the stopping criterion for the nonlinear least squares problem is satisfied. But in this application we are more interested in the performance of the classifier, and not minimizing the objective of the nonlinear least squares problem. Figure 18.19 shows the *classification error* of the classifier (on the training and test data sets) with parameter $\theta^{(k)}$, the kth iterate of the Levenberg–Marquardt algorithm. We can see that the classification errors reach their final values of 0.7% after just a few iterations. This phenomenon is very typical in nonlinear data fitting problems. Well before convergence, the Levenberg–Marquardt algorithm finds model parameters that are just as good (as judged by test error) as the parameters obtained when the algorithm converges.

Feature engineering.   After adding the 5000 random features used in chapter 14, we obtain the training and test classification errors shown in figure 18.20. The error on the training set is zero for small λ. For $\lambda = 1000$, the error on the test set is 0.24%, with the confusion matrix in table 18.2. The distribution of $\tilde{f}(x^{(i)})$ on the training set in figure 18.21 shows why the training error is zero.

Figure 18.22 shows the classification errors versus Levenberg–Marquardt iteration, if we start the Levenberg–Marquardt algorithm with $\beta = 0$, $v = 0$. (This implies that the values computed in the first iteration are the coefficients of the linear least squares classifier.) The error on the training set is exactly zero at iteration 5. The error on the test set is almost equal to its final value after one iteration.

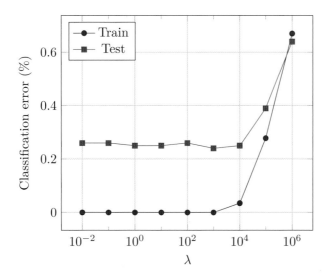

Figure 18.20 Boolean classification error in percent versus λ, after adding 5000 random features.

Outcome	Prediction		
	$\hat{y} = +1$	$\hat{y} = -1$	Total
$y = +1$	967	13	980
$y = -1$	11	9009	9020
All	978	9022	10000

Table 18.2 Confusion matrix on the test set for the Boolean classifier to recognize the digit zero after addition of 5000 new features.

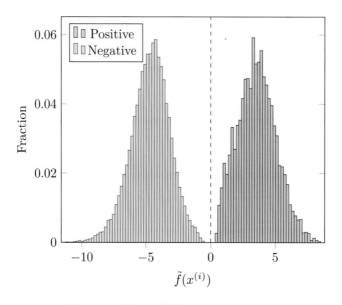

Figure 18.21 The distribution of the values of $\tilde{f}(x^{(i)})$ used in the Boolean classifier (14.1) for recognizing the digit zero, after addition of 5000 new features. The function $\tilde{f}$ was computed by solving the nonlinear least squares problem (18.17).

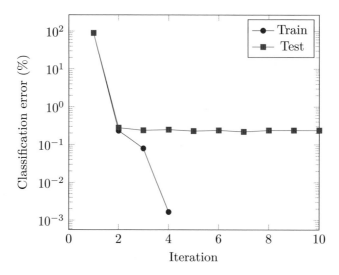

Figure 18.22 Training and test error versus Levenberg–Marquardt iteration for $\lambda = 1000$.

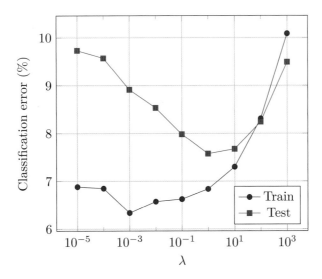

Figure 18.23 Multiclass classification error in percent versus λ.

Multi-class classifier.   Next we apply the nonlinear least squares method to the multi-class classification of recognizing the ten digits in the MNIST data set. For each digit k, we compute a Boolean classifier $\tilde{f}_k(x) = x^T \beta_k + v_k$ by solving a regularized nonlinear least squares problem (18.18). The same value of λ is used in the ten nonlinear least squares problems. The Boolean classifiers are combined into a multi-class classifier

$$\hat{f}(x) = \operatorname*{argmax}_{k=1,\dots,10} \left(x^T \beta_k + v_k \right).$$

Figure 18.23 shows the classification errors versus λ. The test set confusion matrix (for $\lambda = 1$) is given in table 18.3. The classification error on the test set is 7.6%, down from the 13.9% error we obtained for the same set of features with the least squares method of chapter 14.

Feature engineering.   Figure 18.24 shows the error rates when we add the 5000 randomly generated features. The training and test error rates are now 0.02% and 2%. The test set confusion matrix for $\lambda = 1000$ is given in table 18.4. This classifier has matched human performance in classifying digits correctly. Further, or more sophisticated, feature engineering can bring the test performance well below what humans can achieve.

| Digit | \multicolumn{10}{c}{Prediction} | Total |
	0	1	2	3	4	5	6	7	8	9	
0	964	0	0	2	0	2	5	3	3	1	980
1	0	1112	4	3	0	1	4	1	10	0	1135
2	5	5	934	13	7	3	13	10	38	4	1032
3	3	0	19	926	1	21	2	8	21	9	1010
4	1	2	4	2	917	0	7	1	10	38	982
5	10	2	2	31	10	782	17	7	23	8	892
6	8	3	3	1	5	20	910	1	7	0	958
7	2	6	25	5	11	5	0	947	4	23	1028
8	13	10	4	18	16	27	8	9	865	4	974
9	8	6	0	12	43	11	1	19	23	886	1009
All	1014	1146	995	1013	1010	872	967	1006	1004	973	10000

Table 18.3 Confusion matrix for test set. The error rate is 7.6%.

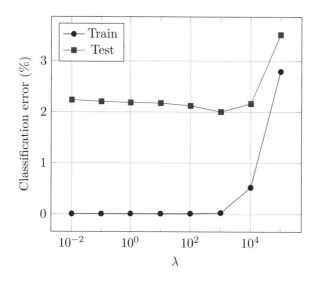

Figure 18.24 Multiclass classification error in percent versus λ after adding 5000 random features.

Digit	Prediction										Total
	0	1	2	3	4	5	6	7	8	9	
0	972	1	1	1	0	1	1	1	2	0	980
1	0	1124	2	2	0	0	3	1	3	0	1135
2	5	0	1006	1	3	0	2	6	9	0	1032
3	0	0	3	986	0	5	0	3	7	6	1010
4	0	0	4	1	966	0	4	1	0	6	982
5	2	0	2	5	2	875	5	0	1	0	892
6	7	2	0	1	3	2	941	0	2	0	958
7	1	7	6	1	2	0	0	1003	3	5	1028
8	3	0	0	4	4	5	3	4	949	2	974
9	2	5	0	5	6	4	1	6	2	978	1009
All	992	1139	1024	1007	986	892	960	1025	978	997	10000

Table 18.4 Confusion matrix for test set after adding 5000 features. The error rate is 2.0%.

Exercises

18.1 *Lambert W-function.* The *Lambert W-function*, denoted $W : [0, \infty) \to \mathbf{R}$, is defined as $W(u) = x$, where x is the unique number $x \geq 0$ for which $xe^x = u$. (The notation just means that we restrict the argument x to be nonnegative.) The Lambert function arises in a variety of applications, and is named after the mathematician Johann Heinrich Lambert. There is no analytical formula for $W(u)$; it must be computed numerically. In this exercise you will develop a solver to compute $W(u)$, given a nonnegative number u, using the Levenberg–Marquardt algorithm, by minimizing $f(x)^2$ over x, where $f(x) = xe^x - u$.

   (a) Give the Levenberg–Marquardt update (18.13) for f.

   (b) Implement the Levenberg–Marquardt algorithm for minimizing $f(x)^2$. You can start with $x^{(1)} = 1$ and $\lambda^{(1)} = 1$ (but it should work with other initializations). You can stop the algorithm when $|f(x)|$ is small, say, less than 10^{-6}.

18.2 *Internal rate of return.* Let the n-vector c denote a cash flow over n time periods, with positive entries meaning cash received and negative entries meaning payments. We assume that its NPV (net present value; see page 22) with interest rate $r \geq 0$ is given by

$$N(r) = \sum_{i=1}^{n} \frac{c_i}{(1 + r)^{i-1}}.$$

The *internal rate of return* (IRR) of the cash flow is defined as the smallest positive value of r for which $N(r) = 0$. The Levenberg–Marquardt algorithm can be used to compute the IRR of a given cash flow sequence, by minimizing $N(r)^2$ over r.

   (a) Work out a specific formula for the Levenberg–Marquardt update for r, *i.e.*, (18.13).

   (b) Implement the Levenberg–Marquardt algorithm to find the IRR of the cash flow sequence

$$c = (-\mathbf{1}_3, \ 0.3\,\mathbf{1}_5, \ 0.6\,\mathbf{1}_6),$$

   where the subscripts give the dimensions. (This corresponds to three periods in which you make investments, which pay off at one rate for 5 periods, and a higher rate for the next 6 periods.) You can initialize with $r^{(0)} = 0$, and stop when $N(r^{(k)})^2$ is small. Plot $N(r^{(k)})^2$ versus k.

18.3 *A common form for the residual.* In many nonlinear least squares problems the residual function $f : \mathbf{R}^n \to \mathbf{R}^m$ has the specific form

$$f_i(x) = \phi_i(a_i^T x - b_i), \quad i = 1, \dots, m,$$

where a_i is an n-vector, b_i is a scalar, and $\phi_i : \mathbf{R} \to \mathbf{R}$ is a scalar-valued function of a scalar. In other words, $f_i(x)$ is a scalar function of an affine function of x. In this case the objective of the nonlinear least squares problem has the form

$$\|f(x)\|^2 = \sum_{i=1}^{m} \left(\phi_i(a_i^T x - b_i) \right)^2 . \quad \cdot$$

We define the $m \times n$ matrix A to have rows $a_1^T, \dots, a_m^T$, and the m-vector b to have entries $b_1, \dots, b_m$. Note that if the functions ϕ_i are the identity function, *i.e.*, $\phi_i(u) = u$ for all u, then the objective becomes $\|Ax - b\|^2$, and in this case the nonlinear least squares problem reduces to the linear least squares problem. Show that the derivative matrix $Df(x)$ has the form

$$Df(x) = \mathbf{diag}(d)A,$$

where $d_i = \phi_i'(r_i)$ for $i = 1, \dots, m$, with $r = Ax - b$.

Remark. This means that in each iteration of the Gauss–Newton method we solve a weighted least squares problem (and in the Levenberg–Marquardt algorithm, a regularized weighted least squares problem); see exercise 12.4. The weights change in each iteration.

18.4 *Fitting an exponential to data.* Use the Levenberg–Marquardt algorithm to fit an exponential function of the form $\hat{f}(x;\theta) = \theta_1 e^{\theta_2 x}$ to the data

$$0, 1, \ldots, 5, \qquad 5.2, 4.5, 2.7, 2.5, 2.1, 1.9.$$

(The first list gives $x^{(i)}$; the second list gives $y^{(i)}$.) Plot your model $\hat{f}(x;\hat{\theta})$ versus x, along with the data points.

18.5 *Mechanical equilibrium.* A mass m, at position given by the 2-vector x, is subject to three forces acting on it. The first force F^{grav} is gravity, which has value $F^{\mathrm{grav}} = -mg(0,1)$, where $g = 9.8$ is the acceleration of gravity. (This force points straight down, with a force that does not depend on the position x.) The mass is attached to two cables, whose other ends are anchored at (2-vector) locations a_1 and a_2. The force F_i applied to the mass by cable i is given by

$$F_i = T_i(a_i - x)/\|a_i - x\|,$$

where T_i is the cable tension. (This means that each cable applies a force on the mass that points from the mass to the cable anchor, with magnitude given by the tension T_i.) The cable tensions are given by

$$T_i = k\frac{\max\{\|a_i - x\| - L_i, 0\}}{L_i},$$

where k is a positive constant, and L_i is the natural or unloaded length of cable i (also positive). In words: The tension is proportional to the fractional stretch of the cable, above its natural length. (The max appearing in this formula means the tension is not a differentiable function of the position, when $\|a_i - x\| = L_i$, but we will simply ignore this.) The mass is in equilibrium at position x if the three forces acting on it sum to zero,

$$F^{\mathrm{grav}} + F_1 + F_2 = 0.$$

We refer to the left-hand side as the residual force. It is a function of mass position x, and we write it as $f(x)$.

Compute an equilibrium position for

$$a_1 = (3,2), \qquad a_2 = (-1,1), \qquad L_1 = 3, \qquad L_2 = 2, \qquad m = 1, \qquad k = 100,$$

by applying the Levenberg–Marquardt algorithm to the residual force $f(x)$. Use $x^{(1)} = (0,0)$ as starting point. (Note that it is important to start at a point where $T_1 > 0$ and $T_2 > 0$, because otherwise the derivative matrix $Df(x^{(1)})$ is zero, and the Levenberg–Marquardt update gives $x^{(2)} = x^{(1)}$.) Plot the components of the mass position and the residual force versus iterations.

18.6 *Fitting a simple neural network model.* A neural network is a widely used model of the form $\hat{y} = \hat{f}(x;\theta)$, where the n-vector x is the feature vector and the p-vector θ is the model parameter. In a neural network model, the function $\hat{f}$ is *not* an affine function of the parameter vector θ. In this exercise we consider a very simple neural network, with two layers, three internal nodes, and two inputs (*i.e.*, $n = 2$). This model has $p = 13$ parameters, and is given by

$$\hat{f}(x;\theta) = \theta_1\phi(\theta_2 x_1 + \theta_3 x_2 + \theta_4) + \theta_5\phi(\theta_6 x_1 + \theta_7 x_2 + \theta_8)$$
$$+ \theta_9\phi(\theta_{10} x_1 + \theta_{11} x_2 + \theta_{12}) + \theta_{13}$$

where $\phi : \mathbf{R} \to \mathbf{R}$ is the sigmoid function defined in (18.16). This function is shown as a *signal flow graph* in figure 18.25. In this graph each edge from an input to an internal node, or from an internal node to the output node, corresponds to multiplication by one of the parameters. At each node (shown as the small filled circles) the incoming values and the constant offset are added together, then passed through the sigmoid function, to become the outgoing edge value.

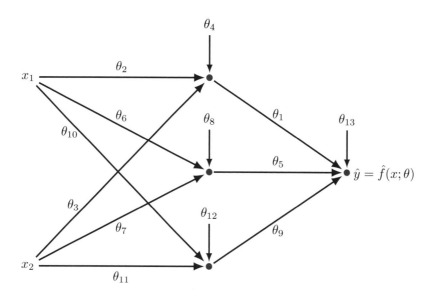

Figure 18.25 Signal flow graph of a simple neural network.

Fitting such a model to a data set consisting of the n-vectors $x^{(1)}, \ldots, x^{(N)}$ and the associated scalar outcomes $y^{(1)}, \ldots, y^{(N)}$ by minimizing the sum of the squares of the residuals is a nonlinear least squares problem with objective (18.4).

(a) Derive an expression for $\nabla_\theta \hat{f}(x; \theta)$. Your expression can use ϕ and ϕ', the sigmoid function and its derivative. (You do not need to express these in terms of exponentials.)

(b) Derive an expression for the derivative matrix $Dr(\theta)$, where $r : \mathbf{R}^p \to \mathbf{R}^N$ is the vector of model fitting residuals,

$$r(\theta)_i = \hat{f}(x^{(i)}; \theta) - y^{(i)}, \quad i = 1, \ldots, N.$$

Your expression can use the gradient found in part (a).

(c) Try fitting this neural network to the function $g(x_1, x_2) = x_1 x_2$. First generate $N = 200$ random points $x^{(i)}$ and take $y^{(i)} = (x^{(i)})_1 (x^{(i)})_2$ for $i = 1, \ldots, 200$. Use the Levenberg–Marquardt algorithm to try to minimize

$$f(\theta) = \|r(\theta)\|^2 + \gamma \|\theta\|^2$$

with $\gamma = 10^{-5}$. Plot the value of f and the norm of its gradient versus iteration. Report the RMS fitting error achieved by the neural network model. Experiment with choosing different starting points to see the effect on the final model found.

(d) Fit the same data set with a (linear) regression model $\hat{f}^{\text{lin}}(x; \beta, v) = x^T \beta + v$ and report the RMS fitting error achieved. (You can add regularization in your fitting, but it won't improve the results.) Compare the RMS fitting error with the neural network model RMS fitting error from part (c).

Remarks. Neural networks used in practice employ many more regressors, layers, and internal modes. Specialized methods and software are used to minimize the fitting objective, and evaluate the required gradients and derivatives.

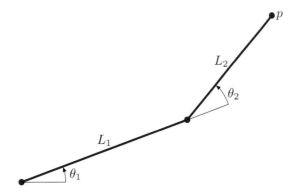

Figure 18.26 Two-link robot manipulator in a plane.

18.7 *Robot manipulator.* Figure 18.26 shows a two-link robot manipulator in a plane. The robot manipulator endpoint is at the position

$$p = L_1 \begin{bmatrix} \cos \theta_1 \\ \sin \theta_1 \end{bmatrix} + L_2 \begin{bmatrix} \cos(\theta_1 + \theta_2) \\ \sin(\theta_1 + \theta_2) \end{bmatrix},$$

where L_1 and L_2 are the lengths of the first and second links, θ_1 is the first joint angle, and θ_2 is second joint angle. We will assume that $L_2 < L_1$, *i.e.*, the second link is shorter than the first. We are given a desired endpoint position p^{des}, and seek joint angles $\theta = (\theta_1, \theta_2)$ for which $p = p^{\text{des}}$.

This problem can be solved analytically. We find θ_2 using the equation

$$\begin{aligned} \|p^{\text{des}}\|^2 &= (L_1 + L_2 \cos \theta_2)^2 + (L_2 \sin \theta_2)^2 \\ &= L_1^2 + L_2^2 + 2L_1 L_2 \cos \theta_2. \end{aligned}$$

When $L_1 - L_2 < \|p^{\text{des}}\| < L_1 + L_2$, there are two choices of θ_2 (one positive and one negative). For each solution θ_2 we can then find θ_1 using

$$\begin{aligned} p^{\text{des}} &= L_1 \begin{bmatrix} \cos \theta_1 \\ \sin \theta_1 \end{bmatrix} + L_2 \begin{bmatrix} \cos(\theta_1 + \theta_2) \\ \sin(\theta_1 + \theta_2) \end{bmatrix} \\ &= \begin{bmatrix} L_1 + L_2 \cos \theta_2 & -L_2 \sin \theta_2 \\ L_2 \sin \theta_2 & L_1 + L_2 \cos \theta_2 \end{bmatrix} \begin{bmatrix} \cos \theta_1 \\ \sin \theta_1 \end{bmatrix}. \end{aligned}$$

In this exercise you will use the Levenberg–Marquardt algorithm to find joint angles, by minimizing $\|p - p^{\text{des}}\|^2$.

(a) Identify the function $f(\theta)$ in the nonlinear least squares problem, and give its derivative $Df(\theta)$.

(b) Implement the Levenberg–Marquardt algorithm to solve the nonlinear least squares problem. Try your implementation on a robot with $L_1 = 2$, $L_2 = 1$, and the desired endpoints

$$(1.0, 0.5), \quad (-2.0, 1.0), \quad (-0.2, 3.1).$$

For each endpoint, plot the cost function $\|f(\theta^{(k)})\|^2$ versus iteration number k.

Note that the norm of the last endpoint exceeds $L_1 + L_2 = 3$, so there are no joint angles for which $p = p^{\text{des}}$. Explain the angles your algorithm finds in this case.

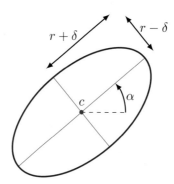

Figure 18.27 Ellipse with center (c_1, c_2), and radii $r + \delta$ and $r - \delta$. The largest semi-axis makes an angle α with respect to horizontal.

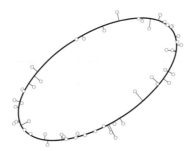

Figure 18.28 Ellipse fit to 50 points in a plane.

18.8 *Fitting an ellipse to points in a plane.* An ellipse in a plane can be described as the set of points

$$\hat{f}(t; \theta) = \left[\begin{array}{c} c_1 + r \cos(\alpha + t) + \delta \cos(\alpha - t) \\ c_2 + r \sin(\alpha + t) + \delta \sin(\alpha - t) \end{array} \right],$$

where t ranges from 0 to 2π. The vector $\theta = (c_1, c_2, r, \delta, \alpha)$ contains five parameters, with geometrical meanings illustrated in figure 18.27. We consider the problem of fitting an ellipse to N points $x^{(1)}, \dots, x^{(N)}$ in a plane, as shown in figure 18.28. The circles show the N points. The short lines connect each point to the nearest point on the ellipse. We will fit the ellipse by minimizing the sum of the squared distances of the N points to the ellipse.

(a) The squared distance of a data point $x^{(i)}$ to the ellipse is the minimum of $\|\hat{f}(t^{(i)}; \theta) - x^{(i)}\|^2$ over the scalar $t^{(i)}$. Minimizing the sum of the squared distances of the data points $x^{(1)}, \dots, x^{(N)}$ to the ellipse is therefore equivalent to minimizing

$$\sum_{i=1}^{N} \|\hat{f}(t^{(i)}; \theta) - x^{(i)}\|^2$$

over $t^{(1)}, \dots, t^{(N)}$ and θ. Formulate this as a nonlinear least squares problem. Give expressions for the derivatives of the residuals.

(b) Use the Levenberg–Marquardt algorithm to fit an ellipse to the 10 points:

$$(0.5, 1.5), \quad (-0.3, 0.6), \quad (1.0, 1.8), \quad (-0.4, 0.2), \quad (0.2, 1.3)$$

$$(0.7, 0.1), \quad (2.3, 0.8), \quad (1.4, 0.5), \quad (0.0, 0.2), \quad (2.4, 1.7).$$

To select a starting point, you can choose parameters θ that describe a circle with radius one and center at the average of the data points, and initial values of $t^{(i)}$ that minimize the objective function for those initial values of θ.

Chapter 19

Constrained nonlinear least squares

In this chapter we consider an extension of the nonlinear least squares problem that includes nonlinear constraints. Like the problem of solving a set of nonlinear equations, or finding a least squares approximate solution to a set of nonlinear equations, the constrained nonlinear least squares problem is in general hard to solve exactly. We describe a heuristic algorithm that often works well in practice.

19.1 Constrained nonlinear least squares

In this section we consider an extension of the nonlinear least squares problem (18.2) that includes equality constraints:

$$
\begin{array}{ll}
\text{minimize} & \|f(x)\|^2 \\
\text{subject to} & g(x) = 0,
\end{array}
\tag{19.1}
$$

where the n-vector x is the variable to be found. Here $f(x)$ is an m-vector, and $g(x)$ is a p-vector. We sometimes write out the components of $f(x)$ and $g(x)$, to express the problem as

$$
\begin{array}{ll}
\text{minimize} & f_1(x)^2 + \cdots + f_m(x)^2 \\
\text{subject to} & g_i(x) = 0, \quad i = 1, \ldots, p.
\end{array}
$$

We refer to $f_i(x)$ as the ith (scalar) residual, and $g_i(x) = 0$ as the ith (scalar) equality constraint. When the functions f and g are affine, the equality constrained nonlinear least squares problem (19.1) reduces to the (linear) least squares problem with equality constraints from chapter 16.

We say that a point x is feasible for the problem (19.1) if it satisfies $g(x) = 0$. A point $\hat{x}$ is a solution of the problem (19.1) if it is feasible and has the smallest objective among all feasible points, i.e., if whenever $g(x) = 0$, we have $\|f(x)\|^2 \geq \|f(\hat{x})\|^2$.

Like the nonlinear least squares problem, or solving a set of nonlinear equations, the constrained nonlinear least squares problem is in general hard to solve exactly. But the Levenberg–Marquardt algorithm for solving the (unconstrained) nonlinear least squares problem (18.2) can be leveraged to handle the problem with equality constraints. We will describe a basic algorithm below, the *penalty algorithm*, and a variation on it that works much better in practice, the *augmented Lagrangian algorithm*. These algorithms are heuristics for (approximately) solving the nonlinear least squares problem (19.1).

Linear equality constraints.   One special case of the constrained nonlinear least squares problem (19.1) is when the constraint function g is affine, in which case the constraints $g(x) = 0$ can be written $Cx = d$ for some $p \times n$ matrix C and a p-vector d. In this case the problem (19.1) is called a nonlinear least squares problem with linear equality constraints. It can be (approximately) solved by the Levenberg–Marquardt algorithm described in chapter 18, by simply adding the linear equality constraints to the linear least squares problem that is solved in step 2. The more challenging problem is the case when g is not affine.

19.1.1   Optimality condition

Using Lagrange multipliers (see §C.3) we can derive a condition that any solution of the constrained nonlinear least squares problem (19.1) must satisfy. The Lagrangian for the problem (19.1) is

$$L(x, z) = \|f(x)\|^2 + z_1 g_1(x) + \cdots + z_p g_p(x) = \|f(x)\|^2 + g(x)^T z, \qquad (19.2)$$

where the p-vector z is the vector of Lagrange multipliers. The method of Lagrange multipliers tells us that for any solution $\hat{x}$ of (19.1), there is a set of Lagrange multipliers $\hat{z}$ that satisfy

$$\frac{\partial L}{\partial x_i}(\hat{x}, \hat{z}) = 0, \quad i = 1, \ldots, n, \qquad \frac{\partial L}{\partial z_i}(\hat{x}, \hat{z}) = 0, \quad i = 1, \ldots, p$$

(provided the rows of $Dg(\hat{x})$ are linearly independent). The p-vector $\hat{z}$ is called an *optimal Lagrange multiplier*.

The second set of equations can be written as $g_i(\hat{x}) = 0$, $i = 1, \ldots, p$, in vector form

$$g(\hat{x}) = 0, \qquad (19.3)$$

i.e., $\hat{x}$ is feasible, which we already knew. The first set of equations can be written in vector form as

$$2Df(\hat{x})^T f(\hat{x}) + Dg(\hat{x})^T \hat{z} = 0. \qquad (19.4)$$

This equation is the extension of the condition (18.3) for the unconstrained nonlinear least squares problem (18.2). The equation (19.4), together with (19.3), *i.e.*, $\hat{x}$ is feasible, form the optimality conditions for the problem (19.1).

If $\hat{x}$ is a solution of the constrained nonlinear least squares problem (19.1), then it satisfies the optimality condition (19.4) for some Lagrange multiplier vector $\hat{z}$

(provided the rows of $Dg(\hat{x})$ are linearly independent). So $\hat{x}$ and $\hat{z}$ satisfy the optimality conditions.

These optimality conditions are not sufficient, however; there can be choices of x and z that satisfy them, but x is not a solution of the constrained nonlinear least squares problem.

19.2 Penalty algorithm

We start with the observation (already made on page 340) that the equality constrained problem can be thought of as a limit of a bi-objective problem with objectives $\|f(x)\|^2$ and $\|g(x)\|^2$, as the weight on the second objective increases to infinity. Let μ be a positive number, and consider the composite objective

$$\|f(x)\|^2 + \mu\|g(x)\|^2. \tag{19.5}$$

This can be (approximately) minimized using the Levenberg–Marquardt algorithm applied to

$$\left\| \begin{bmatrix} f(x) \\ \sqrt{\mu}g(x) \end{bmatrix} \right\|^2. \tag{19.6}$$

By minimizing the composite objective (19.5), we do not insist that $g(x)$ is zero, but we assess a cost or *penalty* $\mu\|g(x)\|^2$ on the residual. If we solve this for large enough μ, we should obtain a choice of x for which $g(x)$ is very small, and $\|f(x)\|^2$ is small, *i.e.*, an approximate solution of (19.1). The second term $\mu\|g(x)\|^2$ is a penalty imposed on choices of x with nonzero $g(x)$.

Minimizing the composite objective (19.5) for an increasing sequence of values of μ is known as the *penalty algorithm*.

Algorithm 19.1 PENALTY ALGORITHM FOR CONSTRAINED NONLINEAR LEAST SQUARES

given differentiable functions $f : \mathbf{R}^n \to \mathbf{R}^m$ and $g : \mathbf{R}^n \to \mathbf{R}^p$, and an initial point $x^{(1)}$. Set $\mu^{(1)} = 1$.

For $k = 1, 2, \ldots, k^{\max}$

 1. *Solve unconstrained nonlinear least squares problem.* Set $x^{(k+1)}$ to be the (approximate) minimizer of

$$\|f(x)\|^2 + \mu^{(k)}\|g(x)\|^2$$

 using the Levenberg–Marquardt algorithm, starting from initial point $x^{(k)}$.
 2. *Update $\mu^{(k)}$:* $\mu^{(k+1)} = 2\mu^{(k)}$.

The penalty algorithm is stopped early if $\|g(x^{(k)})\|$ is small enough, *i.e.*, the equality constraint is almost satisfied.

The penalty algorithm is simple and easy to implement, but has an important drawback: The parameter $\mu^{(k)}$ rapidly increases with iterations (as it must, to drive

$g(x)$ to zero). When the Levenberg–Marquardt algorithm is used to minimize (19.5) for very high values of μ, it can take a large number of iterations or simply fail. The augmented Lagrangian algorithm described below gets around this drawback, and gives a much more reliable algorithm.

We can connect the penalty algorithm iterates to the optimality condition (19.4). The iterate $x^{(k+1)}$ (almost) satisfies the optimality condition for minimizing (19.5),

$$2Df(x^{(k+1)})^T f(x^{(k+1)}) + 2\mu^{(k)} Dg(x^{(k+1)})^T g(x^{(k+1)}) = 0.$$

Defining

$$z^{(k+1)} = 2\mu^{(k)} g(x^{(k+1)})$$

as our estimate of a suitable Lagrange multiplier in iteration $k + 1$, we see that the optimality condition (19.4) (almost) holds for $x^{(k+1)}$ and $z^{(k+1)}$. (The feasibility condition $g(x^{(k)}) = 0$ only holds in the limit as $k \to \infty$.)

19.3   Augmented Lagrangian algorithm

The augmented Lagrangian algorithm is a modification of the penalty algorithm that addresses the difficulty associated with the penalty parameter $\mu^{(k)}$ becoming very large. It was proposed by Magnus Hestenes and Michael Powell in the 1960s.

Augmented Lagrangian.   The *augmented Lagrangian* for the problem (19.1), with parameter $\mu > 0$, is defined as

$$L_\mu(x, z) = L(x, \mu) + \mu\|g(x)\|^2 = \|f(x)\|^2 + g(x)^T z + \mu\|g(x)\|^2. \qquad (19.7)$$

This is the Lagrangian, augmented with the new term $\mu\|g(x)\|^2$; alternatively, it can be interpreted as the composite objective function (19.5) used in the penalty algorithm, with the Lagrange multiplier term $g(x)^T z$ added.

The augmented Lagrangian (19.7) is also the ordinary Lagrangian associated with the problem

$$\begin{aligned} \text{minimize} \quad & \|f(x)\|^2 + \mu\|g(x)\|^2 \\ \text{subject to} \quad & g(x) = 0. \end{aligned}$$

This problem is equivalent to the original constrained nonlinear least squares problem (19.1): A point x is a solution of one if and only if it is a solution of the other. (This follows since the term $\mu\|g(x)\|^2$ is zero for any feasible x.)

Minimizing the augmented Lagrangian.   In the augmented Lagrangian algorithm we minimize the augmented Lagrangian over the variable x for a sequence of values of μ and z. We show here how this can be done using the Levenberg–Marquardt algorithm. We first establish the identity

$$L_\mu(x, z) = \|f(x)\|^2 + \mu\|g(x) + z/(2\mu)\|^2 - \mu\|z/(2\mu)\|^2. \qquad (19.8)$$

We expand the second term on the right-hand side to get

$$
\begin{aligned}
&\mu\|g(x) + z/(2\mu)\|^2 \\
&= \mu\|g(x)\|^2 + 2\mu g(x)^T(z/(2\mu)) + \mu\|z/(2\mu)\|^2 \\
&= g(x)^T z + \mu\|g(x)\|^2 + \mu\|z/(2\mu)\|^2.
\end{aligned}
$$

Substituting this into the right-hand side of (19.8) verifies the identity.

When we minimize $L_\mu(x, z)$ over the variable x, the term $-\mu\|z/(2\mu)\|^2$ in (19.8) is a constant (*i.e.*, does not depend on x), and does not affect the choice of x. It follows that we can minimize $L_\mu(x, z)$ over x by minimizing the function

$$
\|f(x)\|^2 + \mu\|g(x) + z/(2\mu)\|^2, \tag{19.9}
$$

which in turn can be expressed as

$$
\left\| \begin{bmatrix} f(x) \\ \sqrt{\mu}g(x) + z/(2\sqrt{\mu}) \end{bmatrix} \right\|^2. \tag{19.10}
$$

This can be be (approximately) minimized using the Levenberg–Marquardt algorithm.

Any minimizer $\tilde{x}$ of $L_\mu(x, z)$ (or equivalently, (19.9)) satisfies the optimality condition

$$
\begin{aligned}
0 &= 2Df(\tilde{x})^T f(\tilde{x}) + 2\mu Dg(\tilde{x})^T(g(\tilde{x}) + z/(2\mu)) \\
&= 2Df(\tilde{x})^T f(\tilde{x}) + Dg(\tilde{x})^T(2\mu g(\tilde{x}) + z).
\end{aligned}
$$

From this equation we can observe that if $\tilde{x}$ minimizes the augmented Lagrangian and is also feasible (*i.e.*, $g(\tilde{x}) = 0$), then it satisfies the optimality condition (19.4) with the vector z as the Lagrange multiplier. The bottom equation also suggests a good choice for updating the Lagrange multiplier vector z if $\tilde{x}$ is not feasible. In this case the choice

$$
\tilde{z} = z + 2\mu g(\tilde{x}) \tag{19.11}
$$

satisfies the optimality condition (19.4) with $\tilde{x}$ and $\tilde{z}$.

The augmented Lagrangian algorithm alternates between minimizing the augmented Lagrangian (approximately, using the Levenberg–Marquardt algorithm), and updating the parameter z (our estimate of a suitable Lagrange multiplier) using the suggestion (19.11) above. The penalty parameter μ is increased only when needed, when $\|g(x)\|$ does not sufficiently decrease.

Algorithm 19.2 AUGMENTED LAGRANGIAN ALGORITHM

given differentiable functions $f : \mathbf{R}^n \to \mathbf{R}^m$ and $g : \mathbf{R}^n \to \mathbf{R}^p$, and an initial point $x^{(1)}$. Set $z^{(1)} = 0$, $\mu^{(1)} = 1$.

For $k = 1, 2, \ldots, k^{\max}$

1. *Solve unconstrained nonlinear least squares problem.* Set $x^{(k+1)}$ to be the (approximate) minimizer of

$$\|f(x)\|^2 + \mu^{(k)}\|g(x) + z^{(k)}/(2\mu^{(k)})\|^2$$

   using Levenberg–Marquardt algorithm, starting from initial point $x^{(k)}$.

2. *Update $z^{(k)}$.*

$$z^{(k+1)} = z^{(k)} + 2\mu^{(k)}g(x^{(k+1)}).$$

3. *Update $\mu^{(k)}$.*

$$\mu^{(k+1)} = \begin{cases} \mu^{(k)} & \|g(x^{(k+1)})\| < 0.25\|g(x^{(k)})\| \\ 2\mu^{(k)} & \|g(x^{(k+1)})\| \geq 0.25\|g(x^{(k)})\|. \end{cases}$$

The augmented Lagrangian algorithm is stopped early if $g(x^{(k)})$ is very small. Note that due to our particular choice of how $z^{(k)}$ is updated, the iterate $x^{(k+1)}$ (almost) satisfies the optimality condition (19.4) with $z^{(k+1)}$.

The augmented Lagrangian algorithm is not much more complicated than the penalty algorithm, but it works much better in practice. In part this is because the penalty parameter $\mu^{(k)}$ does not need to increase as much as the algorithm proceeds.

Example. We consider an example with two variables and

$$f(x_1, x_2) = \begin{bmatrix} x_1 + \exp(-x_2) \\ x_1^2 + 2x_2 + 1 \end{bmatrix}, \qquad g(x_1, x_2) = x_1 + x_1^3 + x_2 + x_2^2.$$

Figure 19.1 shows the contour lines of the cost function $\|f(x)\|^2$ (solid lines) and the constraint function $g(x)$ (dashed lines). The point $\hat{x} = (0, 0)$ is optimal with corresponding Lagrange multiplier $\hat{z} = -2$. One can verify that $g(\hat{x}) = 0$ and

$$2Df(\hat{x})^T f(\hat{x}) + Dg(\hat{x})^T \hat{z} = 2\begin{bmatrix} 1 & 0 \\ -1 & 2 \end{bmatrix}\begin{bmatrix} 1 \\ 1 \end{bmatrix} - 2\begin{bmatrix} 1 \\ 1 \end{bmatrix} = 0.$$

The circle at $x = (-0.666, -0.407)$ indicates the position of the unconstrained minimizer of $\|f(x)\|^2$.

The augmented Lagrangian algorithm is started from the point $x^{(1)} = (0.5, -0.5)$. Figure 19.2 illustrates the first six iterations. The solid lines are the contour lines for $L_\mu(x, z^{(k)})$, the augmented Lagrangian with the current value of the Lagrange multiplier. For comparison, we also show in figure 19.3 the first six iterations of the penalty algorithm, started from the same point. The solid lines are the contour lines of $\|f(x)\|^2 + \mu^{(k)}\|g(x)\|^2$.

In figure 19.4 we show how the algorithms converge. The horizontal axis is the cumulative number of Levenberg–Marquardt iterations. Each of these requires

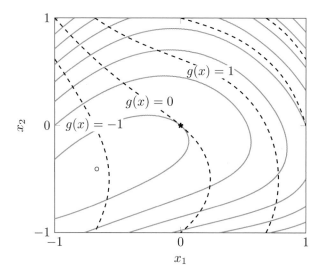

Figure 19.1 Contour lines of the cost function $\|f(x)\|^2$ (solid line) and the constraint function $g(x)$ (dashed line) for a nonlinear least squares problem in two variables with one equality constraint.

the solution of one linear least squares problem (minimizing (19.10) and (19.6), respectively). The two lines show the absolute value of the feasibility residual $|g(x^{(k)})|$, and the norm of the optimality condition residual,

$$\|2Df(x^{(k)})^T f(x^{(k)}) + Dg(x^{(k)})^T z^{(k)}\|.$$

The vertical jumps in the optimality condition norm occur in steps 2 and 3 of the augmented Lagrangian algorithm, and in step 2 of the penalty algorithm, when the parameters μ and z are updated.

Figure 19.5 shows the value of the penalty parameter μ versus the cumulative number of Levenberg–Marquardt iterations in the two algorithms.

19.4   Nonlinear control

A nonlinear dynamical system has the form of an iteration

$$x_{k+1} = f(x_k, u_k), \quad k = 1, 2, \ldots, N,$$

where the n-vector x_k is the state, and the m-vector u_k is the input or control, at time period k. The function $f : \mathbf{R}^{n+m} \to \mathbf{R}^n$ specifies what the next state is, as a function of the current state and the current input. When f is an affine function, this reduces to a linear dynamical system.

In nonlinear control, the goal is to choose the inputs $u_1, \ldots, u_{N-1}$ to achieve some goal for the state and input trajectories. In many problems the initial state

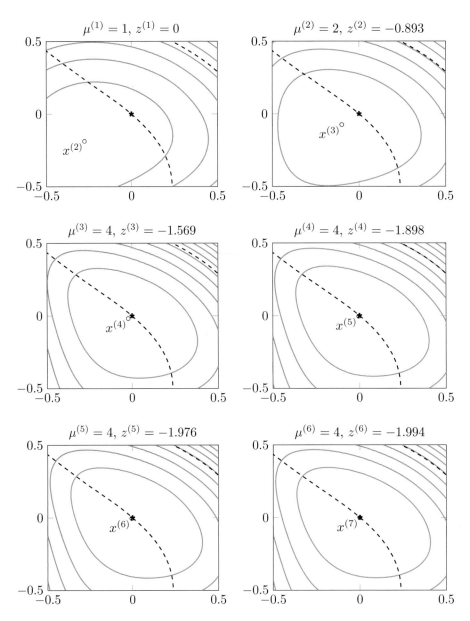

Figure 19.2 First six iterations of the augmented Lagrangian algorithm.

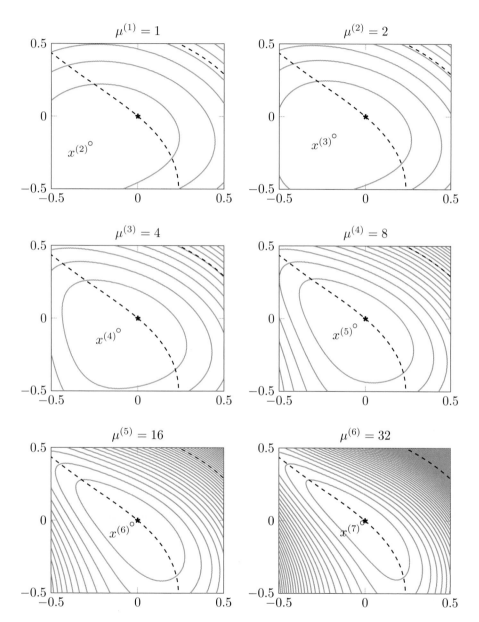

Figure 19.3 First six iterations of the penalty algorithm.

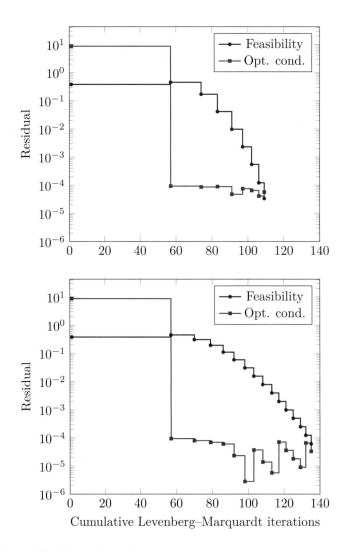

Figure 19.4 Feasibility and optimality condition errors versus the cumulative number of Levenberg–Marquardt iterations in the augmented Lagrangian algorithm (top) and the penalty algorithm (bottom).

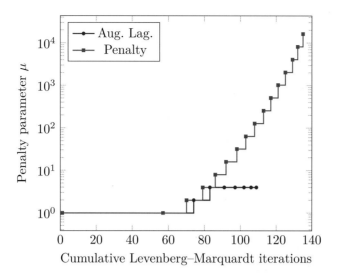

Figure 19.5 Penalty parameter μ versus cumulative number of Levenberg–Marquardt iterations in the augmented Lagrangian algorithm and the penalty algorithm.

x_1 is given, and the final state x_N is specified. Subject to these constraints, we may wish the control inputs to be small and smooth, which suggests that we minimize

$$\sum_{k=1}^{N} \|u_k\|^2 + \gamma \sum_{k=1}^{N-1} \|u_{k+1} - u_k\|^2,$$

where $\gamma > 0$ is a parameter used to trade off input size and smoothness. (In many nonlinear control problems the objective also involves the state trajectory.)

We can formulate the nonlinear control problem, with a norm squared objective that involves the state and input, as a large constrained least problem, and then solve it using the augmented Lagrangian algorithm. We illustrate this with a specific example.

Control of a car.   Consider a car with position $p = (p_1, p_2)$ and orientation (angle) θ. The car has wheelbase (length) L, steering angle ϕ, and speed s (which can be negative, meaning the car moves in reverse). This is illustrated in figure 19.6.

The wheelbase L is a known constant; all of the other quantities p, θ, ϕ, and s are functions of time. The dynamics of the car motion are given by the differential equations

$$\frac{dp_1}{dt}(t) = s(t) \cos \theta(t),$$

$$\frac{dp_2}{dt}(t) = s(t) \sin \theta(t),$$

$$\frac{d\theta}{dt}(t) = (s(t)/L) \tan \phi(t).$$

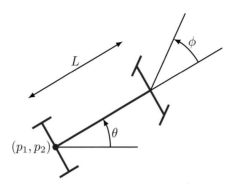

Figure 19.6 Simple model of a car.

Here we assume that the steering angle is always less than $90°$, so the tangent term in the last equation makes sense. The first two equations state that the car is moving in the direction $\theta(t)$ (its orientation) at speed $s(t)$. The last equation gives the change in orientation as a function of the car speed and the steering angle. For a fixed steering angle and speed, the car moves in a circle.

We can control the speed s and the steering angle ϕ; the goal is to move the car over some time period from a given initial position and orientation to a specified final position and orientation.

We now discretize the equations in time. We take a small time interval h, and obtain the approximations

$$
\begin{aligned}
p_1(t + h) &\approx p_1(t) + h s(t) \cos \theta(t), \\
p_2(t + h) &\approx p_2(t) + h s(t) \sin \theta(t), \\
\theta(t + h) &\approx \theta(t) + h(s(t)/L) \tan \phi(t).
\end{aligned}
$$

We will use these approximations to derive nonlinear state equations for the car motion, with state $x_k = (p_1(kh), p_2(kh), \theta(kh))$ and input $u_k = (s(kh), \phi(kh))$. We have

$$
x_{k+1} = f(x_k, u_k),
$$

with

$$
f(x_k, u_k) = x_k + h(u_k)_1 \begin{bmatrix} \cos(x_k)_3 \\ \sin(x_k)_3 \\ (\tan(u_k)_2)/L \end{bmatrix}.
$$

We now consider the nonlinear optimal control problem

$$
\begin{aligned}
\text{minimize} \quad & \sum_{k=1}^{N} \|u_k\|^2 + \gamma \sum_{k=1}^{N-1} \|u_{k+1} - u_k\|^2 \\
\text{subject to} \quad & x_2 = f(0, u_1) \\
& x_{k+1} = f(x_k, u_k), \quad k = 2, \ldots, N-1 \\
& x_{\text{final}} = f(x_N, u_N),
\end{aligned}
\tag{19.12}
$$

with variables $u_1, \ldots, u_N$, and $x_2, \ldots, x_N$.

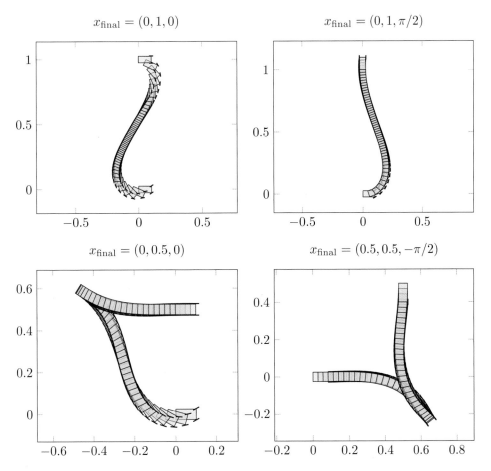

Figure 19.7 Solution trajectories of 19.12 for different end states x_{final}. The outline of the car shows the position $(p_1(kh), p_2(kh))$, orientation $\theta(kh)$, and the steering angle $\phi(kh)$ at time kh.

Figure 19.7 shows solutions for

$$L = 0.1, \qquad N = 50, \qquad h = 0.1, \qquad \gamma = 10,$$

and different values of x_{final}. They are computed using the augmented Lagrangian algorithm. The algorithm is started at the same starting point for each example. The starting point for the input variables u_k is randomly chosen, the starting point for the states x_k is zero.

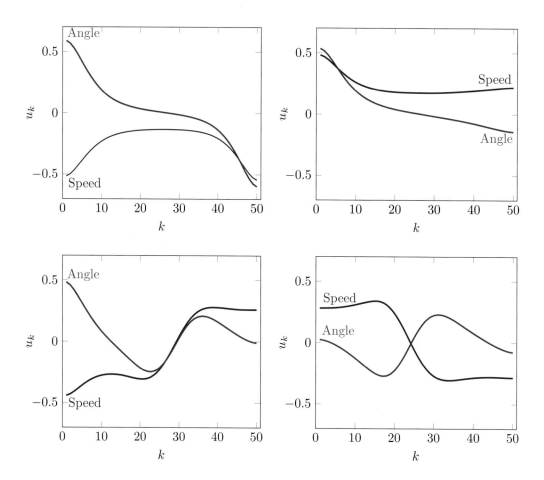

Figure 19.8 The two inputs (speed and steering angle) for the trajectories in figure 19.7.

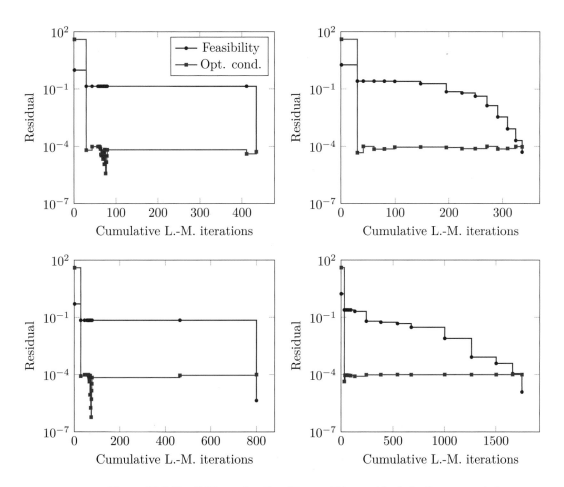

Figure 19.9 Feasibility and optimality condition residuals in the augmented Lagrangian algorithm for computing the trajectories in figure 19.7.

Exercises

19.1 *Projection on a curve.* We consider a constrained nonlinear least squares problem with three variables $x = (x_1, x_2, x_3)$ and two equations:

$$\begin{array}{ll} \text{minimize} & (x_1 - 1)^2 + (x_2 - 1)^2 + (x_3 - 1)^2 \\ \text{subject to} & x_1^2 + 0.5x_2^2 + x_3^2 - 1 = 0 \\ & 0.8x_1^2 + 2.5x_2^2 + x_3^2 + 2x_1x_3 - x_1 - x_2 - x_3 - 1 = 0. \end{array}$$

The solution is the point closest to $(1, 1, 1)$ on the nonlinear curve defined by the two equations.

(a) Solve the problem using the augmented Lagrangian method. You can start the algorithm at $x^{(1)} = 0$, $z^{(1)} = 0$, $\mu^{(1)} = 1$, and start each run of the Levenberg–Marquardt method with $\lambda^{(1)} = 1$. Stop the augmented Lagrangian method when the feasibility residual $\|g(x^{(k)})\|$ and the optimality condition residual

$$\|2Df(x^{(k)})^T f(x^{(k)}) + Dg(x^{(k)})^T z^{(k)}\|$$

are less than 10^{-5}. Make a plot of the two residuals and of the penalty parameter μ versus the cumulative number of Levenberg–Marquardt iterations.

(b) Solve the problem using the penalty method, started at $x^{(1)} = 0$ and $\mu^{(1)} = 1$, and with the same stopping condition. Compare the convergence and the value of the penalty parameter with the results for the augmented Lagrangian method in part (a).

19.2 *Portfolio optimization with downside risk.* In standard portfolio optimization (as described in §17.1) we choose the weight vector w to achieve a given target mean return, and to minimize deviations from the target return value (*i.e.*, the risk). This leads to the constrained linear least squares problem (17.2). One criticism of this formulation is that it treats portfolio returns that exceed our target value the same as returns that fall short of our target value, whereas in fact we should be delighted to have a return that exceeds our target value. To address this deficiency in the formulation, researchers have defined the *downside risk* of a portfolio return time series T-vector r, which is sensitive only to portfolio returns that fall short of our target value ρ^{tar}. The downside risk of a portfolio return time series (T-vector) r is given by

$$D = \frac{1}{T}\sum_{t=1}^{T}\left(\max\left\{\rho^{\text{tar}} - r_t, 0\right\}\right)^2.$$

The quantity $\max\{\rho^{\text{tar}} - r_t, 0\}$ is the shortfall, *i.e.*, the amount by which the return in period t falls short of the target; it is zero when the return exceeds the target value. The downside risk is the mean square value of the return shortfall.

(a) Formulate the portfolio optimization problem, using downside risk in place of the usual risk, as a constrained nonlinear least squares problem. Be sure to explain what the functions f and g are.

(b) Since the function g is affine (if you formulate the problem correctly), you can use the Levenberg–Marquardt algorithm, modified to handle linear equality constraints, to approximately solve the problem. (See page 420.) Find an expression for $Df(x^{(k)})$. You can ignore the fact that the function f is not differentiable at some points.

(c) Implement the Levenberg–Marquardt algorithm to find weights that minimize downside risk for a given target annualized return. A very reasonable starting point is the solution of the standard portfolio optimization problem with the same target return. Check your implementation with some simulated or real return data (available online). Compare the weights, and the risk and downside risk, for the minimum risk and the minimum downside risk portfolios.

19.3 *Boolean least squares.* The *Boolean least squares problem* is a special case of the constrained nonlinear least squares problem (19.1), with the form

$$\begin{array}{ll} \text{minimize} & \|Ax - b\|^2 \\ \text{subject to} & x_i^2 = 1, \quad i = 1, \ldots, n, \end{array}$$

where the n-vector x is the variable to be chosen, and the $m \times n$ matrix A and the m-vector b are the (given) problem data. The constraints require that each entry of x is either -1 or $+1$, *i.e.*, x is a Boolean vector. Since each entry can take one of two values, there are 2^n feasible values for the vector x. The Boolean least squares problem arises in many applications.

One simple method for solving the Boolean least squares problem, sometimes called the *brute force method*, is to evaluate the objective function $\|Ax - b\|^2$ for each of the 2^n possible values, and choose one that has the least value. This method is not practical for n larger than 30 or so. There are many heuristic methods that are much faster to carry out than the brute force method, and approximately solve it, *i.e.*, find an x for which the objective is small, if not the smallest possible value over all 2^n feasible values of x. One such heuristic is the augmented Lagrangian algorithm 19.2.

(a) Work out the details of the update step in the Levenberg–Marquardt algorithm used in each iteration of the augmented Lagrangian algorithm, for the Boolean least squares problem.

(b) Implement the augmented Lagrangian algorithm for the Boolean least squares problem. You can choose the starting point $x^{(1)}$ as the minimizer of $\|Ax - b\|^2$. At each iteration, you can obtain a feasible point $\tilde{x}^{(k)}$ by rounding the entries of $x^{(k)}$ to the values ± 1, *i.e.*, $\tilde{x}^{(k)} = \mathbf{sign}(x^{(k)})$. You should evaluate and plot the objective value of these feasible points, *i.e.*, $\|A\tilde{x}^{(k)} - b\|^2$. Your implementation can return the best rounded value found during the iterations. Try your method on some small problems, with $n = m = 10$ (say), for which you can find the actual solution by the brute force method. Try it on much larger problems, with $n = m = 500$ (say), for which the brute force method is not practical.

Appendices

Appendix A

Notation

Vectors

$$\begin{bmatrix} x_1 \\ \vdots \\ x_n \end{bmatrix}$$
n-vector with entries $x_1, \ldots, x_n$.   3

$(x_1, \ldots, x_n)$	n-vector with entries $x_1, \ldots, x_n$.	3
x_i	The ith entry of a vector x.	3
$x_{r:s}$	Subvector with entries from r to s.	4
0	Vector with all entries zero.	5
$\mathbf{1}$	Vector with all entries one.	5
e_i	The ith standard unit vector.	5
$x^T y$	Inner product of vectors x and y.	19
$\|x\|$	Norm of vector x.	45
$\mathbf{rms}(x)$	Root-mean-square value of a vector x.	46
$\mathbf{avg}(x)$	Average of entries of a vector x.	20
$\mathbf{std}(x)$	Standard deviation of a vector x.	52
$\mathbf{dist}(x, y)$	Distance between vectors x and y.	48
$\angle(x, y)$	Angle between vectors x and y.	56
$x \perp y$	Vectors x and y are orthogonal.	58

Matrices

$$\begin{bmatrix} X_{11} & \cdots & X_{1n} \\ \vdots & & \vdots \\ X_{m1} & \cdots & X_{mn} \end{bmatrix}$$
$m \times n$ matrix with entries $X_{11}, \ldots, X_{mn}$.   107

X_{ij}	The i, jth entry of a matrix X.	107
$X_{r:s, p:q}$	Submatrix with rows $r, \ldots, s$ and columns $p \ldots, q$.	109
0	Matrix with all entries zero.	113
I	Identity matrix.	113
X^T	Transpose of matrix X.	115

$\|X\|$	Norm of matrix X.	117
X^k	(Square) matrix X to the kth power.	186
X^{-1}	Inverse of (square) matrix X.	202
X^{-T}	Inverse of transpose of matrix X.	205
$X^{\dagger}$	Pseudo-inverse of matrix X.	215
$\mathbf{diag}(x)$	Diagonal matrix with diagonal entries $x_1, \ldots, x_n$.	114

Functions and derivatives

$f : A \to B$	f is a function on the set A into the set B.
$\nabla f(z)$	Gradient of function $f : \mathbf{R}^n \to \mathbf{R}$ at z.
$Df(z)$	Derivative (Jacobian) matrix of function $f : \mathbf{R}^n \to \mathbf{R}^m$ at z.

Ellipsis notation

In this book we use standard mathematical ellipsis notation in lists and sums. We write $k, \ldots, l$ to mean the list of all integers from k to l. For example, $3, \ldots, 7$ means $3, 4, 5, 6, 7$. This notation is used to describe a list of numbers or vectors, or in sums, as in $\sum_{i=1,\ldots,n} a_i$, which we also write as $a_1 + \cdots + a_n$. Both of these mean the sum of the n terms $a_1, a_2, \ldots, a_n$.

Sets

In a few places in this book we encounter the mathematical concept of sets. The notation $\{a_1, \ldots, a_n\}$ refers to a *set* with elements $a_1, \ldots, a_n$. This is not the same as the vector with entries $a_1, \ldots, a_n$, which is denoted $(a_1, \ldots, a_n)$. For sets the order does not matter, so, for example, we have $\{1, 2, 6\} = \{6, 1, 2\}$. Unlike a vector, a set cannot have repeated elements. We can also specify a set by giving conditions that its entries must satisfy, using the notation $\{x \mid \text{condition}(x)\}$, which means the set of x that satisfy the condition, which depends on x. We say that a set contains its elements, or that the elements are in the set, using the symbol $\in$, as in $2 \in \{1, 2, 6\}$. The symbol $\notin$ means not in, or not an element of, as in $3 \notin \{1, 2, 6\}$.

We can use sets to describe a sum over some elements in a list. The notation $\sum_{i \in S} x_i$ means the sum over all x_i for which i is in the set S. As an example, $\sum_{i \in \{1,2,6\}} a_i$ means $a_1 + a_2 + a_6$.

A few sets have specific names: $\mathbf{R}$ is the set of real numbers (or scalars), and $\mathbf{R}^n$ is the set of all n-vectors. So $\alpha \in \mathbf{R}$ means that α is a number, and $x \in \mathbf{R}^n$ means that x is an n-vector.

Appendix B

Complexity

Here we summarize approximate complexities or flop counts of various operations and algorithms encountered in the book. We drop terms of lower order. When the operands or arguments are sparse, and the operation or algorithm has been modified to take advantage of sparsity, the flop counts can be dramatically lower than those given here.

Vector operations

In the table below, x and y are n-vectors and a is a scalar.

ax	n
$x + y$	n
$x^T y$	$2n$
$\|x\|$	$2n$
$\|x - y\|$	$3n$
$\mathbf{rms}(x)$	$2n$
$\mathbf{std}(x)$	$4n$
$\angle(x, y)$	$6n$

The convolution $a * b$ of an n-vector a and m-vector b can be computed by a special algorithm that requires $5(m + n) \log_2(m + n)$ flops.

Matrix operations

In the table below, A and B are $m \times n$ matrices, C is an $m \times p$ matrix, x is an n-vector, and a is a scalar.

aA	mn
$A + B$	mn
Ax	$2mn$
AC	$2mnp$
$A^T A$	mn^2
$\|A\|$	$2mn$

Factorization and inverses

In the table below, A is a tall or square $m \times n$ matrix, R is an $n \times n$ triangular matrix, and b is an n-vector. We assume the factorization or inverses exist; in particular in any expression involving A^{-1}, A must be square.

QR factorization of A	$2mn^2$
$R^{-1}b$	n^2
$A^{-1}b$	$2n^3$
A^{-1}	$3n^3$
$A^\dagger$	$3mn^2$

The pseudo-inverse $A^\dagger$ of a wide $m \times n$ matrix (with linearly independent rows) can be computed in $3m^2n$ flops.

Solving least squares problems

In the table below, A is an $m \times n$ matrix, C is a wide $p \times n$ matrix, and b is an m-vector. We assume the associated independence conditions hold.

minimize $\|Ax - b\|^2$	$2mn^2$
minimize $\|Ax - b\|^2$ subject to $Cx = d$	$2(m + p)n^2 + 2np^2$
minimize $\|x\|^2$ subject to $Cx = d$	$2np^2$

Big-times-small-squared mnemonic

Many of the complexities listed above that involve two dimensions can be remembered using a simple mnemonic: The cost is order

$$(\text{big}) \times (\text{small})^2 \text{ flops},$$

where 'big' and 'small' refer to big and small problem dimensions. We list some examples below.

- Computing the Gram matrix of a tall $m \times n$ matrix requires mn^2 flops. Here m is the big dimension and n is the small dimension.
- In the QR factorization of an $m \times n$ matrix, we have $m \geq n$, so m is the big dimension and n is the small dimension. The complexity is $2mn^2$ flops.
- Computing the pseudo-inverse $A^\dagger$ of an $m \times n$ matrix A when A is tall (and has independent columns) costs $3mn^2$ flops. When A is wide (and has independent rows), it is $3nm^2$ flops.
- For least squares, we have $m \geq n$, so m is the big dimension and n is the small dimension. The cost of computing the least squares approximate solution is $2mn^2$ flops.
- For the least norm problem, we have $p \leq n$, so n is the big dimension and p is the small dimension. The cost is $2np^2$ flops.
- The constrained least squares problem involves two matrices A and C, and three dimensions that satisfy $m + p \geq n$. The numbers $m + p$ and n are the big and small dimensions of the stacked matrix $\begin{bmatrix} A \\ C \end{bmatrix}$. The cost of solving the constrained least squares problem is $2(m + p)n^2 + 2np^2$ flops, which is between $2(m + p)n^2$ flops and $4(m + p)n^2$ flops, since $n \leq m + p$.

Appendix C

Derivatives and optimization

Calculus does not play a big role in this book, except in chapters 18 and 19 (on nonlinear least squares and constrained nonlinear least squares), where we use derivatives, Taylor approximations, and the method of Lagrange multipliers. In this appendix we collect some basic material about derivatives and optimization, focusing on the few results and formulas we use.

C.1 Derivatives

C.1.1 Scalar-valued function of a scalar

Definition. Suppose $f : \mathbf{R} \to \mathbf{R}$ is a real-valued function of a real (scalar) variable. For any number x, the number $f(x)$ is the *value* of the function, and x is called the *argument* of the function. The number

$$\lim_{t \to 0} \frac{f(z + t) - f(z)}{t},$$

(if the limit exists) is called the *derivative* of the function f at the point z. It gives the slope of the graph of f at the point $(z, f(z))$. We denote the derivative of f at z as $f'(z)$. We can think of f' as a scalar-valued function of a scalar variable; this function is called the derivative (function) of f.

Taylor approximation. Let us fix the number z. The (first order) *Taylor approximation* of the function f at the point z is defined as

$$\hat{f}(x) = f(z) + f'(z)(x - z)$$

for any x. Here $f(z)$ is the value of f at z, $x - z$ is the deviation of x from z, and $f'(z)(x - z)$ is the approximation of the change in value of the function due to the deviation of x from z. Sometimes the Taylor approximation is shown with a second argument, separated by a semicolon, to denote the point z where the approximation is made. Using this notation, the left-hand side of the equation above is written $\hat{f}(x; z)$. The Taylor approximation is sometimes called the *linearized approximation* of f at z. (Here linear uses informal mathematical language, where affine is sometimes called linear.) The Taylor approximation function $\hat{f}$ is an affine function of x, *i.e.*, a linear function of x plus a constant.

The Taylor approximation $\hat{f}$ satisfies $\hat{f}(z; z) = f(z)$, *i.e.*, at the point z it agrees with the function f. For x near z, $\hat{f}(x; z)$ is a very good approximation of $f(x)$. For x not close enough to z, however, the approximation can be poor.

Finding derivatives. In a basic calculus course, the derivatives of many common functions are worked out. For example, with $f(x) = x^2$, we have $f'(z) = 2z$, and for $f(x) = e^x$, we have $f'(z) = e^z$. Derivatives of more complex functions can be found using these known derivatives of common functions, along with a few rules for finding the derivative of various combinations of functions. For example, the *chain rule* gives the derivative of a composition of two functions. If $f(x) = g(h(x))$, where g and h are scalar-valued functions of a scalar variable, we have

$$f'(z) = g'(h(z))h'(z).$$

Another useful rule is the derivative of product rule, for $f(x) = g(x)h(x)$, which is

$$f'(z) = g'(z)h(z) + g(z)h'(z).$$

The derivative operation is linear, which means that if $f(x) = ag(x) + bh(x)$, where a and b are constants, we have

$$f'(z) = ag'(z) + bh'(z).$$

Knowledge of the derivative of just a few common functions, and a few combination rules like the ones above, is enough to determine the derivatives of many functions.

C.1.2 Scalar-valued function of a vector

Suppose $f : \mathbf{R}^n \to \mathbf{R}$ is a scalar-valued function of an n-vector argument. The number $f(x)$ is the value of the function f at the n-vector (argument) x. We sometimes write out the argument of f to show that it can be considered a function of n scalar arguments, $x_1, \ldots, x_n$:

$$f(x) = f(x_1, \ldots, x_n).$$

Partial derivative. The *partial derivative* of f at the point z, with respect to its ith argument, is defined as

$$\begin{aligned}
\frac{\partial f}{\partial x_i}(z) &= \lim_{t \to 0} \frac{f(z_1, \ldots, z_{i-1}, z_i + t, z_{i+1}, \ldots, z_n) - f(z)}{t} \\
&= \lim_{t \to 0} \frac{f(z + te_i) - f(z)}{t},
\end{aligned}$$

(if the limit exists). Roughly speaking the partial derivative is the derivative with respect to the ith argument, with all other arguments fixed.

Gradient.   The partial derivatives of f with respect to its n arguments can be collected into an n vector called the *gradient* of f (at z):

$$\nabla f(z) = \begin{bmatrix} \frac{\partial f}{\partial x_1}(z) \\ \vdots \\ \frac{\partial f}{\partial x_n}(z) \end{bmatrix}.$$

Taylor approximation.   The (first-order) Taylor approximation of f at the point z is the function $\hat{f} : \mathbf{R}^n \to \mathbf{R}$ defined as

$$\hat{f}(x) = f(z) + \frac{\partial f}{\partial x_1}(z)(x_1 - z_1) + \cdots + \frac{\partial f}{\partial x_n}(z)(x_n - z_n)$$

for any x. We interpret $x_i - z_i$ as the deviation of x_i from z_i, and the term $\frac{\partial f}{\partial x_i}(z)(x_i - z_i)$ as an approximation of the change in f due to the deviation of x_i from z_i. Sometimes $\hat{f}$ is written with a second vector argument, as $\hat{f}(x; z)$, to show the point z at which the approximation is developed. The Taylor approximation can be written in compact form as

$$\hat{f}(x; z) = f(z) + \nabla f(z)^T (x - z).$$

The Taylor approximation $\hat{f}$ is an affine function of x.

   The Taylor approximation $\hat{f}$ agrees with the function f at the point z, *i.e.*, $\hat{f}(z; z) = f(z)$. When all x_i are near the associated z_i, $\hat{f}(x; z)$ is a very good approximation of $f(x)$. The Taylor approximation is sometimes called the linear approximation or linearized approximation of f (at z), even though in general it is affine, and not linear.

Finding gradients.   The gradient of a function can be found by evaluating the partial derivatives using the common functions and rules for derivatives of scalar-valued functions, and assembling the result into a vector. In many cases the result can be expressed in a more compact matrix-vector form. As an example let us find the gradient of the function

$$f(x) = \|x\|^2 = x_1^2 + \cdots + x_n^2,$$

which is the sum of squares of the arguments. The partial derivatives are

$$\frac{\partial f}{\partial x_i}(z) = 2z_i, \quad i = 1, \ldots, n.$$

This leads to the very simple vector formula

$$\nabla f(z) = 2z.$$

(Note the resemblance to the formula for the derivative of the square of a scalar variable.)

   There are rules for the gradient of a combination of functions similar to those for functions of a scalar. For example if $f(x) = ag(x) + bh(x)$, we have

$$\nabla f(z) = a\nabla g(z) + b\nabla h(z).$$

C.1.3   Vector-valued function of a vector

Suppose $f : \mathbf{R}^n \to \mathbf{R}^m$ is a vector-valued function of a vector. The n-vector x is the argument; the m-vector $f(x)$ is the value of the function f at x. We can write out the m components of f as

$$f(x) = \begin{bmatrix} f_1(x) \\ \vdots \\ f_m(x) \end{bmatrix},$$

where f_i is a scalar-valued function of $x = (x_1, \ldots, x_n)$.

Jacobian.   The partial derivatives of the components of $f(x)$ with respect to the components of x, evaluated at z, are arranged into an $m \times n$ matrix denoted $Df(z)$, called the *derivative matrix* or *Jacobian* of f at z. (In the notation $Df(z)$, the D and f go together; Df does not represent, say, a matrix-vector product.) The derivative matrix is defined by

$$Df(z)_{ij} = \frac{\partial f_i}{\partial x_j}(z), \quad i = 1, \ldots, m, \quad j = 1, \ldots, n.$$

The rows of the Jacobian are $\nabla f_i(z)^T$, for $i = 1, \ldots, m$. For $m = 1$, *i.e.*, when f is a scalar-valued function, the derivative matrix is a row vector of size n, the transpose of the gradient of the function. The derivative matrix of a vector-valued function of a vector is a generalization of the derivative of a scalar-valued function of a scalar.

Taylor approximation.   The (first-order) Taylor approximation of f near z is given by

$$\begin{aligned} \hat{f}(x)_i &= f_i(z) + \frac{\partial f_i}{\partial x_1}(z)(x_1 - z_1) + \cdots + \frac{\partial f_i}{\partial x_n}(z)(x_n - z_n) \\ &= f_i(z) + \nabla f_i(z)^T(x - z), \end{aligned}$$

for $i = 1, \ldots, m$. We can express this approximation in compact notation as

$$\hat{f}(x) = f(z) + Df(z)(x - z).$$

For x near z, $\hat{f}(x)$ is a very good approximation of $f(x)$. As in the scalar case, the Taylor approximation is sometimes written with a second argument as $\hat{f}(x; z)$ to show the point z around which the approximation is made. The Taylor approximation $\hat{f}$ is an affine function of x, sometimes called a linear approximation of f, even though it is not, in general, a linear function.

Finding Jacobians.   We can always find the derivative matrix by calculating partial derivatives of the entries of f with respect to the components of the argument vector. In many cases the result simplifies using matrix-vector notation. As an example, let us find the derivative of the (scalar-valued) function

$$h(x) = \|f(x)\|^2 = f_1(x)^2 + \cdots + f_m(x)^2,$$

where $f : \mathbf{R}^n \to \mathbf{R}^m$. The partial derivative with respect to x_j, at z, is

$$\frac{\partial h}{\partial x_j}(z) = 2f_1(z)\frac{\partial f_1}{\partial x_j}(z) + \cdots + 2f_m(z)\frac{\partial f_m}{\partial x_j}(z).$$

Arranging these to form the row vector $Dh(z)$, we see we can write this using matrix multiplication as

$$Dh(z) = 2f(z)^T Df(z).$$

The gradient of h is the transpose of this expression,

$$\nabla h(z) = 2Df(z)^T f(z). \tag{C.1}$$

(Note the analogy to the formula for the scalar-valued function of a scalar variable $h(x) = f(x)^2$, which is $h'(z) = 2f'(z)f(z)$.)

Many of the formulas for derivatives in the scalar case also hold for the vector case, with scalar multiplication replaced with matrix multiplication (provided the order of the terms is correct). As an example, consider the composition function $f(x) = g(h(x))$, where $h : \mathbf{R}^n \to \mathbf{R}^k$ and $g : \mathbf{R}^k \to \mathbf{R}^m$. The Jacobian or derivative matrix of f at z is given by

$$Df(z) = Dg(h(z))Dh(z).$$

(This is matrix multiplication; compare it to composition formula for scalar-valued functions of scalars given above.) This chain rule is described on page 184.

C.2   Optimization

Derivative condition for minimization.   Suppose h is a scalar-valued function of a scalar argument. If $\hat{x}$ minimizes $h(x)$, we must have $h'(\hat{x}) = 0$. This fact is easily understood: If $h'(\hat{x}) \neq 0$, then by taking a point $\tilde{x}$ slightly less than $\hat{x}$ (if $h'(\hat{x}) > 0$) or slightly more than $\hat{x}$ (if $h'(\hat{x}) < 0$), we would obtain $h(\tilde{x}) < h(\hat{x})$, which shows that $\hat{x}$ does not minimize $h(x)$. This leads to the classic calculus-based method for finding a minimizer of a function f: Find the derivative, and set it equal to zero. One subtlety here is that there can be (and generally are) points that satisfy $h'(z) = 0$, but are not minimizers of h. So we generally need to check which of the solutions of $h'(z) = 0$ are in fact minimizers of h.

Gradient condition for minimization.   This basic calculus-based method for finding a minimizer of a scalar-valued function can be generalized to functions with vector arguments. If the n-vector $\hat{x}$ minimizes $h : \mathbf{R}^n \to \mathbf{R}$, then we must have

$$\frac{\partial h}{\partial x_i}(\hat{x}) = 0, \quad i = 1, \ldots, n.$$

In vector notation, we must have

$$\nabla h(\hat{x}) = 0.$$

Like the case of a scalar argument, this is easily seen to hold if $\hat{x}$ minimizes h. Also as in the case of a scalar argument, there can be points that satisfy $\nabla h(z) = 0$ but are not minimizers of h. So we need to check if points found this way are in fact minimizers of h.

Nonlinear least squares. As an example, consider the nonlinear least squares problem, with objective $h(x) = \|f(x)\|^2$, where $f : \mathbf{R}^n \to \mathbf{R}^m$. The optimality condition $\nabla h(\hat{x}) = 0$ is

$$2Df(\hat{x})^T f(\hat{x}) = 0$$

(using the expression (C.1) for the gradient, derived above). This equation will hold for a minimizer, but there can be points that satisfy the equation, but are not solutions of the nonlinear least squares problem.

C.3 Lagrange multipliers

Constrained optimization. We now consider the problem of minimizing a scalar-valued function $h : \mathbf{R}^n \to \mathbf{R}$, subject to the requirements, or constraints, that

$$g_1(x) = 0, \ \ldots, \ g_p(x) = 0$$

must hold, where $g_i : \mathbf{R}^n \to \mathbf{R}$ are given functions. We can write the constraints in compact vector form $g(x) = 0$, where $g(x) = (g_1(x), \ldots, g_p(x))$, and express the problem as

$$\begin{array}{ll} \text{minimize} & f(x) \\ \text{subject to} & g(x) = 0. \end{array}$$

We seek a solution of this optimization problem, *i.e.*, a point $\hat{x}$ that satisfies $g(\hat{x}) = 0$ (*i.e.*, is feasible) and, for any other x that satisfies $g(x) = 0$, we have $h(x) \geq h(\hat{x})$.

The *method of Lagrange multipliers* is an extension of the derivative or gradient conditions for (unconstrained) minimization, that handles constrained optimization problems.

Lagrange multipliers. The *Lagrangian* function associated with the constrained problem is defined as

$$\begin{aligned} L(x,z) &= h(x) + z_1 g_1(x) + \cdots + z_p g_p(x) \\ &= h(x) + g(x)^T z, \end{aligned}$$

with arguments x (the original variable to be determined in the optimization problem), and a p-vector z, called the (vector of) *Lagrange multipliers*. The Lagrangian function is the original objective, with one term added for each constraint function. Each term is the constraint function value multiplied by z_i, hence the name multiplier.

KKT conditions.   The *KKT conditions* (named for Karush, Kuhn, and Tucker) state that if $\hat{x}$ is a solution of the constrained optimization problem, then there is a vector $\hat{z}$ that satisfies

$$\frac{\partial L}{\partial x_i}(\hat{x}, \hat{z}) = 0, \quad i = 1, \ldots, n, \qquad \frac{\partial L}{\partial z_i}(\hat{x}, \hat{z}) = 0, \quad i = 1, \ldots, p.$$

(This is provided the rows of $Dg(\hat{x})$ are linearly independent, a technical condition we ignore.) As in the unconstrained case, there can be pairs x, z that satisfy the KKT conditions but $\hat{x}$ is not a solution of the constrained optimization problem.

The KKT conditions give us a method for solving the constrained optimization problem that is similar to the approach for the unconstrained optimization problem. We attempt to solve the KKT equations for $\hat{x}$ and $\hat{z}$; then we check to see if any of the points found are really solutions.

We can simplify the KKT conditions, and express them compactly using matrix notation. The last p equations can be expressed as $g_i(\hat{x}) = 0$, which we already knew. The first n can be expressed as

$$\nabla_x L(\hat{x}, \hat{z}) = 0,$$

where ∇_x denotes the gradient with respect to the x_i arguments. This can be written as

$$\nabla h(\hat{x}) + \hat{z}_1 \nabla g_1(\hat{x}) + \cdots + \hat{z}_p g_p(\hat{x}) = \nabla h(\hat{x}) + Dg(\hat{x})^T \hat{z} = 0.$$

So the KKT conditions for the constrained optimization problem are

$$\nabla h(\hat{x}) + Dg(\hat{x})^T \hat{z} = 0, \qquad g(\hat{x}) = 0.$$

This is the extension of the gradient condition for unconstrained optimization to the constrained case.

Constrained nonlinear least squares.   As an example, consider the constrained least squares problem

$$\begin{array}{ll} \text{minimize} & \|f(x)\|^2 \\ \text{subject to} & g(x) = 0, \end{array}$$

where $f : \mathbf{R}^n \to \mathbf{R}^m$ and $g : \mathbf{R}^n \to \mathbf{R}^p$. Define $h(x) = \|f(x)\|^2$. Its gradient at $\hat{x}$ is $2Df(\hat{x})^T f(\hat{x})$ (see above) so the KKT conditions are

$$2Df(\hat{x})^T f(\hat{x}) + Dg(\hat{x})^T \hat{z} = 0, \qquad g(\hat{x}) = 0.$$

These conditions will hold for a solution of the problem (assuming the rows of $Dg(\hat{x})$ are linearly independent). But there can be points that satisfy them and are not solutions.

Appendix D

Further study

In this appendix we list some further topics of study that are closely related to the material in this book, give a different perspective on the same material, complement it, or provide useful extensions. The topics are organized into groups, but the groups overlap, and there are many connections between them.

Mathematics

Probability and statistics. In this book we do not use probability and statistics, even though we cover multiple topics that are traditionally addressed using ideas from probability and statistics, including data fitting and classification, control, state estimation, and portfolio optimization. Further study of many of the topics in this book requires a background in basic probability and statistics, and we strongly encourage you to learn this material. (We also urge you to remember that topics like data fitting can be discussed without ideas from probability and statistics.)

Abstract linear algebra. This book covers some of the most important basic ideas from linear algebra, such as linear independence. In a more abstract course you will learn about vector spaces, subspaces, nullspace, and range. Eigenvalues and singular values are useful topics that we do not cover in this book. Using these concepts you can analyze and solve linear equations and least squares problems when the basic assumption used in this book (*i.e.*, the columns of some matrix are linearly independent) does not hold. Another more advanced topic that arises in the solution of linear differential equations is the matrix exponential.

Mathematical optimization. This book focuses on just a few optimization problems: Least squares, linearly constrained least squares, and their nonlinear extensions. In an optimization course you will learn about more general optimization problems, for example ones that include inequality constraints. Convex optimization is a particularly useful generalization of the linearly constrained least squares problem. Convex optimization problems can be solved efficiently and nonheuristically, and include a wide range of practically useful problems that arise in many application areas, including all of the ones we have seen in this book. We would strongly encourage you to learn convex optimization, which is widely used in many applications. It is also useful to learn about methods for general non-convex optimization problems.

Computer science

Languages and packages for linear algebra. We hope that you will actually use the ideas and methods in this book in practical applications. This requires a good knowledge and understanding of at least one of the computer languages and packages that support linear algebra computations. In a first introduction you can use one of these packages to follow the material of this book, carrying out numerical calculations to verify our assertions and experiment with methods. Developing more fluency in one or more of these languages and packages will greatly increase your effectiveness in applying the ideas in this book.

Computational linear algebra. In a course on computational or numerical linear algebra you will learn more about floating point numbers and how the small round-off errors made in numerical calculations affect the computed solutions. You will also learn about methods for sparse matrices, and iterative methods that can solve linear equations, or compute least squares solutions, for extremely large problems such as those arising in image processing or in the solution of partial differential equations.

Applications

Machine learning and artificial intelligence. This book covers some of the basic ideas of machine learning and artificial intelligence, including a first exposure to clustering, data fitting, classification, validation, and feature engineering. In a further course on this material, you will learn about unsupervised learning methods (like k-means) such as principal components analysis, nonnegative matrix factorization, and more sophisticated clustering methods. You will also learn about more sophisticated regression and classification methods, such as logistic regression and the support vector machine, as well as methods for computing model parameters that scale to extremely large scale problems. Additional topics might include feature engineering and deep neural networks.

Linear dynamical systems, control, and estimation. We cover only the basics of these topics; entire courses cover them in much more detail. In these courses you will learn about continuous-time linear dynamical systems (described by systems of differential equations) and the matrix exponential, more about linear quadratic control and state estimation, and applications in aerospace, navigation, and GPS.

Finance and portfolio optimization. Our coverage of portfolio optimization is basic. In a further course you would learn about statistical models of returns, factor models, transaction costs, more sophisticated models of risk, and the use of convex optimization to handle constraints, for example a limit on leverage, or the requirement that the portfolio be long-only.

Signal and image processing. Traditional signal processing, which is used throughout engineering, focuses on convolution, the Fourier transform, and the so-called frequency domain. More recent approaches use convex optimization, especially in non-real-time applications, like image enhancement or medical image reconstruction. (Even more recent approaches use neural networks.) You will find whole courses on signal processing for a specific application area, like communications, speech, audio, and radar; for image processing, there are whole courses on microscopy, computational photography, tomography, and medical imaging.

Time series analysis. Time series analysis, and especially prediction, plays an important role in many applications areas, including finance and supply chain optimization. It is typically taught in a statistics or operations research course, or as a specialty course in a specific area such as econometrics.

Index